IONOSPHERIC TECHNIQUES AND PHENOMENA

GEOPHYSICS AND ASTROPHYSICS MONOGRAPHS

AN INTERNATIONAL SERIES OF FUNDAMENTAL TEXTBOOKS

VOLUME 13

IONOSPHERIC TECHNIQUES AND PHENOMENA

by

ALAIN GIRAUD and MICHEL PETIT
Centre National d'Etudes des Télécommunications,
Centre National de la Recherche Scientifique,
92131 Issy les Moulineaux, France

D. REIDEL PUBLISHING COMPANY
DORDRECHT : HOLLAND / BOSTON : U.S.A.
LONDON : ENGLAND

Library of Congress Cataloging in Publication Data

Giraud, Alain, 1938–
Ionospheric techniques and phenomena.
(Geophysics and astrophysics monographs; v. 13)
Published in 1975 under title: Physique de l'ionosphère.
Bibliography: p.
Includes index.
1. Ionosphere. 1. Petit, Michel, 1935– joint author. II. Title. III. Series.
QC881.2.16G57 1978 551.5′145 78–16478
ISBN 90–277–0499–6

Published by D. Reidel Publishing Company,
P.O. Box 17, Dordrecht, Holland

Sold and distributed in the U.S.A., Canada and Mexico
by D. Reidel Publishing Company, Inc.
Lincoln Building, 160 Old Derby Street, Hingham,
Mass. 02043, U.S.A.

Printed in The Netherlands

'I make out two ways of cultivating Science: the first is to enlarge the stock of knowledge by making discoveries; the other is to bring together discoveries and to tidy them up, in order that more men be enlightened, and that every one take part, according to his reach, in his century's outlook.'

Denis Diderot

This photograph was taken at Chatanika, Alaska, on March 24, 1973. It pictures a vivid synthesis of the modern (post-IGY) outlook on the ionosphere. The 'dish' antenna in the foreground is that of an incoherent scatter radar system pulse sounding the plasma in the upper atmosphere; the distant trail is that of a rocket launched with a payload of instruments about to telemeter back to Earth real-time data on the ionizing radiation. These experiments exemplify the powerful techniques of investigation that have superseded the original tools of ionospheric research used for almost half a century. The background lights in the sky are those of an intense auroral display, symbolizing contemporary interest in ionization phenomena as the moving picture which solar and magnetospheric activity project on the screen of the terrestrial atmosphere. (Courtesy of M. Baron, Stanford Research Institute)

TABLE OF CONTENTS

PART THREE / THE INTERPRETATION OF IONOSPHERIC PHENOMENA

LIST OF PLATES

PREFACE

If our eyes were radio rather than optical wide-band detectors it is well known that for us the brightest object in the sky would still be the Sun; that planets, stars and the Milky Way would still shine feebly (and that we would still occasionally be blinded by man-made sources). What is less well known is that quite a different earthbound overcast would hover about us, with its climatic zones, its seasonal changes, its unpredictable storms and scintillating transparence. To be sure, we can get a sort of glimpse of this peculiar type of weather when we tune our receiver to radio broadcasting from some remote spot, or photograph the Earth from space at certain specific wavelengths. Nevertheless no one has ever looked at the ionized shroud of the Earth without the help of sophisticated apparatus, and this is one of the reasons why in this domain the *phenomena* are not easily abstracted from the use of specific *techniques.*

For generations, the study of the ionosphere has been deeply interwoven with the practice of radio communication and detection. Today however, ionospheric physics is best thought of as a branch of space physics; that part of physics which deals with processes at work in the solar system and methods developed for its exploration. For some years, the interest of ionospherists has shifted: it rests no longer in the 'electronosphere' – that of radio engineers – as in the coupling between the neutral atmosphere on one side and the magnetosphere on the other, within the larger frame of solar-terrestrial relationships. In a sense the ionosphere, which has seen its role as 'tracer' of the properties of the air and the electromagnetic field at very high altitudes become more and more valuable at the same time as its importance as a privileged support of long range communications declined, has in the process somewhat lost its identity, but not its interest. In fact, ionospheric techniques and phenomena have become *bonafide* methods to study quantitatively such otherwise elusive beings as neutral atmospheric composition; tidal and gravity waves; and magnetospheric convection or precipitation patterns.

This evolution has affected the point of view we have adopted in this work. In contrast to earlier writers on the subject, we have not emphasized the rich heritage concerning the morphology of electron concentration but stress instead the more recent data e.g. from incoherent scatter radars and mass spectrometers, which have 'put the *ion* back in *ion*osphere'.

The backbone of this book evolved from lectures delivered at the University of Paris and at the International School of Atmospheric Physics (Erice, Sicily). The material was prepared for a French edition called 'Physique de l'ionosphère' (Presses Universitaires de France, 1975). An early translation was made by Dr. Paul Koch.

While the French version was being completed and the English version prepared in manuscript form, S. J. Bauer's excellent introduction to the 'Physics of Planetary

Ionospheres' (Springer Verlag, 1973) appeared, and P. Banks and G. Kockarts' encyclopedic reference opus magnum, 'Aeronomy' (Academic Press, 1973) whose second volume is almost entirely devoted to ionospheric processes. Although each followed a different path on not quite overlapping territory, these authors obviously had much of the same ideas in mind as we had, and naturally fed on the same literature; so we pondered for a while whether their work did not render ours superfluous. We finally felt that, in addition to witnessing the lasting vigor of ionospheric research (one third of the referenced articles are post-1970), the publication in English of an extended and more up to date version of our monograph was nevertheless justified by our particular emphasis on the dual theoretical and experimental approach to the subject matters.

In fact the title and the plan, distinguishing between techniques and phenomena, are meant to underline this approach. This is in no way to be construed as putting up any barrier between experiment and theory; on the contrary, the spirit which guided us implies that both theory and experiment are prerequisites for an understanding of both the techniques used and the phenomena observed. It is our deep conviction that if a lasting value is to emerge from the otherwise frightening hold of science on human affairs, it lies in this sobering dialectics between theory and practice. Whether we have succeeded in conveying more of the experience on which this conviction is based is for the reader to judge.

In preparing a work of this type, it is inevitable to glean, from here and there in the literature, facts and presentations whose original authors may not all have been acknowledged individually.

Therefore let us avow our debt to the international collectivity of our colleagues and ask, at the same time, the indulgence of those who shall criticize the manner in which their speciality is treated. We must thank especially our friends at the Centre National d'Etudes des Télécommunications (C.N.E.T.) and Centre National de la Recherche Scientifique (C.N.R.S.) who have played a role in the inspiration, writing and editing of this book, in particular: François Du Castel with whom we formed the plan; Michel Blanc, Rudolf Burke, Roger Gendrin, Owen Storey, Jacques Testud, Philippe Waldteufel, who have contributed to some sections; Nicole Adane who has typed in French and in English most of the chapters; E. Jamin, whose careful reading of the proofs was extremely helpful; and the craftsmen who have done the 'art' work.

Issy les Moulineaux

HISTORICAL INTRODUCTION

As Max Born once remarked, 'an active scientist has no time to read about the history of science'. The authors of this book are no exception, and the following notes are second or third hand. Interested readers can refer to Chapman's 'Historical Introduction to Aurora and Magnetic Storms' (1968), and to 'Fifty years of the Ionosphere' edited by Ratcliffe (1974), where fairly complete bibliographies of the original articles are compiled.

The men whose portraits appear in these historical notes, Gauss, Marconi and Fabry, are not particularly famous for their work in ionospheric research. Yet they have indirectly made outstanding contributions to ionospheric science. They were selected to illustrate the three roots (and the three routes) of speculative and practical interest in the ionosphere before our generation of space explorers (cf. Waynick, 1974):

– the study of terrestrial magnetism,
– the development of radio communications, and
– the analysis of interactions between light and gases.

Had we prefered instead to honor the memory of physicists who have devoted most of their life work to the ionosphere, there is no doubt that the best choices would have been Edward Appleton (1872–1965) and Sydney Chapman (1888–1970).

1. Geomagnetism and Aurorae

During a long first phase in the history of our science, up to the end of the 19th century, the objects of study were the fugitive and indirect effects of the vaguely hinted existence of the ionosphere itself. Luminous phenomena in the arctic sky were called 'Aurora Borealis' by Gassendi in 1623, of which very ancient accounts were kept, and slow variations of the terrestrial magnetic field, noticed by Gellibrand in 1634, awakened the curiosity of scholars and suggested all sorts of hypotheses.

It was a clockmaker of London, Graham, who in 1722 observed for the first time microscopic fluctuations in the direction of compass needles. Hiortier and Celsius, nearly 20 yrs later, in Uppsala, made the great discovery that some of these fluctuations accompanied displays of the Aurora, seemingly confirming Halley's 1716 hypothesis that these were due to 'magnetic effluvia'.

Humboldt in 1808 coined the term 'magnetic storms' to describe these sporadic phenomena. Under his leadership, a systematic study of geomagnetism was laid out by Gauss, Weber, and later Airy, who verified that the superposed components of the variations observed, including the regular solar quiet diurnal variation, were world-wide effects due to external causes.

In spite of the discovery of the 11 yr. solar cycle by Schwabe (1844), and that of the relation between the frequency of magnetic storms and the number of sunspots by Sabine (1852) and Wolf (1852), (cf. Figure 0), it will take some fifty more years to correlate the 27 days period of recurrence of magnetic storms to that of the Sun's rotation.

Although the first coherent ideas on the existence of an ionosphere are generally

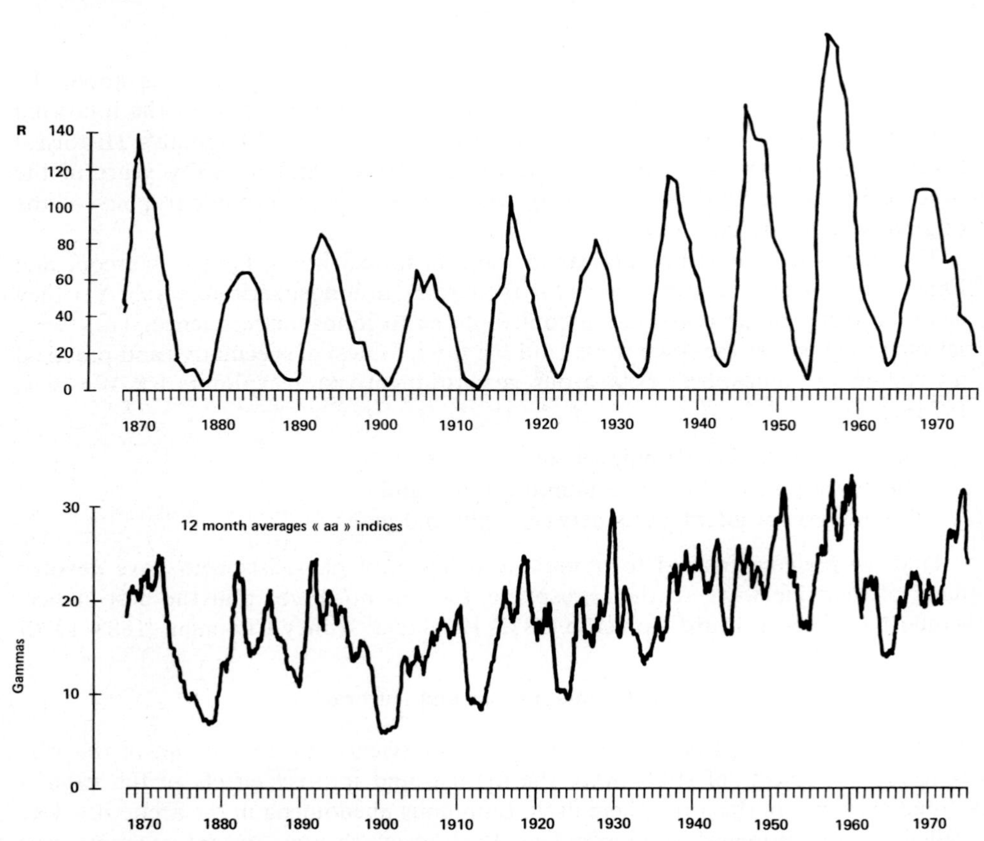

Fig. 0. A century of solar and geomagnetic activity (Courtesy of P.-N. Mayaud). The Wolf number R is $k\,(10\ g + f)$. where f is the total number of observed sunspots, g is the number of spot groups, and k is a normalizing factor to bring the counts of different observatories into agreement. The aa index is the average of the amplitude in γ (10^{-5} G), over three-hour intervals, of the irregular variations of the horizontal components of the Earth's magnetic field measured at two antipodal sites (Europe and Australia) with about 50 degrees latitude. A nearly perfectly regular 11 years cycle is the dominant feature, but its amplitude is modulated by a longer term variation where a 90 years cycle can be detected (Waldmeier, 1966). Eddy (1976) has reviewed convincing evidence that within the past, solar and geomagnetic activity have been considerably attenuated and probably enhanced with respect to their present level. In particular, the last prolonged calm (Maunder's minimum) which apparently occurred between 1645 and 1715 during the reign of Louis XIV, might have caused the concurrent 'Little Ice Age' (Le Roy Ladurie, 1967). It may also explain why a long time elapsed between the telescopic discovery of sunspots by Galileo and Scheiner and the revelation of their now obvious cyclic behavior, and why the scientific curiosity for geomagnetic and auroral phenomena, which had started in the first half of the XVIIth century, did not resume until the beginning of the XVIIIth century.

Karl Friedrich Gauss (1777–1855) on the terrace of the Göttingen observatory. The prince of mathematicians' fascination for metrical fields, and his virtuosity with their spherical harmonics expression put him at the heart of the first international venture in geophysics: a network of geomagnetic recording stations. He designed a precision magnetometer with Wilhelm Weber and introduced electromagnetic measure into the first absolute unit system, in recognition for which their names were later given to magnetic and electric induction in the CGS and MKSA systems, respectively. In his 'General Theory of Terrestrial Magnetism', published in Leipzig in 1839, after having demonstrated from the network data that the source of the regular and irregular components of the Earth's magnetic field variations was external, Gauss discussed the possibility that they might be due to airborne 'galvanic currents'. He mentioned as support for this hypothesis 'the enigmatic phenomena of the Aurora Borealis, in which there is every appearance that electricity in motion performs a principal part' (cf. Green, 1946). (Lithography from the Stadtmuseum in Göttingen)

attributed to Stewart, whose theory that the daily oscillation of the compass needle was due to a system of electric currents circulating in the upper atmosphere appeared in the Encyclopedia Britannica in 1878, it is remarkable that Gauss in 1839, and Lord Kelvin in 1860 had already conceived similar theories.

Meanwhile, the arctic regions had been explored and the oval zone surrounding the northern geomagnetic pole, wherein Aurorae occur with the greatest frequency, delimited (Loomis, 1860; Fritz, 1874). Cook in 1773, and other occidental travelers since then, had reported on the existence of a similar Auroral zone in the austral hemisphere. The green line in the Auroral light was observed spectroscopically by Angström in 1867; the mystery of its origin has led to one of the famous long lasting wrong hypotheses of science: the existence of an unknown gas in the upper atmosphere, 'geokoronium'. By 1881, Goldstein had reached the conclusion that the agents of field aligned Auroral rays were corpuscular and not electromagnetic.

The first international polar year, in 1882–1883, witnessing the coordinated effort which, from then on, the scientific community applied to these problems, marked the end of this first historical phase.

2. The Birth of Ionospheric Science

The turn of the century saw the beginning of a very active period of some thirty years, marked by the work of eminent scientific personalities who, for the first time, came to specialize in these problems. Stimulated by spectacular progress in the experimental field, their research was able in a short time to lay down a solid physical foundation for the phenomena, which appeared more and more to be bound up with the action of the Sun on the terrestrial environment. The agents of this action, ultraviolet light and 'cathode rays', nevertheless remained hypothetical as they could not be observed through the screen of the atmosphere.

Before 1900 Schuster developed the 'dynamo theory' of the diurnal variation of the magnetic field, attributing the driving force of the electric currents proposed by Stewart to the tidal motions of the atmosphere, theorized since the days of Laplace.

In 1901 Birkeland measured the altitude of the Aurora, and defined the Auroral 'electrojet' responsible for the stormtime variations of the geomagnetic field at high latitudes.

The same year Marconi transmitted a radio signal over the Atlantic, a feat which was not in agreement with the very exact mathematical theory of the diffraction of electromagnetic waves by the Earth's surface. Kenelly at Harvard in the U.S.A., Heaviside in Great Britain, and Nagaoka in Japan, independently furnished, if in vague terms, the correct explanation of the success of Marconi's demonstration, positing a permanent conductive layer in the rarefied air above. This layer took on the name 'Kenelly-Heaviside layer'. Kenelly, in his original paper, seems to have foreseen the future of ionospheric physics: 'As soon as long distance wireless waves come under the sway of accurate measurements, we may hope to find, from the observed attenuations, data for computing the electrical conditions of the upper atmosphere'.

Eccles in 1912, then Larmor (1924a, b), Appleton (1925), Nichols and Schelleng (1925a, b), Hartree (1929), elaborated the theory of the propagation of radio-

With Hertz's spark gap transmitter, Popov's aerial and Branly's detector, Guglielmo Marconi, at the age of 22 in Rome, had made the first radio telegraph over a distance of 100 m. Four years later, working in Great Britain, he had progressively increased the range to the point where he boldly attempted to surmount the 'wall of ocean 160 km high' between Europe and America. He is seen here on the left directing on Signal Hill (Newfoundland) the launching of the kite antenna with which he hoped to receive the message transmitted from Poldhu (Cornwall). Ratcliffe (1974) gave a modern account of the experiment. He reports: 'The signals were heard on headphones under very difficult conditions, and there have even been suggestions that Marconi deceived himself in thinking that he had heard them'. Whether or not he did, he said he had heard them, and this turned out to be the right thing to say. What Marconi had in mind was the dream of trans-continental communications, and for the dream to become reality it took exactly what is required to hear a weak signal drowned in a large noise: an unordinary amount of faith! Marconi, who received the Nobel prize in 1909, was never much interested in the scientific developments of radio soundings. Just as most people, as long as the ionosphere worked, he did not otherwise bother about its existence. (Courtesy of Marconi Co., Chelmsford)

frequency waves in ionized gases. The experiments which indirectly measured the height of the reflecting layer were developed by De Forest (1913) and Fuller (1915) in the U.S.A., and by Appleton and Barnett (1924a, b) in Great Britain (interference between the 'ground' and 'sky' waves). In 1925 Tuve and Breit published the results of the first direct measurement by pulse sounding. This method, giving target distance from echo delay, in which 'radar' (Radio Detection And Ranging) first appeared, was the forerunner of the network of 'ionosondes' which for thirty years became practically the only source of all the data used for the description of the structure and forecasting of the variations of the ionosphere, the necessity for which became progressively clearer as radiocommunications developed.

The idea that the ionosphere was produced by the ionizing action of the ultraviolet radiation from the Sun on the upper atmosphere was extremely current after it had been proposed by Taylor, and Fleming, but it was Hulburt in 1928 and Chapman in 1931 who were the first to formulate it quantitatively. Because the theory required a very energetic agent to ionize the atmosphere down to 60 mile heights, and because the solar far ultraviolet radiation was underestimated, Chapman however thought that corpuscular ionization was also involved.

The correlation between magnetic storms and the presence of large spots on the Sun, as shown by their 27 days period of recurrence, had been definitely proved by Chree in 1922. Chapman, with Ferraro, starting in 1930, discussed the detailed consequences of the earlier hypotheses of Birkeland and Störmer, which had been improved by the criticism of Schuster (1911) and Lindemann (1919), on the role of clouds of solar corpuscular radiation on magnetic and auroral activity. They inferred the existence of a geomagnetic 'cavity' and of a 'ring current' circling the Earth during disturbed conditions.

The second International Polar Year (1932–1933), fifty years after the first, was coincident with the installation of permanent arctic geophysical observatories (Sodankyla, Trömso).

By the time of the second world war, which gave a strong impetus to radar and radio communication technology, the overall structure and the variations of the ionospheric 'layers' and 'currents' had been already well described, but remained, in general, poorly understood.

3. Solar-Terrestrial Physics

Following the observation during the 1933 solar eclipse in Canada that the *E* region was under direct solar control, Appleton and Naismith had established in 1935 the basic theory of ionospheric formation by solar X.U.V. radiation. But the understanding of the charge budget processes remained confused, and it was not before 1947 that Bates and Massey did away with the then popular electron attachment and ion-ion recombination mechanisms, suggesting ion-molecule and dissociative recombination schemes as the main operating channels for chemical equilibrium in the ionosphere. The role of plasma drifts in the *F* region behaviour was pointed out by Martyn in 1947, who also showed, with Baker (1953), and independently Maeda, that ionospheric conductivities were adequate to sustain the 'dynamo' variations of the geomagnetic field.

The name of Charles Fabry (1867–1945) is one of a famous lineage of French opticians from Fresnel to Kastler. His most remarkable achievements were in the field of interferometry, and his students paved the way for the invention of the laser and Fourier transform spectroscopy. His interest in photometry led him to pioneer work in aeronomy: he proved the existence of the ozone layer and of the airglow. This work was pursued by Vassy, Dufay, Cabannes, Barbier and the Service d'Aéronomie of the Centre National de la Recherche Scientifique. In a 1928 study on the scattering of light by ionized gases, Fabry hit upon the understanding that the scattered radiation was bearing information on the electron temperature, not going so far as suggesting this as a new remote sounding technique for observing the upper atmosphere; the implementation of the method was beyond technological capacity. By the time it became feasible, Fabry's ideas had been forgotten, and had to be rediscovered by Gordon thirty years later to turn into one of the most fruitful experimental innovations in the field of external geophysics. (Courtesy of Madame Coppens, Institut d'Optique, Orsay)

While the fundamentals of the main ionospheric regions phenomena described by ionosondes and magnetometers were thus well within grasp, the stage was set for new experimental developments to open the lower and higher regions of the ionosphere to investigation. The study of whistler propagation by Storey (1952), of partial reflections by Gardner and Pawsley (1953), of wave interactions by Fejer (1955) were the first successful endeavours in this domain. At about the same time, the discoveries by Barbier, in France and Africa, of the latitudinal structure of the red airglow (transequatorial arcs and subauroral arcs), and the work of Akasofu and Chapman in Alaska on auroral morphology, were renewing interest in the geomagnetic control of aeronomical phenomena.

The International Geophysical Year (1957–1958), 25 years after the Second International Polar Year, coincided with the installation of permanent antarctic observatories, with the beginning of the operation on a large scale of spatial means of observation (following the launching of the first artificial satellite in October 1957), and with the implementation of a new and powerful means of observation from the ground: incoherent scatter sounding (Gordon, 1958; Bowles 1958, 1961). It inaugurated the era of maturity for external geophysics. An unprecedented number of researchers, experimentalists and theoreticians, benefiting from an exceptional financial effort by scientific policy makers, rapidly came to make the upper atmosphere one of the most studied and best known of the natural media, and to discover and then explore the vast domain of permanent interaction between the terrestrial magnetic field and the solar corona, named the 'magnetosphere' by Gold in 1959, and which had been previewed by the work of Alfvén, Van Allen, and Virnov among others.

A fundamental contribution to these researches, owing to equipment carried by rockets or satellites, has been the direct measurement of the spectrum of the ionizing energy, photonic or corpuscular, not observable from the ground but postulated for a long time. Another aspect has evidently been the considerable progress made in the description and comprehension of the structure and variations of the neutral atmosphere and geomagnetic field. In addition, modern ionospheric research has furnished the motivation for much laboratory work, needed as well for the interpretation of the observations as for the establishment of quantitative models of the medium.

By 1969 (moon landing), the topside of the ionosphere had been mapped with its transequatorial arch, its midlatitude trough; the concept of the plasmasphere and of the polar wind had been formulated; the composition, the temperature and the motions of ionospheric plasma had been surveyed from the bottom to the top and on the way to being elucidated (cf. Rishbeth and Gariott, 1969).

By that same year, the fervor and favor for big science in general, and space science in particular, had already reached its apogee. The outward look that had been directed to the cosmos was being reflected back as an anxious sight on our own planet. The key words were becoming ecology, environment, and pollution. For ionospherists this was not a farfetched turnabout: whether from a cosmic vantage, or from a more humble human point of view, the ionosphere is a frontier of the terrestrial world, and its study, standing halfway between meteorology and astrophysics, has earned a deserved recognition as a cross-road of scientific research.

PART ONE

THE IONOSPHERIC ENVIRONMENT

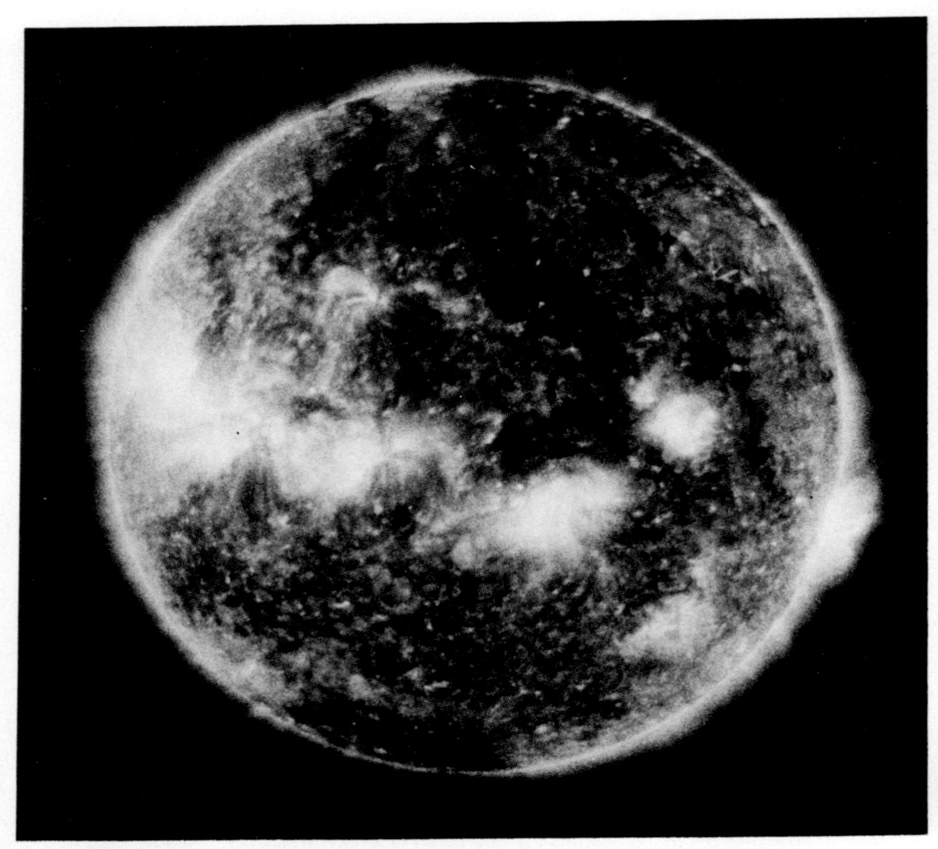

This image of the far UV Sun was recorded onboard a rocket payload on January 15, 1974, with broad band filters letting in the wavelength range 150 to 600Å absorbed in and responsible for the main ionospheric layers of the Earth. Most of the emission originates in highly ionized atoms Mg^{9+}, Mg^{10+}, Si^{11+}, Fe^{13+}, Fe^{14+} and Fe^{15+} which are produced in collisions at temperatures of 1 to 2.5 million degrees K in the hottest and densest portions of the star's corona, concentrated at the feet of 'loop' structures, above active regions, connecting areas of opposite magnetic polarity. It can be seen that there are no regions of uniform coronal emission, a strong indicator of the essential variability of terrestrial UV environment. (Courtesy of G. Brueckner, U.S. Naval Research Laboratory)

CHAPTER I

THE ATMOSPHERE AND THE VERTICAL STRUCTURE OF THE IONOSPHERIC PLASMA

While the origin and bulk composition of the Earth's atmosphere have been traced to crustal outgassing (N_2, H_2O, CO_2), nuclear radioactive decay (Ar, He), and biophotosynthesis (O_2) over the past several billion years, its structure is essentially governed by the pull of gravity on the one hand, and the action of sunlight on the other hand. The former would have the gases sink to the planet's surface, layering according to weight. The latter would have them heated away into space. The resulting equilibrium 'barometric' stratification of the atmosphere has been more or less understood since the XVIIIth century (Pascal, Bernoulli, Dalton) but it remained for our century to explore it fully and to analyze the complex photochemical and dynamical processes responsible for the composition, the temperature and the motions of upper air.

1.1. Hydrostatic Equilibrium in the Earth's Gravitational Field

1.1.1. TERRESTRIAL GRAVITY

The terrestrial gravitational force per unit mass or *gravitational acceleration*, can be considered with good approximation to be directed towards the center of the Earth and to have a magnitude $g' = GM/R^2$, where R is the distance from the Earth's center, M the total mass of the solid Earth, and G the universal gravitational constant.

Expressed in terms of the local altitude z, the Earth's radius R_0, and its sea level value g'_0, this gravitational acceleration becomes

$$g' = g'_0/(1 + z/R_0)^2 \tag{I.1}$$

Because a reference frame spinning with the Earth is not an inertial system of reference, the force felt in such an earthbound frame is the sum of the above radial gravitational force, and of a centrifugal force directed perpendicularly to the Earth's axis of rotation (Figure I.1). The resulting apparent gravitational acceleration, or

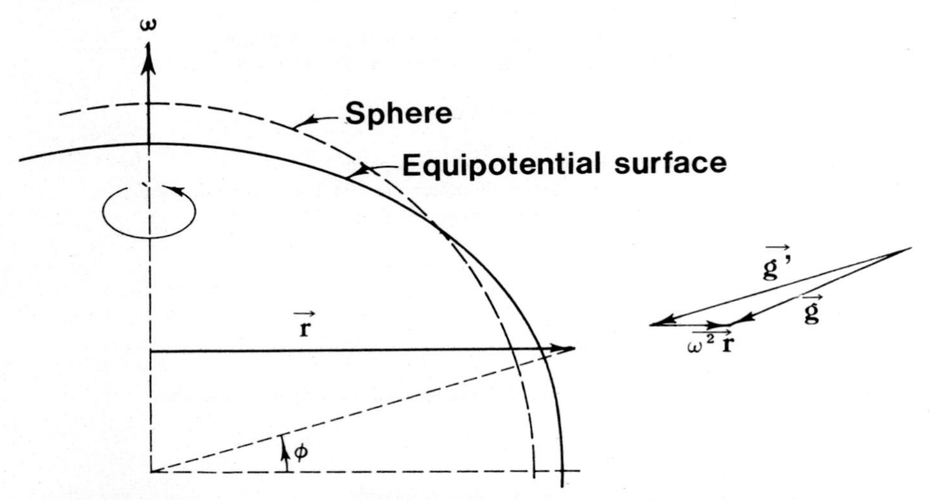

Fig. I.1. In a reference frame spinning with the Earth, the apparent gravitational acceleration or *gravity* g is the sum of the spherically symmetric gravitational acceleration g and of the cylindrically symmetrical centrifugal acceleration $\omega^2\vec{r}$. These two forces become equal and opposite—and hence gravity vanishes within this frame—6.6 R_0 away from the Earth's center in the equatorial plane (geostationary orbit).

gravity g, is no longer directed exactly towards the center of the Earth, except at the geographical poles and equator, as the Earth's flattening witnesses. Neglecting however the small deviation between the vertical (plumb line) and the radial (Earth's center) directions, it is easily computed that

$$g = g' - \Omega^2 (R_0 + z) \cos^2\varphi \tag{I.2}$$

where Ω is the angular velocity of the Earth's rotation in inertial (stellar) space, and φ is the geographical latitude.

With $R_0 = 6356.9$ km at the pole ($\varphi = 90°$), this gives $g_0 = 9.832$ m s^{-2}, and with $R_0 = 6378.4$ km at the equator ($\varphi = 0°$), $g_0 = 9.78$ m s^{-2}, for the extreme sea level ($z = 0$) values. The centrifugal term, which eventually equilibrates gravitational acceleration 6.6 R_0 away from the Earth's center at the equator (the 'geostationary' altitude), only becomes significant at a height of several hundred km, so that expression (I.2) above is too exact for most atmospheric applications.

If we are satisfied with a precision of about 1%, up to 400 km altitude we can neglect the centrifugal term and use Equation (I.1) as for a non-rotating atmosphere. Furthermore, we can replace Equation (I.1) with the linear approximation

$$g = g_0 (1 - 2z/R_0) \tag{I.3}$$

using for g_0 a mean sea level value of 9.81 m s^{-2} and for R_0 a mean radius of 6370 km. Finally, when considering an atmospheric layer of thickness smaller than a few tens of km, we can use a constant mean value for g throughout the layer's depth.

1.1.2. THE BAROMETRIC DISTRIBUTION OF GASES

If ρ be the density (mass per unit volume) of air, the force of gravity on a unit volume of the atmosphere is ρg. The weight of the atmosphere above a horizontal unit surface at altitude z is therefore $\int_z^\infty \rho g \, dz$, which, under equilibrium static conditions exactly balances the gas pressure $p(z)$. For instance the mean sea level pressure over the whole Earth is 1013 mbar (1.013×10^6 dyn cm^{-2}), corresponding to a total mass for the atmosphere of about 5×10^{19} kg, a figure first computed by Pascal in the first half of the XVIIth century.

Air pressure thus diminishes with altitude as the weight of the overlying atmosphere, a statement best expressed by the derived equation (the 'barometric' equation of Bernoulli)

$$-\frac{\partial p}{\partial z} = \rho g. \tag{I.4}$$

In words: pressure decreases upwards at a rate proportional to air density. Air however being a compressible fluid, its density is itself decreasing with pressure. According to the perfect gas state equation, density is in fact proportional to pressure, the coefficient of proportionality being itself proportional to the air mean molecular mass m and inversely proportional to the absolute temperature T:

$$p = nkT, \ \rho = nm = \sum_i n_i m_i$$

where n_i is the molecular numerical density, or *concentration*, per unit volume, of the different components of air with molecular mass m_i, and k is Boltzmann's constant.

The barometric equation can therefore be explicited as

$$\frac{\partial p}{p}\left(= \frac{\partial n}{n} + \frac{\partial T}{T} = \frac{\partial \rho}{\rho} + \frac{\partial T}{T} - \frac{\partial m}{m} \right) = -\frac{\partial z}{H} \tag{I.5}$$

with $H = kT/mg$.

The parameter H, which lumps the local gravitational and thermodynamical characteristics of a given atmospheric volume, has the dimension of length, and is known as the *scale height*. It can receive several useful interpretations:

—If we assume air to be incompressible, we can identify $H(z)$, from $p = \rho g H$, as the height above z an overlying homogeneous atmosphere would have to extend to exert the pressure $p(z)$. For instance, at the lowermost level, with $m_0 = 4.85 \times 10^{-23}$ g and $T = 273°$K, the scale height is computed to be approximately 8 km. This means that while a mercury column of 760 mm height, or a water column of 10 m height are required, it would take an S.T.P. air column almost as high as Mount Everest to equilibrate atmospheric pressure.

– From $mgH = kT$, we can relate $H(z)$ to the height above z where an air molecule has increased its gravitational potential energy by an amount of the order of its thermal kinetic energy at z. Equivalently we can relate $H(z)$ to the distance that would reach on the average air molecules leaving in vacuo the altitude z vertically upwards with their thermal speed. Indeed, as the air rarefies with increasing altitude, there comes a level where the molecule mean free path becomes so long that such a situation becomes realistic (the so-called 'exobase', situated around 600 km), and the scale height there is some measure of the actual extension of the atmosphere. Because as we shall see the very high atmosphere or 'exosphere' is hot and composed of light gases, this extension is quite large, resulting in the *geocorona*, a quasispherical halo of primarily H atoms on ballistic paths about the Earth. (It must be noted however that the 'hottest' H atoms are shot out on hyperbolic orbits, as their thermal speed exceeds the escape velocity $(2GM/R)^{1/2} = 11.2 \times (R_0/R)^{1/2}$ km s^{-1}, so that the atmosphere in a way never ends but to some extent evaporates into space.)

– Finally, from the solution to Equation (I.5)

$$p(z) = p(z_0) \exp \{ (z_0 - z)/H \} \tag{I.6}$$

we can identify H with the altitude interval over which the pressure actually decreases by 1/e, provided H be constant over that interval. For instance, air pressure at the top of Mount Everest, slightly over a scale height up, is about one-third of sea level pressure.

The assumption H = const. can be a useful approximation over limited altitude ranges, but in general since air temperature and composition, as well as gravity, are altitude dependent, the exact altitude dependence of air pressure and density, must be computed from the exact solution to Equation (I.5):

$$\frac{p(z)}{p(z_0)} = \frac{n(z)}{n(z_0)} \frac{T(z)}{T(z_0)} = \exp\left\{ - \int_{z_0}^{z} dz/H(z)\right\} \tag{I.7}$$

requiring knowledge of the altitude dependence of the scale height, that is essentially, of the air temperature and mean molecular mass.

However, since air is a mixture of gases, the total pressure p is in fact the sum of the partial pressures $p_i = n_i kT$ exerted by each constituent. Dalton's law, stating that under equilibrium conditions these partial pressures are the same as the pressures that would be exerted by each constituent if it were alone, should allow the barometric equation to be integrated separately for each constituent, giving

$$\frac{p_i(z)}{p_i(z_0)} = \frac{n_i(z)}{n_i(z_0)} \frac{T(z)}{T(z_0)} = \exp\left\{ - \int_{z_0}^{z} dz/H_i(z)\right\}. \tag{I.8}$$

Equation (I.8) implies that the lighter a constituent, the less its partial pressure decreases with altitude: the mean molecular mass should diminish as the atmosphere gets depleted in heavy constituents at the same time as it rarefies. Boyle, who discovered the compound nature of air, could not understand why 'the azotic gas, being lighter, should not form a distinct stratum on top of the oxygen gas'. That such a situation, called 'gravitational separation' indeed takes place, although it had long been predicted (Maris, 1928), was not confirmed until rocket-borne mass spectrometers were launched into the upper atmosphere in the late 50's: gravitational separation does not occur below a level, called the 'turbopause', situated around 100 km altitude, because up to this height turbulence efficiently mixes the different gases, forcing them to a common scale height, and resulting in a constant mean molecular mass very nearly equal to the sea level value m_0.

The vertical structure of the atmosphere is thus divided into two regions separated by the turbopause (Figure I.2): a lower region, sometimes called the 'turbosphere' or

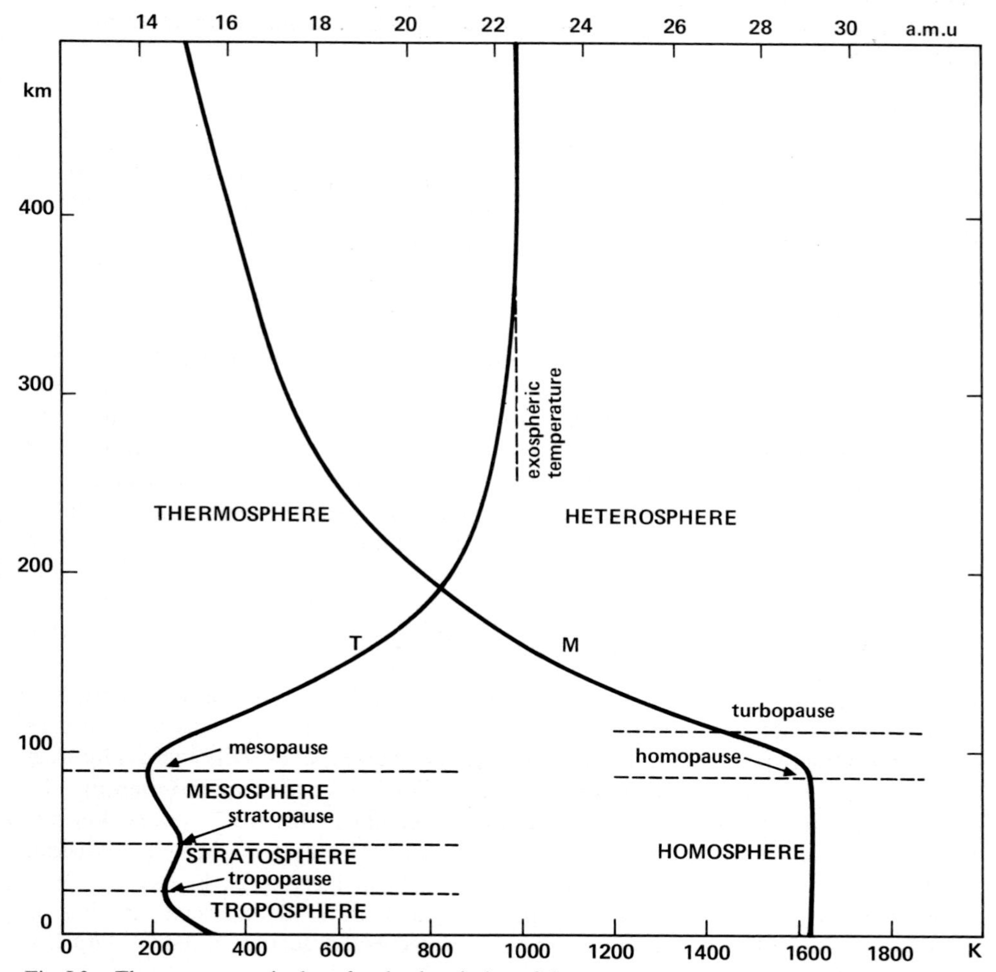

Fig. I.2. The current terminology for the description of the upper atmospheric layers is based either on the temperature (T in deg K) or the mean molecular mass (M in atomic mass units) vertical profiles.

the 'homosphere', where hydrostatic equilibrium of gases is 'mixing equilibrium' as defined by Equation (I.7) with $H(z) = kT(z)/m_0 g(z_0)$; and an upper region, sometimes called the 'diffusosphere' or 'heterosphere', where hydrostatic equilibrium of gases is 'diffusive equilibrium' as defined by Equation (I.8) with $H_i(z) = kT(z)/m_i g(z)$. To be precise we should note that there is a difference in definition, and level, between turbopause and 'homopause', connected with the small decrease in air mean molecular mass due to dissociation of O_2 below the onset of gravitational separation.

As we shall see, the mean temperature throughout the homosphere is about 250°K, with the result that the mean scale height is such that the pressure decreases by a factor 10 about every 16 km up to 96 km, where it is down to 10^{-6} atm., with a corresponding mean free path of 5 cm; from there, the pressure diminishes increasingly slowly under the effect of both decreasing molecular mass and increasing temperature up to the 'exobase', which can be defined as the level where the mean free path exceeds the scale height, and the notions of thermodynamic temperature and pressure break down.

1.1.3. DEPARTURE FROM HYDROSTATIC EQUILIBRIUM

Hydrostatic equilibrium is a good approximation to the vertical distribution of air because the overwhelming forces it experiences are gravity and vertical pressure gradients. Hydrostatic equilibrium in fact should describe exactly the distribution of the 'permanent' gases N_2, O_2, Ar, and CO_2, neither created nor destroyed significantly within the atmosphere, nor escaping at appreciable rates. But it need not apply to those constituents with large sources or sinks, whose proportion may then drastically differ from mixing or diffusive equilibrium. This is the case for H_2O, whose condensation or sublimation point is easily reached, and whose partial pressure then cannot exceed the saturation pressure for equilibrium between its liquid or solid phase on condensation nuclei. It is the case for most gases due or subject to solar induced fast photochemical processes, in particular oxygen allotropes O_3 and O. It is again the case for the lighter gases, H and He, which escape to space.

For those constituents which are not distributed according to hydrostatic equilibrium, the effect of turbulent mixing and molecular diffusion is to transport them with a vertical flux tending to establish their hydrostatic distribution, given by:

$$\varphi_i = -(D_i + K)\left(\frac{dn_i}{dz} + \frac{n_i}{T}\frac{dT}{dz}\right) - n_i\left(\frac{D_i}{H_i} + \frac{K}{H}\right). \tag{I.9}$$

In this expression (cf. Colegrove *et al.*, 1965), φ_i is defined as positive when upwards, D_i is the molecular diffusion coefficient of the given constituent within the others (roughly inversely proportional to air density, and thus exponentially increasing with altitude), and K is a phenomenological 'eddy diffusion' coefficient representing the role of turbulent mixing processes acting equally on all constituents, a poorly known, presumably weakly dependent function of altitude. The role of thermal diffusion, which intervenes only for very light gases, has been neglected (cf. Kockarts, 1963).

We can verify that under equilibrium conditions $\varphi_i = 0$ represents mixing or diffusive equilibrium according to whether molecular diffusion or eddy diffusion dominates.

When $D_i \ll K$ (turbosphere) Equation (I.9) leads to Equation (I.7); for instance, in an isothermal layer, $dn_i/dz = -n_i/H$. When $K \ll D_i$ (diffusosphere) Equation (I.9) leads to Equation (I.8); for an isothermal layer $dn_i/dz = -n_i/H_i$. The turbopause can thus be seen to be the level where $D_i = K$. Note that this can lead to different turbopause levels for the different constituents with different molecular diffusion coefficients. The observed turbopause altitudes of about 100 km, occurring where the D_i have decreased to about 10^6 to 10^7 cm^2 s^{-1} imply that this is the order of magnitude of the eddy diffusion coefficient in the transition region. Note also that changes in the strength of the mixing there could be expected to result in changes in the diffusosphere base level.

Since the D_i keep increasing exponentially with height, the higher the altitude above the turbopause, the smaller the departure from diffusive equilibrium necessary to compensate for a source or sink. Diffusive equilibrium should thus be a good approximation to the permanent distribution of any constituent from some altitude up in the diffusosphere. On the contrary, hydrostatic equilibrium, whether diffusive or mixed, need not be a good approximation of the distribution of photo-sensitive constituents produced and destroyed in the turbosphere or lower diffusosphere, like for instance O. Their concentration at a given altitude results from their budget or continuity equations:

$$\frac{\partial n_i}{\partial t} = q_i - l_i - \frac{d\varphi_i}{dz} \tag{I.10}$$

where φ_i is given by Equation (I.9) and q_i and l_i represent the local source and (concentration dependent) sink terms that must be explicited from the photochemistry (next section). At the limit, in the source region where the production and recombination rates are highest, their concentration profile at equilibrium is determined by the equality $q_i = l_i$ with div φ_i negligible, a situation known as 'chemical equilibrium'.

1.2. The Interaction Between the Ultraviolet Solar Spectrum and the Atmosphere

By far the greatest part of solar radiant energy originates in the Sun's light emitting shell or 'photosphere'. The bulk of its emission approximating a 6000°K black body spectrum peaking in yellow green light, is largely incapable of exciting electronic transitions in the free gaseous molecules of the atmosphere, which is thus quite transparent to it. Most of the solar energy intercepted by the Earth (about 10^6 erg cm^{-2} s^{-1}) reaches the lowermost level, where it is absorbed by the surface to the extent that it is not reflected back into space. It is therefore the ground and seas that constitute the main heat sources for the bulk of the atmosphere: air temperature cools as one goes up, a common knowledge evidence (Figure I.2).

This bottomside heating, through radiative transfer, turbulent transport, and the water evaporation-condensation cycle, is confined to a layer some 10 km thick, the 'turning' layer or *troposphere*, representing some three-fourths of the total air mass, whose polyatomic constituents (mainly H_2O and CO_2) are able at the resulting equilibrium temperature profile to radiate away via their IR band spectra, through the overlying thin atmosphere, the totality of the energy convected from below.

It became known during the first quarter of the XXth century that air does not cool indefinitely with distance from the surface. Experiments with sound propagation (Whipple), and later the first systematic high altitude balloon measurements (Tesseirenc de Bort) suggested the existence of a (local) second heat source above a temperature minimum (*tropopause*) regularly found between 10 and 20 km altitude (Ramanathan, 1929). The behavior of the temperature profile above this 'cold shell of air' remained, however, poorly known until rocket-borne pressure measurements in the late 40's, although Lindemann and Dobson's (1923) work on visible meteor trails, and Humphrey's (1933) ice crystals interpretation of the high-latitude summer night-time sunlit ('noctilucent') clouds, had identified a second deep temperature minimum or cold shell (the *mesopause*) at 80–85 km, implying a third heat source above.

In the same way, upper air departure from sea level composition remained crudely estimated long after Fowler and Strutt (Lord Rayleigh IV) had in 1917 shown the role of an O_3 layer in cutting off the solar spectrum at 2900Å, McLennan and Schrun's (1924) identification of the green line of O (geokoronium!) in the light from the night sky, and the 1925 proof of the Kenelly-Heaviside ion layer hypothesis.

Although, it was quite clear that higher atmospheric heat and reactive constituents were of photochemical origin, the detailed unraveling of these aeronomical problems had to wait for the space era, if only because the photolytic solar radiation being absorbed in the process involved could only be observed at altitudes exceeding balloon capability.

1.2.1. THE HIGH ENERGY TAIL OF THE SOLAR SPECTRUM AND SOLAR ACTIVITY INDICES

In the pioneer years of aeronomy, it was customary to estimate the Sun's invisible radiation by extrapolating to shorter wavelengths the visible 6000°K Planck continuum, although it was suspected that the Sun was radiating more strongly in the far UV below 1000 Å, and that a line emission spectrum corresponding to the principal series of H(1215, 1226, . . . 912), He (584, 537, . . . 506) and He^+ (304, 256, . . . 228 Å) would be present and brightening up during solar 'flares'.

As direct experimental knowledge of the high energy tail of the solar spectrum accumulated with space borne instrumentation from 1946 on, first with optical spectroscopy in the near UV (Baum, *et al.*, 1946), then with counters in the X-ray domain (Friedman, *et al.*, 1951), and finally with photoelectric techniques in the intermediate 100 to 1000Å range (Hinterreger, 1965), the picture which emerged was one of far greater complexity and variability.

Naturally, this work entailed a significant progress in the understanding of the structure of the emitting regions in the Sun's outer envelopes, and of the absorbing layers in the Earth's upper atmosphere as well.

In short, as one considers shorter and shorter wavelengths (Figure I.3), the solar spectrum is observed to cool from the visible photosphere's 6000°K temperature to about 4500°K as the radiation originates from higher and colder levels in the Sun's radiatively heated lower atmosphere or 'chromosphere', until one reaches the far UV range below about 1800Å where the character of the spectrum changes to that of numerous discrete emission lines originating from excited and ionized atoms in the mechanically heated hot upper atmosphere or 'corona' of the Sun. The number and intensity of the far UV and X-ray emitting coronal active regions is related to sunspots below and other visible manifestations of solar activity, and exhibit, like all

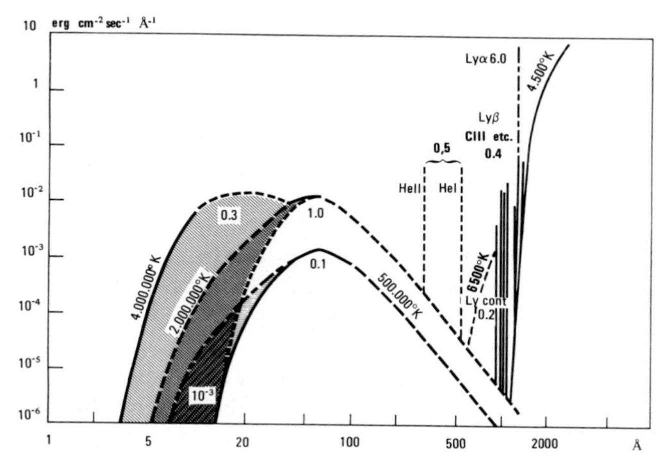

Fig. I.3 Schematic representation of the solar UV spectrum. The figures are estimates in erg cm^{-2} s^{-1} of the energy flux within the main contributions identified here, emission lines and continua. The temperatures shown are those of the emitting regions. The cross hatched areas give an idea of the range of X-rays variability.

these manifestations, irregular as well as periodic variations. A 27 day cycle corresponds to solar rotation (cf. Figure I.6), and an 11 yr cycle (cf. Figure 0) to the waxing and waning of global solar activity.

As a continuous spaceborne watch of the XUV spectrum is not available, indices have been defined which hopefully monitor these fluctuations from ground based observations, like the old Wolf or Zurich number from counting sunspots, the visible singly ionized Ca emission line Ca II, or the radio Sun noise level between 3 and 30 cm wavelength, originating from active regions like the far UV spectrum. The radioflux at 10.7 cm (2830 MHz) is the most generally used index for aeronomical purposes, called S when expressed in units of 10^{-22} W m^{-2} Hz^{-1}. S can vary from minima of about 70 to maxima in excess of 250. It should be clear however that a single index cannot yield too faithful an image of the complexity exhibited by the variable Sun.

1.2.2. ATMOSPHERIC PHOTOCHEMISTRY

1.2.2.1. *The Absorption of UV Solar Photons.*

Photolysis occurs the deeper into the atmosphere the thinner the absorbing gas and the smaller the efficiency of the corresponding photolytic process.

The vertical profile of the rate of photolysis must be calculated in terms of the transmission of the photolysing radiation (determined by the absorptivity of the atmosphere and the zenith angle of the Sun χ) and the yield of the relevant photolytic process. At a given altitude, and for a given wavelength, the significant parameters are:

- the intensity of radiation $I\,(z, \lambda)$ (photons cm^{-2} s^{-1}) transmitted from above,
- the concentration $n_i(z)(cm^{-3})$ of the neutral molecules undergoing photolysis,
- the cross section $\sigma_i(\lambda)$ (cm^2) for photolysis of these molecules.

The frequency of photon molecule interactions is σ I (s^{-1}) and the rate of photolysis is $q(z, \lambda) = n_i(z)\, \sigma\, (\lambda)\, I(z, \lambda)$ ($cm^{-3}\ s^{-1}$).

The intensity of radiation $I(z,\lambda)$ is obtained with the help of the Beer-Lambert law (the intensity decrease is proportional to the number of absorbing molecular along the optical path): $-\ dI = \sec \chi\, n(z)\, \sigma_a dz$ where $\sigma_a(\lambda)$ is the absorption cross sections of air of molecular concentration $n(z)$. Integration of this expression gives:

$$I(z,\lambda) = I_\infty\, e^{-\tau(z,\lambda)} \tag{I.11}$$

where I_∞ is the intensity of radiation at the exterior of the atmosphere, and τ is the optical thickness of the atmosphere above the altitude under consideration

$$\tau(z,\lambda) = \sec \chi\, \sigma_a(\lambda) \textstyle\int_z^\infty n(z)\, dz. \tag{I.12}$$

This neglects the curvature of the Earth and is not exact for $\chi > 80°$, i.e. near sunrise and sunset, for which case sec χ must be replaced by the Chapman function Ch(χ), which has been tabulated (Wilkes, 1954), and can be approximated (Smith and Smith, 1972).

Qualitatively, the situation is simple. Because of the exponential increase of the atmospheric density with depth along the trajectory of the incident photons, in the region of weak absorption ($\tau < 1$, the 'optically thin' part of the atmosphere for the radiation considered) the photolysis rate is proportional to the concentration of the photolysable constituent. But further down, where the absorption becomes strong ($\tau > 1$, the 'optically thick' part of the atmosphere for the radiation considered) the photolysis rate decreases rapidly, along with the photon flux. Therefore, maximum photolysis occurs near the altitude where $\tau = 1$.

Note that since $\displaystyle\int_z^\infty n(z)\, dz = n(z)\, H(z)$, the altitude of unit optical depth is determined by the relation $n(z)\, H(z) = \cos \chi / \sigma_a$.

There are two main mechanisms that entail large absorption rates, responsible for the existence of two overlapping and interacting photochemical layer systems:

– the first and higher operating mechanism is photoionization of all the main atmospheric gases, with cross sections of order 10^{-18} cm^2, thus peaking where $nH \simeq 10^{18}$ cm^{-2}, that is in the lower thermosphere. It requires photons of energy greater than about 10 eV, and thus absorbs the highly variable extreme UV and X radiation averaging a few ergs $cm^{-2}\ s^{-1}$. This gives rise to the existence of the ionosphere.

– the second mechanism is photodissociation of molecular gases, mainly O_2, with cross sections of the order 10^{-20} cm^2, peaking in the upper mesosphere. It requires about 5 eV; and thus absorbs wavelengths shorter than about 2400 Å totalling a few tens of ergs $cm^{-2}\ s^{-1}$. This gives rise essentially to the O layer. Because in the so-called Herzberg continuum (2026 to 2424 Å) the photodissociation cross section of O_2 drops to 10^{-23} cm^2, the radiation in this range can penetrate deep into the stratosphere, where it leads in fact to the formation of an O_3 layer, because at these denser levels the equilibrium $O + O_2 \leftrightharpoons O_3$ is heavily shifted to the right. Photodissociation of O_3, which proceeds most efficiently in the Hartley region of the spectrum, between about 2000 and 3000 Å, absorbs about 2000 ergs $cm^{-2}\ s^{-1}$.

Note that except for O_3, which is both produced and destroyed photochemically,

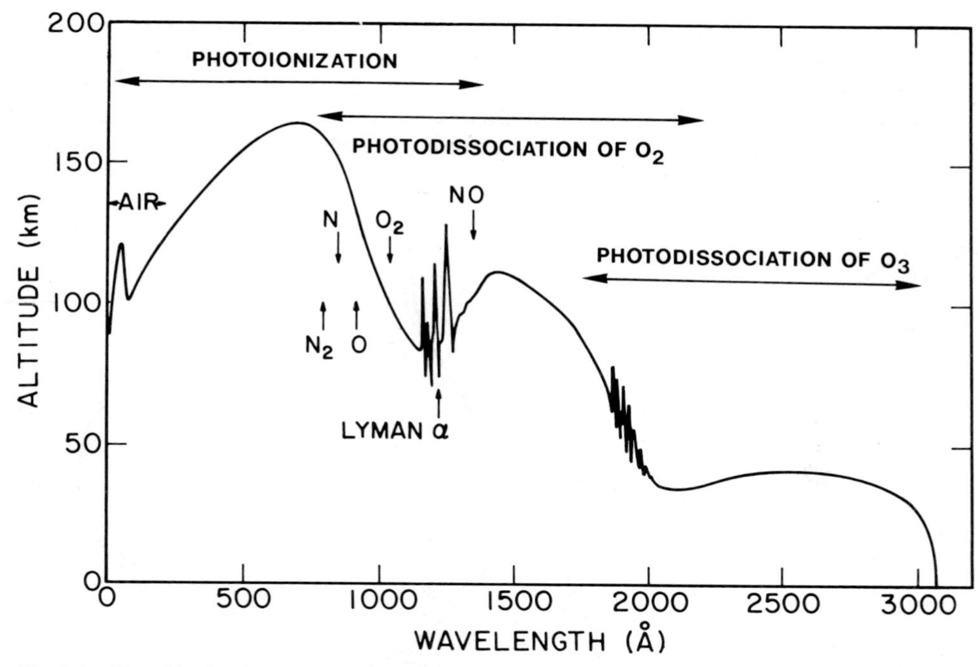

Fig. I.4. The altitude of unit optical depth for an overhead sun reflects essentially the varying cross section of the main absorbing process. It is the level where the maximum energy is dissipated at each wavelength. The arrows indicate the threshold wavelengths for photoionization of the indicated species.

only a small fraction of the main absorbers, O_2 for instance, is affected by these processes, and their hydrostatic distribution is hardly disturbed.

Taking into account the distribution with depth of the absorbers, and the variation with wavelength of the absorption coefficients, it is possible to estimate in a detailed way the penetration spectrum of solar radiation. Figure I.4 indicates the unit optical depth level, where the intensity is reduced to $1/e$ of its free space value for an overhead Sun.

1.2.2.2. *The Photochemical Gases*

While the above rough description accounts for the main features of UV transmission through the upper atmosphere, it must be realized that the number of primary or secondary photochemical reactions initiated by absorption processes is almost unlimited. Models which track all sources and sinks of any particular species, radical or excited states, can become exceedingly complicated. And yet, if we leave aside for the moment the ionized component which we shall study at length in Chapter VI, because of the difficulty in observing minor constituents within the abundant N_2, O_2 background, even after numerous recent investigations both ground-based and *in situ*, our present knowledge about the chemical composition of the photochemical layers between the tropopause and the mesopause (the so-called 'chemosphere') remains scanty, and largely dependent on models based on estimates of the different terms in continuity equations like (I.10).

While early theoretical models neglected all constituents except O_2, and were addressed primarily to a first approximation of O, O_2, O_3 equilibria, later work included the fuller treatment of the so-called oxygen-hydrogen atmosphere then the oxygen-hydrogen-nitrogen atmosphere (Hesstvedt, 1968). A serious difficulty is the lack of precise determination of the eddy diffusion coefficient profile intervening in the estimation of the transport terms.

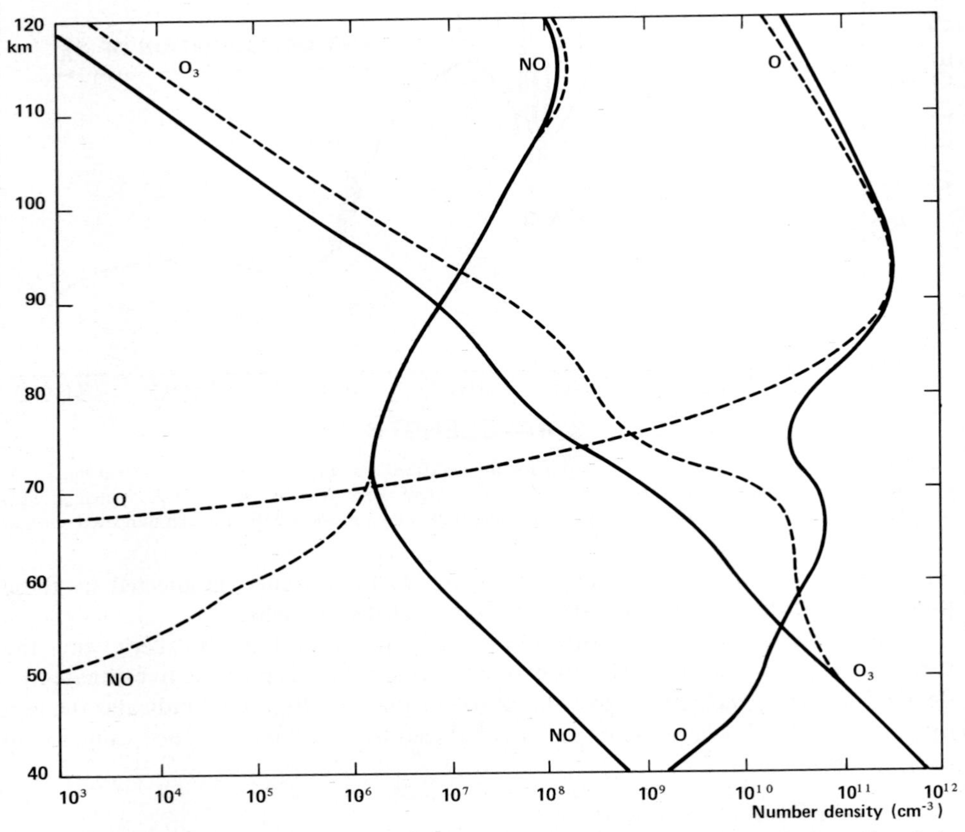

Fig. I.5. A model for the midday (full curves) and midnight (broken curves) vertical profiles of the concentration of the minor neutral constituents O, O_3 and NO in the mesosphere and lower thermosphere. They are the main photolytic gases of interest for ionospheric chemistry. O and O_3 result from the photodissociation of O_2 followed essentially by $O + O_2 \rightarrow O_3$. NO results essentially from the reaction $N(^2D) + O_2 \rightarrow NO + O$, the excited N atoms originating in the photodissociation of N_2 and the dissociative recombination of NO^+ ions with electrons (Ogawa and Shimazaki, 1975).

Figure I.5 shows a tentative model of the main photolytic constituents of interest or ionospheric chemistry, establishing their overall qualitative distribution with height. It is hardly necessary to emphasize the essentially variable character of their proportions ('mixing ratios'), as a consequence of the varying intensity and geometry of solar UV irradiation, and of the probably everchanging strength of turbulent transport (Banks and Kockarts, 1973).

1.2.3. TEMPERATURE AND DYNAMICS OF THE UPPER ATMOSPHERE

A continuity equation can be written for heat just as for the concentration of photochemically produced constituents (I.10). The local source and sink terms express direct (excess energy from photodissociation and photoionization) and indirect (excess energy from chemical recombinations) solar heating, and IR radiative cooling; the transport term includes both molecular diffusion (heat conduction) and turbulent mixing (convection). Just as for photolytic gases, which have a tendency to be in chemical equilibrium in the strongly absorbing homosphere, and to reach diffusive equilibrium in the thinning diffusosphere, so the effect of heat transport is small in the warm upper stratosphere and lower mesosphere which tend to be in radiative equilibrium, but dominates in the thermosphere.

The lack of precise determination of the thermal eddy diffusion profile is again a major obstacle to a theoretical description in the transition region, but the difficulty of the problem is somewhat alleviated by the possibility of inferring precise experimental temperature profiles from rocket pressure or density measurements, since the mean molecular-mass is rather well known. These profiles have indicated that the minimum at the mesopause can vary from as low as 120°K in summer to about 200°K in winter, a mean value compatible with the value of the eddy diffusion coefficient there (Johnson, 1956).

Above the turbopause, the role of convective transport all but ceases, and since the only remaining diatomic and monoatomic gases are poor radiators, as a first approximation it can be assumed that all the heat absorbed must be conducted downwards. The heat conductivity equation becomes

$$\tilde{Q} = \tilde{D}\frac{\partial T}{\partial z}$$

where $\tilde{Q}$ is the total heat input above the altitude considered, and $\tilde{D}$ the local heat conduction coefficient. As $\tilde{Q}$ decreases exponentially with altitude and $\tilde{D}$ is itself proportional to some power of the temperature, this leads to a rapidly diminishing temperature gradient with increasing altitude. There results, as Johnson (1956) showed, that the atmosphere must be nearly isothermal above about 250 km. Because the temperature profile is 'anchored' at the nearly constant mesopause minimum the temperature of the outer isothermal layer, or 'exospheric temperature' (Figure I.6), undergoes huge variations following the variations of Q impressed by the diurnal rotation of the Earth and the intrinsic fluctuations of solar XUV emission (for the treatment of the time dependent heat balance equation, cf. Harris and Priester, 1962).

1.2.3.1 *The Variations of Exospheric Temperature and Thermospheric Models*

As we shall see, direct information on the temperature within the thermosphere has come from recent ionospheric studies (Figure I.6a) but indirect data for the variations of exospheric temperature have been obtained regularly since 1958 from the decay of satellite orbits, an effect of the varying air drag as the thermosphere thermally expands and contracts (Figure I.6b). Under the assumption of diffusive equilibrium (Equation I.8) and with fixed base level conditions, air composition remains

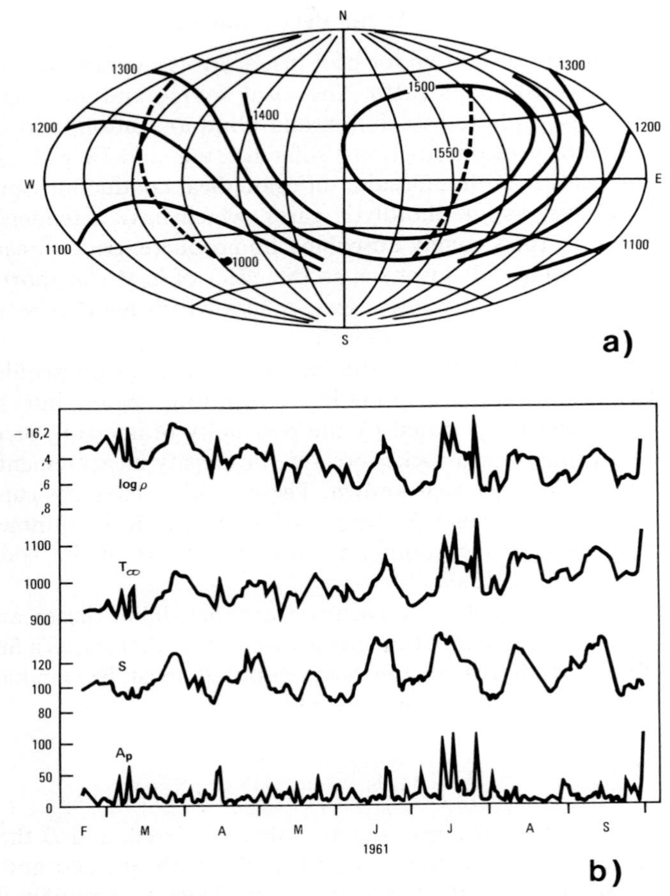

Fig. I.6. (*a*) A map of exospheric temperature on the dayside of the Earth for the June solstice and average solar activity conditions.
(*b*) Day to day variations in air density (ρ in g cm^{-3}) within the thermosphere from the observed drag on the EXPLORER 9 satellite. Interpretation in terms of the variations of exospheric temperature T_∞ (deg. K). Comparison with the indices S of solar activity and A_p of geomagnetic activity. Note that over several months, the dominant variation is the 27 day solar rotation period.

constant at a given pressure as the thermosphere 'breathes'. According to such a model air density in the diffusosphere is then only dependent on the vertical temperature profile; as the main parameter of the temperature profile is the exospheric temperature T_∞, a nearly one to one correspondence can be established between density at any altitude and T_∞.

Harmonic analysis of the satellite perigee drag data revealed six different components of thermospheric density variations, and thus presumably of exospheric temperature (Roemer, 1971):

– the diurnal component,
– a semi-annual component,
– an annual (seasonal/latitudinal) component,

– a 27 day component,
– an 11 yr component, and
– a geomagnetic activity component.

The diurnal variation, the semi-annual variation and the annual variation, which essentially depend on the geometry of solar illumination, can be expressed as empirical functions of time, date, and geographical location.

The 27 day (solar rotation) and 11 yr (solar cycle) variations are best expressed empirically in terms of the solar activity index S.

The geomagnetic activity variation can be expressed in terms of an index of the type to be discussed in the next chapter, and accounts for the input of energy by corpuscular precipitations and the dissipation of ionospheric currents in the auroral zones.

Fitting amplitude and phase coefficients to these components of exospheric temperature variations with well chosen base level conditions (for instance, a mean of mass spectrometric rocket borne measurements in the lower thermosphere, Von Zahn, 1970) form the basis for thermospheric time dependent static 'models' (CIRA, 1975) which as tables or computer subroutines yielding local instantaneous vertical profiles for temperature, density and composition extrapolate in a practical heuristic manner the large amount of data from satellite orbits decay observations (Figure I.7).

Naturally, the assumptions made in such models about the lower boundary conditions for the diffusosphere (for instance a uniform composition and temperature at all times over the whole Earth) are somewhat arbitrary, and lead to distortions in the

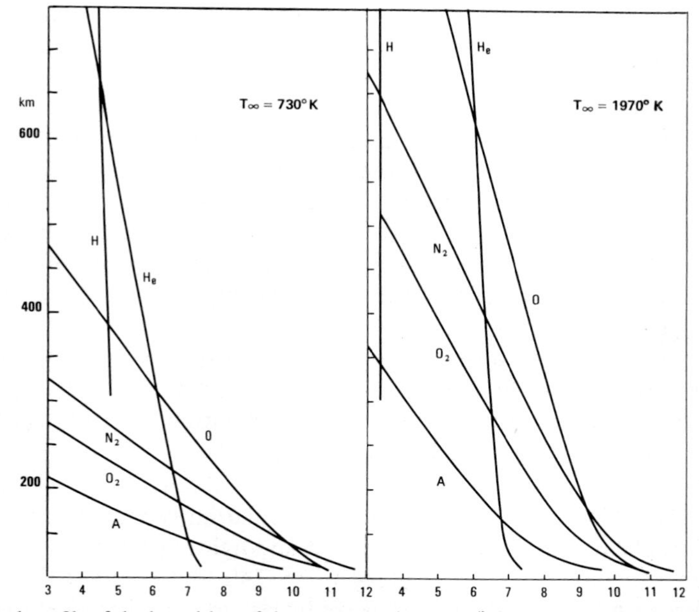

Fig. I.7. Vertical profile of the logarithm of the concentration (cm^{-3}) of the different gases in the thermosphere, for two extreme values of exospheric temperature, under the assumption of diffusive equilibrium. In contrast to other gases, whose concentration at the diffusosphere base level of 120 km is assumed to be fixed, H becomes less abundant when the temperature increases, because of thermal escape.

calculated world wide thermospheric temperature or composition distribution when fitted to the actual density data.

1.2.3.2. *The Spectrum of Upper Atmospheric Motions*

One dimensional static models at any rate provide at best a convenient framework for approximating thermospheric conditions, because the atmosphere never reaches equilibrium under the ever-changing illumination geometry and energy input conditions: adiabatic redistribution of heat and latitudinal and vertical transports are set up which the symmetry of the globe constrains to definite modes. The proper theoretical model should include the full hydrodynamic and thermodynamic coupled equations for a rotating spherical shell, with the proper excitation correctly described, an exceedingly complicated problem (Dickinson, 1972; Volland, 1969a, b).

Actually, the great variety of transport phenomena that occur in the upper atmosphere in response to global or local departure from thermodynamic and hydrostatic equilibrium occupies such a wide range of space and time scales that it can be divided into categories which, although not entirely distinguishable from one another, are the object of quite distinct observational methods and theoretical treatments.

1.2.3.2.1. *Seasonal Circulation.* In the winter hemisphere, throughout the polar night, there is no photodissociation of O_2, and yet no deficiency of O in the thermosphere is observed; indeed, there is a relative excess of O there as the polar N_2 atmosphere cools, lowering its scale height (Hedin, *et al.*, 1973). An even stronger 'winter bulge' exists for He, an effect so large that it was detected from satellite drag studies (Keating and Prior, 1968; Keating, *et al.*, 1970) before it was confirmed by mass spectrometric measurements (Reber, *et al.*, 1971; Von Zahn, *et al.*, 1973).

Johnson and Gottlieb (1969) have proposed a scheme whereby the cross equatorial winds produce an inflow to the polar region of different atmospheric gases above the turbopause which is proportional to the scale height of the constituents, a sevenfold factor from N_2 to He. The return flow in the homosphere is not dependent on scale height and as a result the lighter gases tend to accumulate and build up in concentration in the region of inflow. Due to the opposite deviation of its atomic mass from the mean molecular mass of thermospheric air, Ar presents a corresponding summer bulge (Von Zahn, *et al.*, 1973).

This global differential transport from the warm expanded summer thermosphere to the cool contracted winter thermosphere, together with the variations of temperature, can probably account for the semi-annual variation in thermospheric densities observed in mid-latitudes (Alcayde, *et al.*, 1974).

1.2.3.2.2. *Diurnal Circulation.* The seasonal circulation is embedded within a much stronger diurnal circulation, transferring heat and lighter gases from the hotter (sub-solar region as it goes round the globe every 24h and responsible for the phase difference between the diurnal density wave (maximum at 1400 local time) and temperature wave (maximum at 1700) above a given spot on Earth (Roemer, 1971). These diurnal winds can be tentatively approximated from the pressure gradients described by the empirical drag fitted thermospheric models (Kohl and King, 1967;

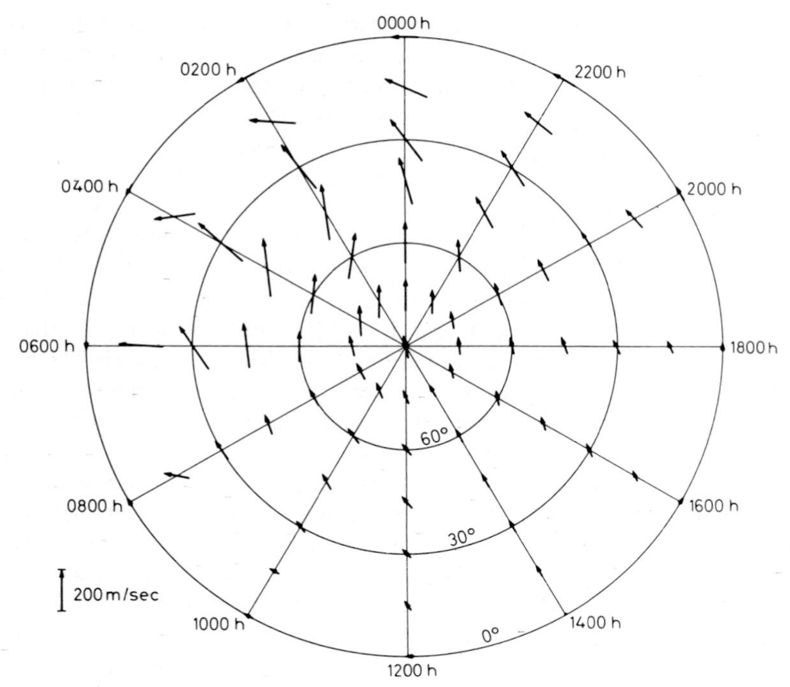

Fig. I.8. A model of the diurnal component of thermospheric circulation over an hemisphere, inferred from static models of pressure distribution for equinox and medium solar activity conditions.

Geisler, 1966, 1967), due account being taken of the Coriolis force and 'ion drag' (cf. Chapter VII). Figure I.8 exemplifies a calculation from such models, which could well bear a small likeness to the actual wind field, especially in the disturbed polar regions, as minor errors on the pressure values may well induce very large errors in the pressure gradients (Amayenc, 1975).

1.2.3.2.3. *Geomagnetic Disturbances.* As mentioned before, there is a substantial increase in thermospheric temperature at high latitudes during geomagnetically disturbed conditions (Blamont and Luton, 1972), sufficient to drive equatorward winds that reverse the average circulation and result in large perturbations of thermospheric composition (Von Zahn, 1975).

1.2.3.2.4. *Tides and Internal Gravity Waves.* In contrast to the preceding components of thermospheric motions, driven by energy absorbed within the thermosphere, an upward energy flux is propagated to the thermosphere from the denser troposphere, stratosphere and mesosphere, where pressure oscillations are generated by a variety of mechanisms.

The atmosphere being generally stably stratified, departures from vertical hydrostatic equilibrium are opposed by restoring forces. Being in neutral equilibrium with respect to horizontal displacements, it can thus sustain propagating oblique waves. The conservation of energy requires that the induced air motions increase exponentially with altitude, that is with rarefying air density, until they can be damped. The damping processes are non-linear interactions between waves leading to turbulent degeneration of the motion, 'ion drag' when air has to carry charged particles across the magnetic field, heat conduction and viscosity when the mean free path becomes comparable to the wavelength. These processes are not efficient below thermospheric levels (Volland and Mayr, 1972).

A particularly important case of atmospheric oscillations is the 'tides' excited by solar heating of the stratosphere (O_3 absorption of near UV radiation). This is a thick source traveling around the globe every 24h, and the resulting waves are constrained to obey closure relations in the vertical, zonal and meridional directions, with the result that the dominant mode is semi-diurnal below 120 to 130 km (Chapman and Lindzen, 1970; Bernard 1974). The diurnal winds in the thermosphere can receive a similar treatment (Volland and Mayr, 1973). However, the corresponding waves are evanescent and their excitation is mainly due to local absorption of the EUV flux; this is nothing more than an alternative description of the diurnal circulation.

Atmospheric waves, called 'internal gravity waves' (to distinguish them from surface waves) can be excited by air flow across mountain ranges, meteorological fronts, explosions (volcanoes; bombs), auroral phenomena, and propagate as free waves (Hines, 1972). Such waves have periods ranging from a few minutes to a few hours. In the *D* and *E* regions, their vertical scale size is typically 5 to 50 km and the corresponding horizontal scale size 10 to 300 km. Only a filtered fraction of the launched waves reaches the thermosphere. This filtering depends upon the general circulation and the background temperature distribution.

1.2.3.2.5. *Turbulence*. We have seen that small scale motions responsible for the phenomenological 'eddy diffusion' process are necessary to explain the vertical transport properties of the atmosphere below the diffusosphere. Such small scale motions have been experimentally detected through radar observations of meteor trails (Roper and Elford, 1963) and optical observations of the structure of artificial trails (Blamont and De Jaeger, 1961; Zimmermann, 1966; Justus, 1966; Blamont and Barat, 1967, 1968), in the 90 km region.

Turbulence is generated by winds when for one reason or another (strong shear, temperature gradient smaller than the superadiabatic lapse rate) instabilities can grow. The resulting vortices can themselves be unstable and break down into smaller eddies to the scale where viscosity damps the motion. Application of the Richardson and Reynolds criteria to atmospheric models permits to conclude qualitatively that the conditions for turbulence to appear can indeed be found in the mesosphere and lower thermosphere. It is difficult however to arrive at precise quantitative results, because of the inadequacy of classical turbulence theory. Questions like the variability of turbulence, or the altitude at which it eventually terminates are still controversial (Elford and Roper, 1967; Zimmerman *et al.*, 1973).

1.3. Stratifications of the Ionosphere

The existence of free electrical charges in the upper atmosphere was hypothesized early when overhead electric currents were hypothesized to explain the regular variations of the Earth's magnetic field. It became evident when radio waves were seen to bounce off the sky. It is an aeronomical necessity in the framework of solar terrestrial relationships.

The name 'ionosphere' is accepted to have two meanings: According to one, it refers to that part of the Earth's atmosphere which is appreciably ionized; according to the other, it relates to the ionized component itself. This distinction is not trivial, because it leads in turn to distinct ways of describing ionospheric stratification into 'layers' or 'regions' on top of one another, with fundamentally different characteristics and behavior, as shown in the next two sections.

1.3.1. THE IONOSPHERE AS A PHOTOCHEMICAL COMPONENT OF THE UPPER ATMOSPHERE

The neutral photochemical layers due to the dissociation of atmospheric gases, and the ion layers due to photoionization, exhibit an overall stratification with a bottomside, the 'lower ionosphere', akin to the chemosphere, near chemical equilibrium;

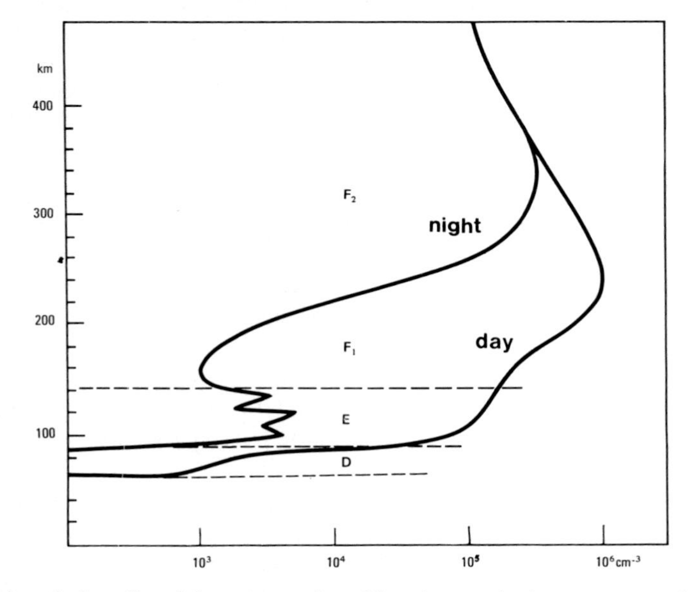

Fig. I.9. Typical vertical profiles of the concentration of free electrons in the upper atmosphere at middle latitudes, and terminology of ionospheric layers. Note the large diurnal variation in the D, E and F_1 layers.

and a topside the 'upper ionosphere', akin to the diffusosphere, near diffusive equilibrium (Figure I.9).

The 'peak' of the ionosphere, at some 300 km altitude, is much higher than the peaks of the photodissociated layers situated about or below 100 km, because photoionization has a larger cross section than photodissociation and so proceeds higher up. The ion-chemical processes are much faster than neutral chemical processes, so that they compete more efficiently with diffusion as explained in Chapter VI.

1.3.1.1. *Charge Neutrality*

Let us denote by n_e the electron concentration and by n_i the positive ion concentration per unit volume. Throughout the ionosphere, we have $n_e = n_i$ (below about 75 km, where negative ions, that is electrons bound to neutral molecules, are present, they must be counted with the electrons). This charge neutrality is maintained on a global scale, and locally in the underlying chemical equilibrium region, by the fact that ionization and recombination processes always affect positive ion-electron pairs. In the overlying diffusive equilibrium region, local charge neutrality is maintained by electrostatic scale height coupling between the electrons and ions gases ('ambipolar' diffusive equilibrium). We shall study this phenomenon in detail in Chapter VII, but it is useful to describe it briefly here: because of their very light mass, electrons tend to distribute themselves with a very large scale height; electrostatic attraction however requires that they must pull up the ions with them and, as a result, the equilibrium situation determining the neutral plasma scale height approximately corresponds to diffusive equilibrium for pseudo-neutral particles consisting of the ion-electron pairs with effective mass $(m_i + m_e)/2 \simeq m_i/2$. The upper ionosphere, with an effective scale height twice as great as its parent neutral gas, thus in effect 'floats' on top of the atmosphere, just as the dissociated O and H layers float on top of their molecular parent gases.

1.3.1.2. *Composition*

Early radio soundings, which first described the bottomside ionosphere, were quite sensitive to vertical gradients in the concentration of free electrons, and mistakenly pictured it as a superposition of distinct strata which were named, after Appleton's notebook, the *D*, *E*, and *F* layers. This division turned out however to be quite meaningful in the light of modern composition studies bearing on the nature of the positive ions. It does describe rather accurately the overlapping strata of dominant chemical species which make up the ion population of the ionosphere (Figure I.10):

– The *D* region (60 to 90 km) corresponds to a sparse layer of polyatomic ion 'clusters' (10^2 to 10^4 cm^{-3}).

– The *E* region (90 to 140 km) corresponds to a moderately dense (10^3 to 10^5 cm^{-3}) layer of molecular ions NO^+ and $O_2{}^+$ in the midst of which fluctuate thin layers of metallic atomic ions occasionally peaking in the so-called 'sporadic *E*' (*Es*) phenomenon.

– The *F* region corresponds to a dense layer of atomic O^+ ions (10^5 to 10^6 cm^{-3}). One still finds a subdivision into the F_1 region, the transition between molecular and atomic ions, and the F_2 region, the 'peak' of O^+ ions.

A further stratum, out of reach before the 1950's is the thinning layer of H ions extending on top of the *F* region, best named the 'protonosphere'.

1.3.1.3. *Temperature*

Drukarev (1946) seems to have been the first in foreseeing that photoionization would produce an electron gas with excess energy, and that its temperature should greatly exceed that of the neutral gas when the rate of ion production is high. This important effect was not confirmed experimentally before the rocket flights of the

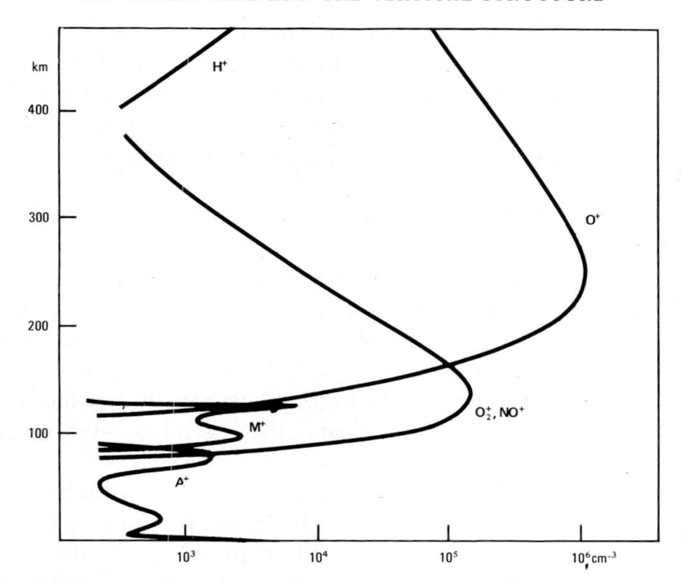

Fig. I.10. Typical vertical profiles of the concentration of the main atmospheric ions, during the day, at middle latitudes. M^+ are atomic metallic ion of meteoric origin, whose concentration can rise to over 10^6 cm^{-3} in sporadic *E* layers.

$A^{\pm}$ are polyatomic ion clusters of positive and negative charge. Omitted in the figure are species like He^+, $N_2{}^+$, N^+, O^{++} etc ... which are always minor components.

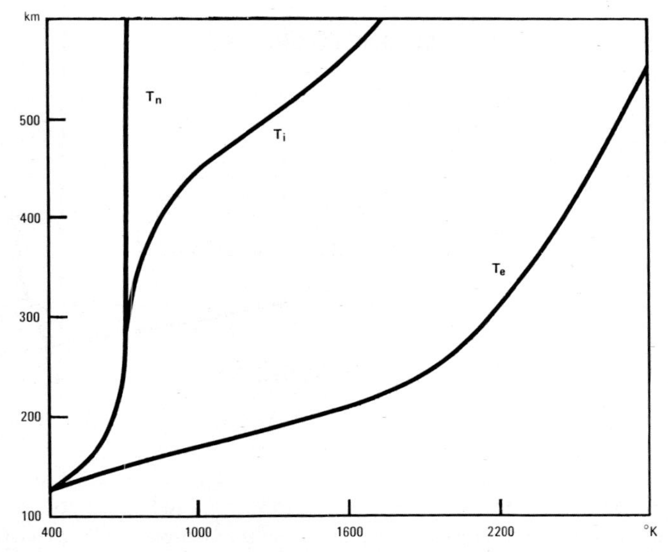

Fig. I.11. Typical vertical profiles of the temperatures of neutrals (T_n), of ions (T_i) and of electrons (T_e), in the daytime middle latitude ionosphere. At night, when photoionization stops, thermal equilibrium is restored, and T_e and T_i become equal to T_n.

1950's. Because the thermalization of the electron gas, and that of the ion gas, proceed much more rapidly than the mutual thermalization of the electrons and the ions, the situation is such that both electrons and ions belong to approximately thermalized populations, but at temperatures which need not be equal, nor equal to the neutral gas temperature (Figure I.11) as explained in Chapter VIII.

1.3.2. THE IONOSPHERE AS A WEAKLY IONIZED MAGNETOPLASMA

In the preceding section, it was shown that the ionosphere, considered as the ionized component of the atmosphere, could be described in parallel and in contrast to other atmospheric photochemical layers. Such a description however does not account for the plasma physical processes which make ionospheric phenomena so complex and interesting. For an alternate, and more complete description, it is necessary then to consider as 'the ionosphere', not just the ionized component, but the whole medium consisting of the charged particles embedded in the neutral gas and permeated by the magnetic field of the Earth.

The interactions between these can be defined with a small number of characteristic parameters for the trajectory of test particles or the response of the medium to a test perturbation. In this section we recall the expressions for these characteristic lengths and frequencies, whose numerical values as a function of altitude are given in Table I.1 and Figure I.12 on the basis of the data from Figures I.10 and I.11, that is of

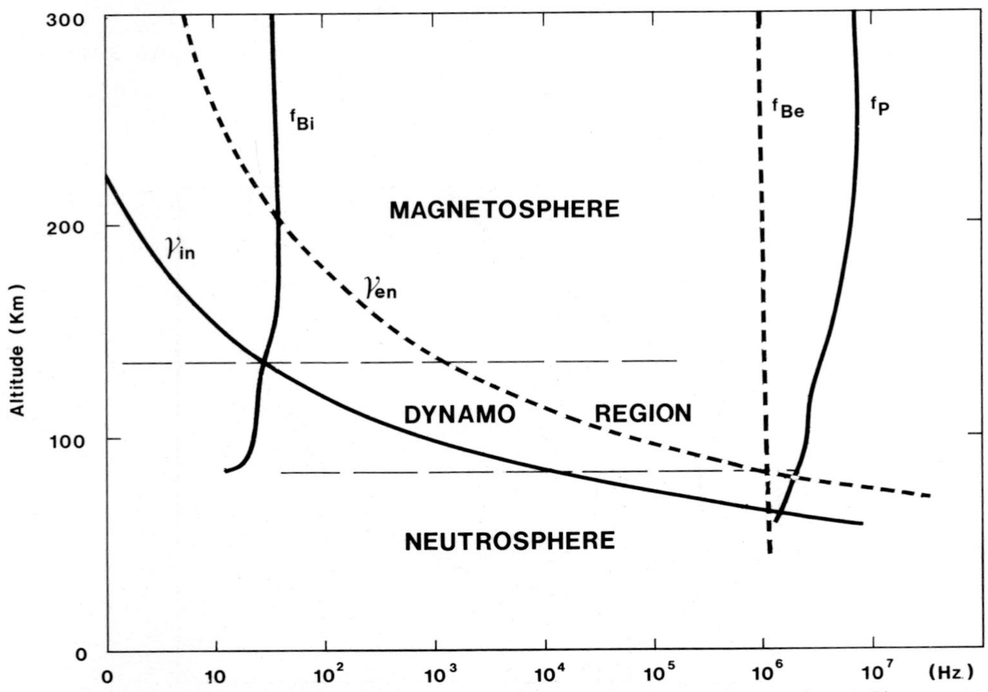

Fig. I.12. Vertical profile of the characteristic frequencies of the ionospheric plasma (cf. text). The exponential decrease with altitude of the collision frequencies with neutral molecules defines the lower lying collision dominated 'neutrosphere', where $\nu_{en} > f_{Be}$, $\nu_{in} > f_{Bi}$, and an overlying geomagnetically controlled region where $\nu_{en} < f_{Be}$. $\nu_{in} < f_{Bi}$.

the number densities (concentration) $n_e(z)$, $n_i(z)$ and $n_n(z)$, of the temperatures $T_e(z)$, $T_i(z)$, $T_n(z)$ (or equivalently, of the mean thermal velocities $v_e(z)$, $v_i(z)$, $v_n(z)$), and of the magnetic field intensity $B(z)$ (cf. next chapter).

TABLE I.1

Typical values for molecular neutral concentration n_n, degree of ionization n_e/n_n, thermal velocities of charged particles v_e and v_i, mean free path of electrons λ_e^n, Debye length λ_e^D and Larmor radii of electrons and ions λ_e^B, λ_i^B, in the ionosphere.

Altitude (km)	75	100	150	200	400	800	1 200	3 000
n_n (cm^{-3})...	10^{15}	10^{13}	5×10^{10}	8×10^9	10^8	10^6	2×10^5	10^4
n_e/n_n	10^{-12}	3×10^{-9}	4×10^{-6}	10^{-4}	4×10^{-3}	4×10^{-2}	10^{-1}	1
v_e (cm s^{-1})	10^7	10^7	2×10^7	2.5×10^7	3×10^7	3.5×10^7	3.7×10^7	4×10^7
v_i (cm s^{-1})	3×10^4	4×10^4	8×10^4	1.2×10^5	1.4×10^5	2×10^5	4×10^5	10^6
λ_e^n (cm)	3	3×10^2	6.4×10^4	4×10^5	3×10^7	3×10^9	1.6×10^{10}	3×10^{11}
λ_e^D (cm)	3	0.6	0.4	0.3	0.5	1.7	2.7	4.4
λ_e^B (cm)	1.2	1.2	2.5	3.3	4.3	6.4	8.0	16.2
λ_i^B (cm)	3.5×10^2	2.8×10^2	4×10^2	5.1×10^2	6.2×10^2	6.6×10^2	6.3×10^2	8.6×10^2

Firstly, there are characteristic parameters associated with 'Coulomb' ion-electron interactions. It is clear that with every deviation from neutrality, electrostatic restoring forces are set up with respect to which the charged particles act as harmonic oscillators. The characteristic frequency of electronic plasma oscillations will be shown to be:

$$f_e^p = \frac{1}{2\pi}\left(\frac{n_e e^2}{m_e \varepsilon_0}\right)^{1/2}$$

where e is the electron charge and ε_0 the permittivity of free space. This is called the (electron) *plasma frequency* (Figure I.12). The corresponding ion plasma frequencies $f_i^p = \frac{1}{2\pi}\left(\frac{n_i e^2}{m_i \varepsilon_0}\right)^{1/2}$ are much smaller, because of the much larger ion masses.

To the plasma frequency, there corresponds a characteristic length, proportional to v_e/f_e^p, and usually defined by:

$$\lambda_e^D = v_e/2\pi f_e^p = \left(\frac{\varepsilon_0 k T_e}{n_e e^2}\right)^{1/2} \qquad \text{(cf. Table I.1)}$$

which is called the *Debye length.* It is basically the distance covered by an electron during one cycle of a plasma oscillation, and represents the distance over which potential differences find themselves naturally 'shielded' by their effect on the charged particles distribution. This means in particular that fluctuations in electron concentration can exist independently of ions only at scales smaller than this Debye shielding length. Moreover, plasma oscillations can develop for wavelengths greater than the Debye length only.

Secondly, there are characteristic parameters associated with collisional charged particles-neutral particles interactions. One can think of them in terms of 'billiard balls' encounters, since the distance of approach necessary for an interaction to take place (of the order of magnitude of the radius r of the larger of the two particles) is much smaller than the mean interparticle distance $n_n{}^{1/3}$; note that up to 3000 km, there are many more neutral particles than charged particles (cf. Table I.1). Elastic cross sections $\sigma^{n}_{i,e} = \pi(r_{i,e} + r_n)^2$ can be defined, with approximate values given by taking $r_n = r_i = 10^{-8}$ cm, $r_e = 0$. The *mean free path* between two collisions of a charged particle with neutral particles is then given by

$$\lambda^{n}_{i,e} = 1/n_n\sigma^{n}_{i,e} \qquad \text{(cf. Table I.1)}$$

Note that the electron mean free path is about four times larger than the ions. A *collision frequency* of charged particles with neutral particles can accordingly be defined by

$$\nu^{n}_{i,e} = v_{i,e}/\lambda^{n}_{i,e} \qquad \text{(Figure I.12)}$$

much larger for electrons because they travel much faster.

Thirdly, there are characteristic parameters associated with the fact that the charged particles are subject to the effect of the geomagnetic field, which permanently deflects them through the Lorentz-Laplace force $\pm e(\vec{\mathbf{v}}_{e,i} \wedge \vec{\mathbf{B}})$, causing their trajectories to spiral around magnetic field lines with an angular frequency, non-velocity dependent:

$$\omega_{e,i} = eB/m_{e,i} \qquad \text{(Figure I.12)}$$

Electrons and ions rotate in opposite directions, electrons some orders of magnitude faster.

To these *gyrofrequencies*, or *cyclotron frequencies*, there correspond mean radii of gyration, or *Larmor radii*, given by

$$\lambda^{B}_{e,i} = v_{e,i}/\omega_{e,i} = v_{e,i}m_{e,i}/eB \qquad \text{(cf. Table I.1)}$$

much larger for the ions.

If we analyze Figure I.12, we can arrive on the basis of the above discussion at a very straightforward conclusion, essentially about the exponentially diminishing role of the neutral atmosphere on charged particles with altitude:

Below about 85 km, the largest characteristic frequencies, whether for ions or for electrons, are the collision frequencies. This means that the behavior of the charged particles is dominated by collisions with the neutral particles. In this region, which can be properly called the 'neutrosphere', ionospheric physics is not much different from classical aeronomy. Note that this ionospheric region coincides roughly with the *D* region.

Above about 130 km on the other hand, whether for electrons or ions, the collision frequencies drop below the other characteristic frequencies. This means that the ionospheric plasma becomes increasingly decoupled from the atmosphere and essentially under electromagnetic control, so that this overlying region, the *F* and topside ionosphere, could legitimately be called the 'magnetosphere' (actually, a true definition of the base of the magnetosphere cannot be found in three dimensional space, but resides in phase space).

In the intermediate region, which coincides roughly with the *E* region, the collision dominated ion population and the electromagnetically controlled electron population have quite different, if coupled, behaviors. This is the 'dynamo region' where ionospheric currents flow.

Note that the approximate coincidence between the regions defined according to the terminologies of this and the preceding section (Figures I.9 and I.12) is largely fortuitous; it would, and probably has been, destroyed if the strength of the Earth's magnetic field, or the composition of the atmosphere, or the solar UV spectrum were to undergo large qualitative changes.

This unusual photograph of the Earth was shot from the Moon with a UV camera-spectrograph during the Apollo 16 mission on April 21, 1972. It shows clearly, on the nightside of our planet, effects at ionospheric heights of the invisible quasi-dipolar magnetospheric structure caging the atmosphere. One must imagine that the North and South luminous belts at low and high latitudes respectively are connected across the equatorial plane by arching toroïdal shells. The outer, higher latitude shell is being filled from the outside with energetic magnetospheric electrons and protons spiraling tightly about magnetic field lines with a wide distribution of 'pitch angles'. At the feet of the field lines, they dip into the atmosphere the lower down the higher their pitch angle and energy. Collisions with the neutral gas excite a number of molecules and atoms to optically radiating states. This is the 'aurora', which was first described scientifically from the ground by De Mairan after a trip to Lapland in the early XVIIIth century. The inner, lower latitude shell is being filled with thermal energy electrons and ions rising up from the equatorial ionosphere below, under the effect of the dynamo electric field set up by winds blowing across the geomagnetic field. As this plasma arches back to the neutral atmosphere, it accumulates into ridges where ion-electron recombination, and associated photo-emission is enhanced. These tropical airglow belts were discovered by Barbier at Tamanrasset (Sahara) in 1959. (Courtesy of G. R. Carruthers, U.S. Naval Research Laboratory)

CHAPTER II

THE LATITUDINAL STRUCTURE OF THE IONOSPHERE AND THE MAGNETOSPHERE

2.1. Terrestrial Magnetism

The weakly ionized, horizontally stratified photochemical plasma pictured in Chapter I accounts for the main features of the local ionosphere anywhere on Earth as coupled to the atmosphere and to itself, and thus shaped on a large scale into a grossly spherical, symmetric shell by the gravitational field, with superimposed diurnal-longitudinal and seasonal-latitudinal variations associated with the varying geometry of solar illumination due to the Earth's rotation and orbital motion.

We have seen in the previous chapter that the influence of the terrestrial magnetic field on the motion of the different species of charged particles is felt starting at the altitude where their frequency of collisions with neutral particles falls below their gyrofrequencies. This lower boundary of the zone of geomagnetic control can be defined more generally by the condition, valid from about the top of the E region upward, $B^2/2\mu_0 > n_n kT_n$, which we interpret as meaning that the magnetic energy density or 'magnetic pressure' on the ionospheric plasma, exceeds the neutral gas energy density, or kinetic pressure, an exponentially decreasing function of altitude. This implies that the peak and upper ionospheric layers are structured by the magnetic field of the Earth. There is indeed a marked latitudinal pattern on all ionospheric phenomena with the dipolar symmetry characteristic of the field's geometry.

The geomagnetic pressure itself however diminishes with distance from the Earth, and is eventually overwhelmed by the kinetic pressure of solar corpuscular radiation. The geomagnetic structure is thus itself modulated by direct solar control, and because of the toroïdal topology which connects the outer magnetospheric regions to the polar caps at atmospheric levels, this modulation is an essential feature of the higher latitudes ionosphere.

2.1.1. THE MAIN FIELD

At the surface of the ground, and throughout the atmosphere, the magnetic field very much resembles that of a dipole located near the center of the Earth, just as if the globe was uniformly magnetized, as Gilbert was the first to remark in the XVIth century (Figure II.1). The origin of this magnetism, which is not yet completely

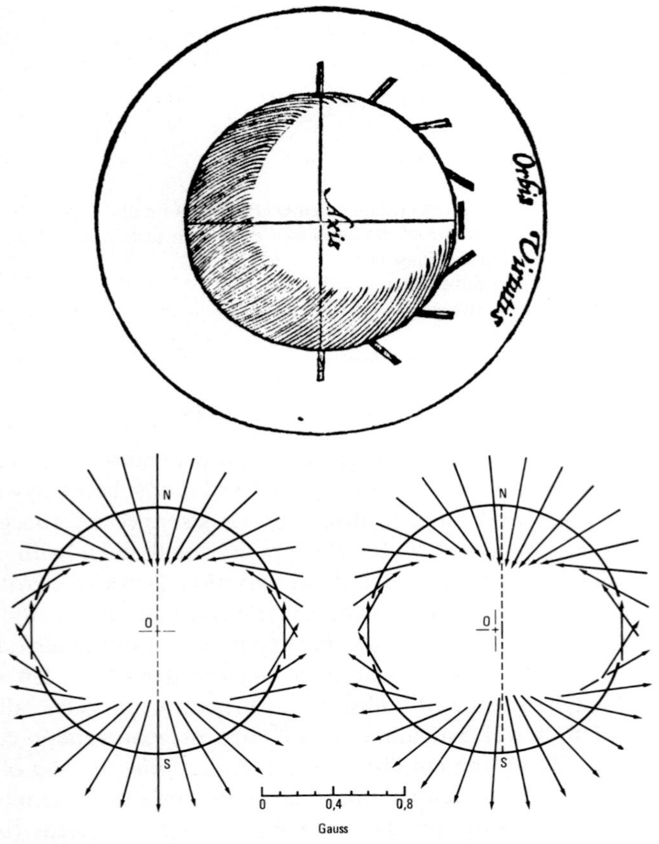

Fig. II.1. Top: William Gilbert's 1600 drawing of his experiment with a 'terella', a polished sphere of magnetite, which helped him demonstrate that 'the Earth is a magnet'. Magnitude and direction of the induction vector for a centered dipole (left), and for an excentric dipole (right), near the surface of the Earth, along a magnetic meridian. The best fit offset is about 350 km towards the Pacific Ocean. Note that the names of the poles are those of the pointing compass needle, reversed with respect to those of the Earth considered as a magnet.

understood, is to be found in an internal electrical current system. Although it might be subject to periodic instabilities of a catastrophic nature (inversion of the North and South poles), as the study of remanent magnetism on the expanding ocean floors suggests, and undergoes secular changes (slow variations in the strength of the dipole moment, slow displacement of the geographical positions of the magnetic poles) (cf. McElhinny and Merrill, 1975), this gross magnetism is perfectly stable on the characteristic time-scale of ionospheric phenomena.

The axis of the best fit dipole is at present inclined about 11.5° with respect to the axis of rotation of the globe. The geographical coordinates of the geomagnetic North Pole are 78.5°N, 291°E (Northern Greenland); the geomagnetic equator is correspondingly inclined with respect to the geographic equator, crossing it Southward over the Pacific Ocean and Northward over the Atlantic.

One usually expresses the properties of the main field in terms of this centered dipole approximation by means of the 'dipole' or 'geomagnetic' coordinates:

– R, the radial distance from the center of the Earth measured in terrestrial radii, $R = (z + R_0)/R_0$;

– Λ, the geomagnetic latitude (or dipole latitude), taken as positive in the Northern hemisphere and negative in the Southern hemisphere.

Within the frame of these approximations and conventions, the components of the terrestrial magnetic induction vector can be written:

$$\text{radial component: } B_r = -2 \sin \Lambda R^{-3} B_{eq}$$

$$\text{tangential component: } B_\lambda = \cos \Lambda R^{-3} B_{eq}. \qquad \text{(II.1)}$$

The minus sign in the expression for B_r signifies that this component is directed downward in the Northern hemisphere. The component B_λ is directed northward in both hemispheres. Both components are expressed in terms of B_{eq}, the best fit average value of B along the geomagnetic equator at ground level equal to 0.31 G, or 31,000 γ, or 31×10^{-6} T.

Starting from these expressions, it is easily found that at a given point the downward angle of inclination of the induction vector from the horizontal plane is:

$$I = \arctan(-B_r/B_\lambda) = \arctan(2 \tan \Lambda). \qquad \text{(II.2)}$$

This 'dip angle' is independent of radial distance (hence of altitude). The magnitude of the induction vector is:

$$B = (B_r^2 + B_\lambda^2)^{1/2} = (1 + 3 \sin^2 \Lambda^{1/2}) R^{-3} B_{eq}. \qquad \text{(II.3)}$$

At a given latitude, this magnitude is inversely proportional to the cube of the radial distance. At a given radial distance (altitude), the induction increases by about a factor of two in going from the equator to the poles.

Because of the excentricity of its actual dipole component (Figure II.1) and of local subterranean inhomogeneities, the real magnetic field exhibits from point to point significant permanent deviations from the values calculated from the centered dipole approximation (Figure II.2). These effects, for which one adjusts in very complete models by fitting a sum of spherical harmonics to the network of observational data, tend to fade with altitude, except in the extreme cases (e.g. the South-Atlantic 'anomaly'). The real magnetic 'dip' equator and poles, where the induction vector is

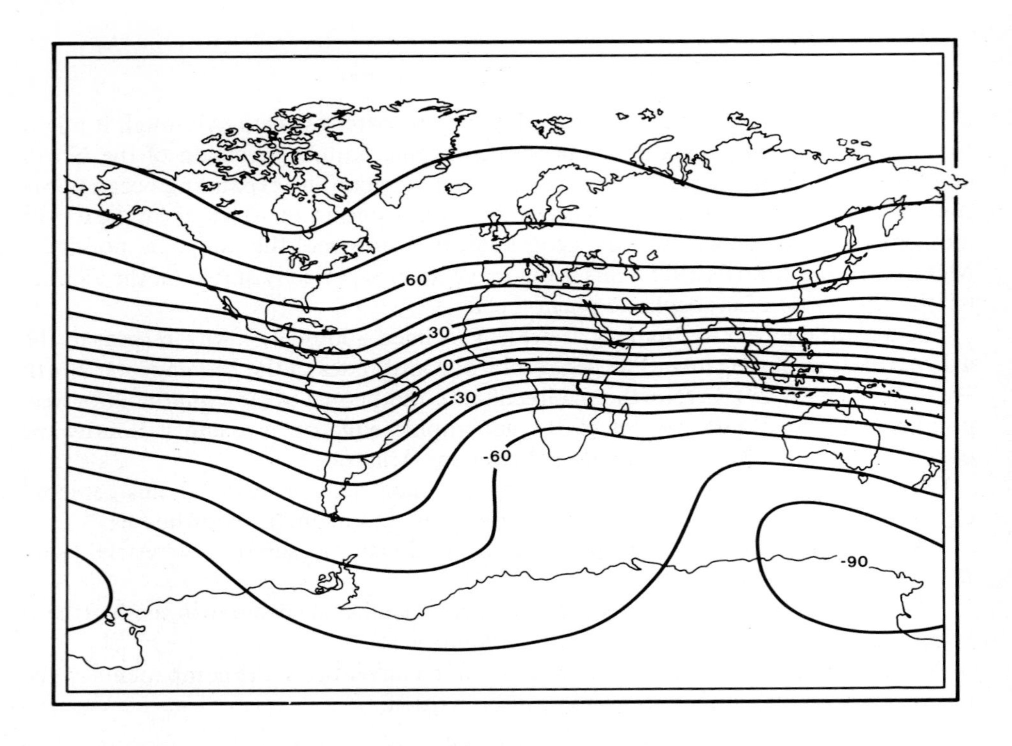

Fig. II.2. Map of the dip I (in degrees, upper panel) and of the total intensity B (in G, lower panel) of the Earth's magnetic field at ground level for 1965. The minimum around 0.238 G is the 'South Atlantic anomaly' (after Wilcox and Ness, 1965).

actually horizontal and vertical, in particular, differ from the dipole equator and poles. Note that this dip equator is the one usually referred to when speaking of the geomagnetic equator in ionospheric terminology. Generally speaking, the dip latitude is defined by Equation (II.2), when I is the actually observed value of inclination.

2.1.2. VARIATIONS AND DISTURBANCES

The measured magnetic induction at ground level is subject to short term time fluctuations, both quasi-diurnal and irregular (with periods of the order of hours, minutes or even seconds and less). These are of small amplitude, in general a fraction of a per cent of the total field; large events can reach 1 or 2%. Empirical indices have been defined which characterize the 'planetary' geomagnetic activity, or degree of agitation of the mean field over the whole Earth (cf. Figure 0). The international K_p index (Table II.1) averaging data from a number of observation stations, is used to

TABLE II.1

(Cf. Bartel's 'The technique of scaling indices K and Q geomagnetic activity,' *IGY Annals* **4**, London, 1957.) Like the 'Beaufort' scale of meteorological winds, the K scale of magnetic unrest indicates with a figure the strength of 'storms'.

K indices	Deflection (γ)
0	0–15
1	15–30
2	30–60
3	60–120
4	120–210
5	210–360
6	360–600
7	600–990
8	990–1500
9	1500 and more

define the 'force' of the irregular perturbations, or storms, along a logarithmic scale rising from 0 to 9, and to choose for each month certain days representing quiet and perturbed conditions. There exist, however, variations localized in amplitude, the most important of which affect the high latitude auroral ovals. This 'auroral' activity is expressed by a special index, AE, computed from measurements of stations at geomagnetic latitude between 55 and 70°.

As Gauss had demonstrated, the larger part of the magnetic fluctuations at the Earth's surface must originate in overhead electric currents, the remainder being due to secondary currents induced within the Earth.

The regular quasi-diurnal components of magnetic fluctuations include the 'solar quiet' (Sq) variation, a diurnal oscillation with a period of 24 hours and an amplitude of about $\pm 25\gamma$, and the lunar variation (L), a semi-diurnal oscillation with a period of 24.8 hours like ocean tides, and an amplitude of about $\pm 5\gamma$.

The irregular component of magnetic fluctuations is usually characterized by a 'sudden commencement' followed by a complicated sequence of more or less repeatable patterns. The 'disturbance' variation *D* is defined for a given location as the departure from the regular quiet behaviour of the field. It can be broken down into the 'storm time' disturbance *Dst*, and the 'solar daily' disturbance *SD*.

The (permanent) *Sq* and (sporadic) *SD* components are organized with a latitudinal and longitudinal structure which suggests that they are due to a current system fixed in solar time, underneath which the Earth rotates, as discussed in Chapter VII (cf. Figures VII.4 and 8).

2.2. The Interaction Between the Solar Wind and the Geomagnetic Field

2.2.1. THE SOLAR WIND

One of the most notable discoveries of the space age has been that of the expansion of the solar corona. Before the 1950's, it had been thought that the interplanetary medium was essentially empty save for rare eruptive events. The study of comet tails by Biermann, followed by the theoretical investigations of Parker, had led however to the idea that the Sun's million degrees hot upper atmosphere could not reach hydrostatic equilibrium and was in a state of perpetual outflow. Subsequent work confirmed that indeed the solar magnetic field is too weak to confine this fully ionized H plasma which instead traps the field away as it escapes. The great complication of the magnetic details observable at photospheric level somehow get smeared in the process and the far field is a fluctuating residue. The combination of the radial coronal expansion with the proper rotation of the Sun every 27 days eventually shapes the solar-interplanetary magnetic field into a spiral structure, which measurements have shown to be divided in wheeling sectors of alternate positive and negative 'polarity' (Figure II.3).

In the neighborhood of the Earth's orbit, the fluctuating but permanent coronal streaming or 'solar wind' has a velocity V_s of a few hundred km s^{-1} on the average and a temperature of the order of 10^5 °K; thus, the proton flow is largely supersonic (their thermal speed is of the order of a few tens km s^{-1}), and forms an almost unidirectional 'jet' coming from the Sun, but the electron flow is subsonic (their velocity distribution is almost isotropic at a few thousand km s^{-1}). The concentration of coronal particles n_s, having decreased during the transit time from Sun to Earth roughly as the square of the distance, fluctuates around a few cm^{-3}.

The trapped coronal magnetic field, whose square has decreased as the plasma pressure, fluctuates around a few γ and as mentioned above periodically reverses its 'polarity', that is from sunward to anti-sunward, with a small fluctuating North–South component.

Aside from this everpresent quasi-thermal magnetoplasma flow of coronal origin, the matter ejected by the sun includes much more energetic components (solar cosmic rays) in the form of sporadic beams of relativistic particles threading their way through the interplanetary electromagnetic background. Their emission is connected to localized short-lived violent events at the surface of the star which are also the source of bursts in its electromagnetic radiation, particularly in the radio and X-ray range. In Chapter VI, we shall see that these solar 'flares' are the cause of ionospheric perturbations (X-ray induced sudden ionospheric disturbances or SID, and proton

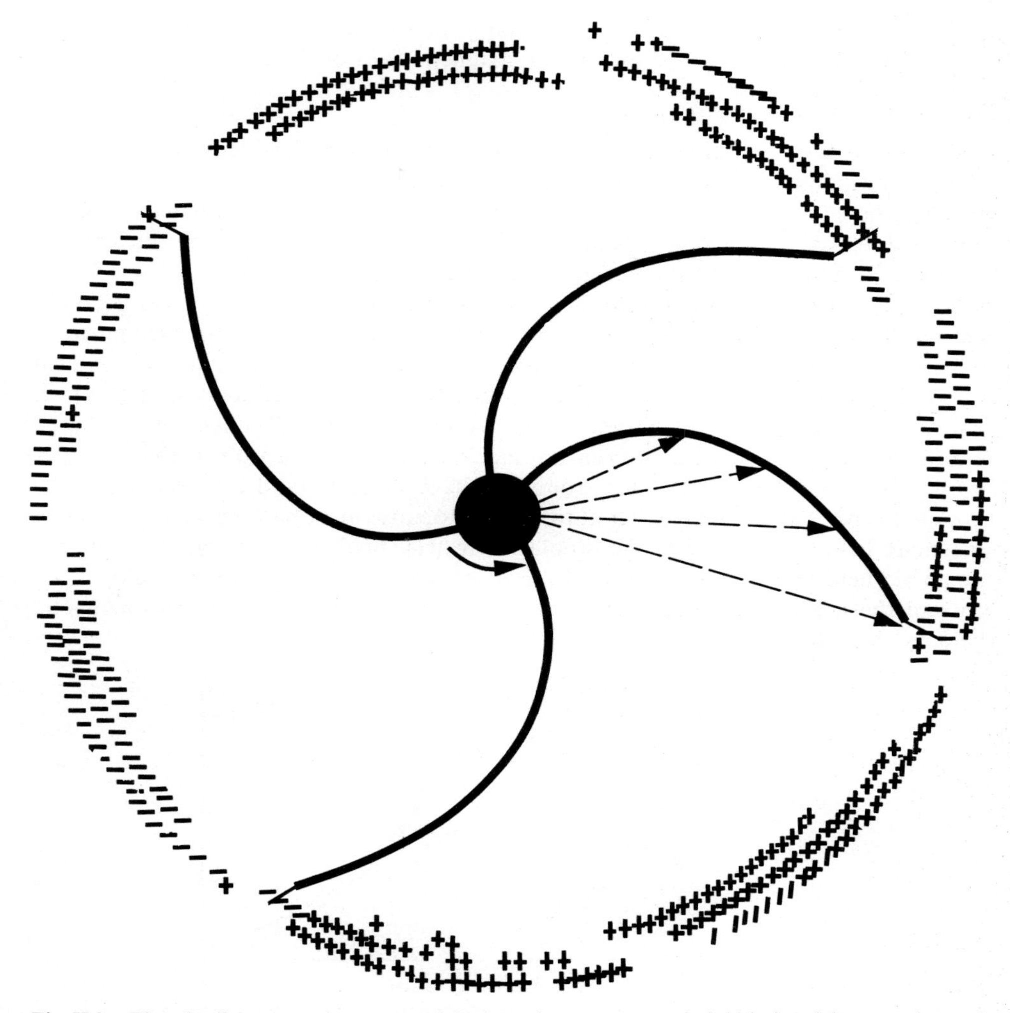

Fig. II.3. The wheeling sectored structure of the interplanetary magnetic field in inertial space, observed during three 27 days rotations of the Sun by the IMP 1 satellite. Plus (minus) sign indicates a sunward (antisunward) field.
Arrows indicate the radial trajectory of plasma ejected from a single emitting region, but on successive days, transporting 'frozen' field lines. The curvature of the sector's boundaries indicated corresponds to an average solar wind speed of a few hundred km s^{-1}.

induced polar cap absorption or PCA) which should not be confused as they have long been with perturbations or 'storms' tied in with the magnetoplasma wind induced geomagnetic activity to be discussed in this chapter.

2.2.2. THE BOUNDARIES OF THE MAGNETOSPHERE

There is a rather sharp outer limit to the magnetosphere of the Earth: *the magnetopause*, roughly situated at a distance from the Earth's center where the geomagnetic pressure becomes less than the kinetic pressure of the solar wind. The magnetosphere has in fact the overall structure of a 'cavity' in this hot and weakly magnetized interplanetary plasma, carved out by the magnetic field of the Earth.

2.2.2.1. *The Magnetopause*

Figuratively, the solar wind 'sweeps away' the geomagnetic field outside the magnetopause but cannot penetrate inside it. The magnetopause is thus the equilibrium surface where the terrestrial magnetic field successfully resists the solar wind; it circumscribes the cavity which the field, 'compressed' by the pressure of the coronal plasma and keeping it from entering, hollows out.

More precisely, one can describe the interaction in the following way: the geomagnetic field curves the trajectories of solar wind electrons and protons incident from without so that they are specularly 'reflected'. The geometrical locus of the focal points of the trajectories of the particles which turn around defines a surface which one can consider as a current sheet. This is because there is an effective flow of charge parallel to the sheet, resulting from the fact that under the action of the Lorentz-Laplace force all the protons turn around in the same sense (and the electrons in the opposite sense). The current sheet at the magnetopause must be such as to generate a magnetic field which cancels the dipolar terrestrial field in the exterior; as for the magnetic field in the interior, it is the sum of the dipole field and the field of the current sheet. This is a more exact phrasing of the 'compression' of the geomagnetic field by the solar wind.

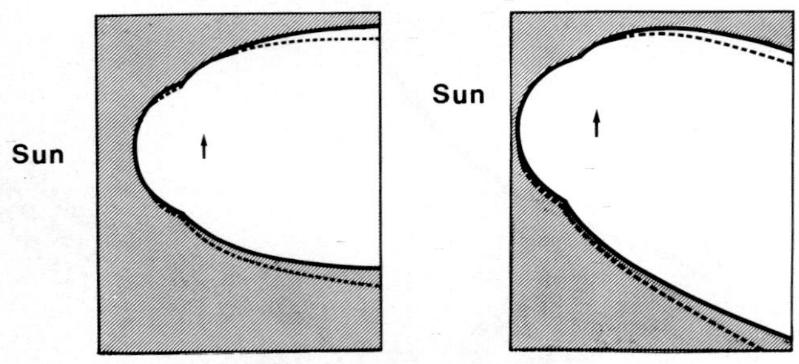

Fig. II.4. Computed shape of the magnetospheric cavity (magnetopause), in the Sun-Earth meridian plane, for various inclinations of the magnetic dipole axis to the sun's direction.

We can understand the magnetopause quantitatively by approximating it locally with its tangent plane (Figure II.4). The solar wind boundary current sheet produces locally a magnetic field whose component normal to the plane must be continuous, but whose tangential components on the two sides are oppositely directed and equal. Since the field of the current sheet must compensate for the dipole field immediately outside, the total B field immediately inside must be equal to twice the projection of the dipole field onto the tangent plane. Furthermore the pressure $B^2/2\mu_0$ of this field must balance the kinetic pressure of the solar wind particles on the exterior (essentially due to the protons, whose mass m_p is very much greater than that of the electrons), itself equal to twice (incident momentum + reflected momentum) the product of the component normal to the tangent plane of the flux $n_s V_s$ of protons and that of their momentum $m_p V_s$. For example, the radial distance R_M of the nose of the magnetosphere, where, for reasons of symmetry, the solar wind must be perpendicularly incident upon the magnetopause, is characterized by the relation $(2\ B_{\text{dipole}})^2/2\mu_0 = 2m_p n_s V_s^2$. If we further assume that the geomagnetic equator is close to the plane of the ecliptic, we have $B_{\text{dipole}} \simeq B_{\text{eq}}/R_M^3$ (R_M being expressed in Earth radii), from which

$$R_M = (\mu_0 m_p n_s V_s^2/B_{\text{eq}})^{-1/6}.$$

Inserting the above mentioned values of B_{eq}, n_s and V_s, we find R_M to be about 10 R_o, say an 'altitude' of about 60,000 km.

The general solution of the problem, for all angles of incidence (differently but correctly posed by Chapman and Ferraro in the 1930s in trying to explain magnetic storms by the advent of 'clouds' of solar plasma) which gives a first approximation for the mean shape of the magnetospheric cavity, was not obtained before the 1960s (Spreiter and Briggs, 1962). Figure II.4 gives an idea of the results for different inclinations of the axis of the geomagnetic dipole with respect to the Sun–Earth axis. There is a relatively good agreement with the observed crossings of the magnetopause by satellites in very eccentric orbits.

Note the existence on the forward face of the magnetopause of two singular points, North and South. At these points, the dipole field is perpendicular to the magnetopause and consequently cancelled by its reflection in the tangent plane. Since the magnetic field offers it no resistance there, the solar wind can penetrate the magnetospheric cavity freely. These neutral regions, which in reality have been observed to spread out across the meridian plane on the morning and evening sides are called 'polar cusps' or 'polar clefts'.

2.2.2.2. *The Magnetotail, Bow Shock, and Magnetosheath*

The above description of the magnetopause is oversimplified and does not take into account the phenomena involving the solar wind magnetic field, which is one of the outstanding problems of magnetospheric physics. First, whereas the position of the magnetopause is roughly determined by the hydrodynamic pressure balance we have described, the interplanetary magnetic field plays a role which cannot be neglected: a southward turning of the solar wind magnetic field has been shown to erode the sunward face of the magnetosphere and to push the magnetopause closer to the Earth, even when the dynamical pressure of the solar wind stays constant (Aubry, *et al.*, 1970).

This phenomenon is linked with field line reconnection occurring along the magnetic null line which appears when the interplanetary magnetic field B_s opposes the equatorial geomagnetic field as proposed early by Dungey (1962). The corresponding magnetic flux is transferred into the back of the magnetosphere. This transfer basically explains the existence of the long tail of the magnetosphere which otherwise should be of a drop-like form. The distance to which the tail, and hence the magnetopause remain well defined, is unknown although some evidence has been obtained for its existence at distances of 500 to 1000 R_0 downstream. A thin sheet of current flows in the equatorial plane transverse to the tail across the neutral magnetic sheet separating the sunward magnetic field in the North lobe from the antisunward magnetic field in the South lobe (Figure II.16).

The presence of the interplanetary magnetic field also causes the solar wind to behave like a continuous fluid, supersonic relative to the Earth, i.e. with bulk velocity greater than the Alfvén velocity $[B/(\mu_0 n m_p)^{1/2}]$ which plays in magnetohydrodynamics (the study of low frequency and large scale behavior of magnetized plasmas) a role similar to sound velocity in ordinary gas dynamics. Thus, a shock wave is generated in front of the magnetopause at some 14 R_0 from the Earth. Within this collisionless bow shock, the bulk velocity of the solar wind is transferred into kinetic and electromagnetic random energy; the processes involved are still poorly understood and constitute one of the fascinating plasma physics problems of the Earth's environment. The region extending between the bow shock and the magnetopause is called the magnetosheath: the plasma there is turbulent and hotter than the solar wind plasma, with a typical temperature of 10^6 K or 100 eV. This magnetosheath plasma flowing around the magnetosphere has direct access to the magnetosphere through the polar clefts and also the neutral sheet across the tail where it is called the 'plasma sheet'. The magnetopause otherwise acts as an efficient barrier to the direct entry of the solar plasma (but not of high energy corpuscules).

2.2.3. THE MAGNETOSPHERIC FIELDS

In the interior of the cavity defined by the magnetopause there are established both a space-time distribution of the electromagnetic field, $\vec{\mathbf{E}}$, $\vec{\mathbf{B}}$, and a distribution in position and velocity of the charged particles of solar and terrestrial origin (essentially protons and electrons). These two distributions are related ('self-consistent') to a large extent, since the E field is affected by the distributions of the particles of both signs while the currents resulting from the velocity distribution of the charges add to the internal currents of the Earth and to the magnetopause (solar wind boundary) currents. Conversely, the distribution of electric and magnetic fields determines the trajectories of the charged particles.

In this section we restrict ourselves to the effect of boundary conditions at the periphery and at the center of the magnetospheric system, which lead to a first approximation for the distribution of the fields. In the next section we shall consider the dynamics of the charged particles. The difficult self-consistent problem of relating the two will be discussed in Chapter VII.

2.2.3.1. *The Magnetic Field*

Once the surface of the magnetopause has been determined, one can, in principle, calculate the magnetic field throughout the interior of the magnetospheric cavity (assumed a vacuum) by adding at each point the terrestrial dipole field and the field due to the currents at the magnetopause, themselves determined, we recall, by the condition that they must nearly cancel the dipole field in the exterior. Figure II.5 shows such a model of the magnetospheric field (Mead, 1964) that is in reasonable agreement with measurements made by magnetometers carried by far-travelling satellites, except in the 'tail' of the magnetosphere.

The dipole field near the Earth is only slightly modified inside a distance of several terrestrial radii. In this region the differential equation of a line of force is as a first approximation, that of the dipole and can be written as $(\mathrm{d}R/\mathrm{d}\Lambda)/R = B_r/B_\Lambda$, which gives after integration $R = \cos^2\Lambda/\cos^2\Lambda_0$, where Λ_0 is the dipole latitude where the line

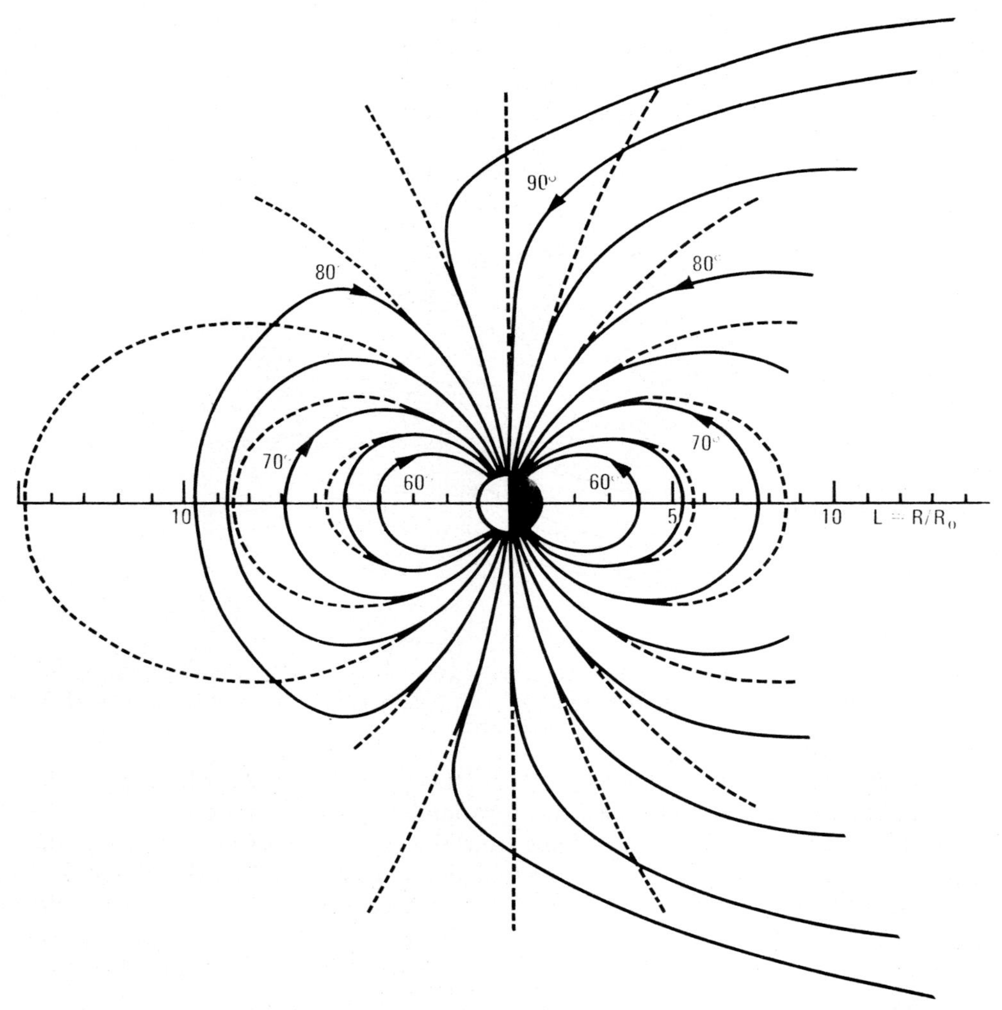

Fig. II.5. Dashed curves are dipolar field lines. Full curves are a magnetic field model in the meridian plane of the magnetosphere, obtained by requiring the magnetopause current sheet to anihilate the dipole field outside. L is distance to the Earth's center counted in Earth's radii R_0.

of force intersects the surface of the Earth. The same line of force crosses the equator at radial distance

$$L = \sec^2 \Lambda_0. \tag{II.4}$$

All the lines of force which have the same L value form a toroidal 'magnetic shell', and this parameter is convenient for characterizing the various shells, as well as their intersection with atmospheric levels. Figure II.6 shows the curves $L =$ constant at 300 km, projected on a planisphere.

Beyond $L = 5$, approximately, the lines of force are distorted with respect to those of the dipole field, but up to about 80° of latitude on the daylight side and about 70° on the nightside, each point on the Southern Hemisphere is still connected by a line of force to a point on the Northern Hemisphere. Two such points are called 'magnetic conjugates'. By an extension based on adiabatic invariants (see 3.3.1.3 below), we continue to identify lines of force by a parameter L (McIllwain, 1961), which no longer represents however the radial distance at which they cross the equatorial plane.

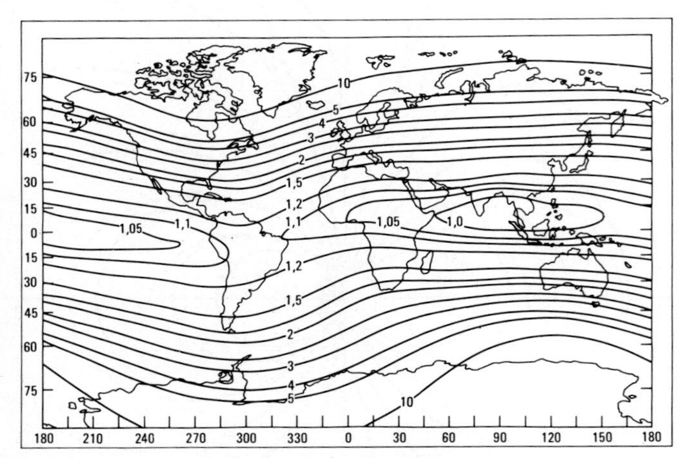

Fig. II.6. Map, in geographic coordinates, of the intersection of equi L shells with a spherical surface located 300 km above ground level. Up to roughly $L = 5$, L is the radial distance in Earth's radii of the shells' apex.

The latitude defined by equation (II.4), that is arc sec$\sqrt{L}$, where L is the McIllwain parameter, is then called the 'invariant' magnetic latitude.

On the other hand, the lines of force from the polar caps open up towards the tail of the magnetosphere. The intersection of the boundary between closed and open lines of force with the atmosphere was suspected and shown to define the poleward borders of the auroral ovals. These borders are magnetic conjugates, as is otherwise demonstrated by the 'mirror' aspect of the Arctic and Antarctic auroras.

Models such as the one in Figure II.5 can furnish only an approximate average field. In reality, one must take account of the inclination of the dipole axis (which recesses daily and annually with respect to the Sun–Earth axis), of significant variations in the location of the magnetopause due to fluctuations in the solar wind parameters, and of additional magnetic fields due to currents which flow within the magnetospheric cavity itself. These last are both ionospheric currents, to which we will return in detail in Chapter VII, and currents corresponding to the drifts of energetic particles populating the magnetospheric cavity, of which the most important are the 'ring' current encircling the Earth, and the tail currents around the North and South lobes. Whatever they are, we know that it is the sum of these currents that determines geomagnetic activity, i.e. small variations in the field measured, in particular, at ground level. These currents are included in more recent models of the geomagnetic field (Olson and Pfister, 1974).

2.2.3.2. *Electric Fields*

When discussing electric fields, one must first of all clearly specify the frame of reference in which the field is expressed, since the relative nature of the $\vec{\mathbf{E}}$ field causes it to transform with an additional term $\vec{\mathbf{V}} \times \vec{\mathbf{B}}$ in switching from a given frame to another moving at velocity $\vec{\mathbf{V}}$ with respect to the former (Lorentz invariance). Secondly, charged particles being free to move in the magnetosphere along the magnetic field lines, as long as collisions can be neglected, potential drops parallel to the magnetic field must remain very small, and one assumes in general that $\vec{\mathbf{E}}_{//}$, the component parallel to $\vec{\mathbf{B}}$, vanishes, save on exceptional occasions, when 'anomalous resistivity' is caused by turbulence or instabilities. Thirdly, if we limit ourselves to stationary cases ($\partial/\partial t = 0$), in the reference frame that moves locally with the bulk velocity of thermal plasma, the electric field vanishes. If the electric field perpendicular to $\vec{\mathbf{B}}$ is not zero, all the particles of the plasma are therefore moving with a velocity $\vec{\mathbf{V}}_c$ given by

$$\vec{\mathbf{E}} = -\vec{\mathbf{V}}_c \times \vec{\mathbf{B}} \tag{II.5}$$

thus representing the electric field in terms of the motion of the plasma in the magnetosphere. (This equivalence led to the amusing consequence that experimenters measuring plasma motion often express their results as electric field data and *vice versa*!) One can also consider $\vec{\mathbf{V}}_c$ as representing the motion of the magnetic tubes of force, as in the hydromagnetic concept of 'field frozen in the plasma'.

A convenient reference frame is that fixed to the Earth: since in a first approximation the lower ionospheric plasma rotates with the Earth, at least in middle latitudes, in this frame $\vec{\mathbf{E}} = 0$ close to the planet. Another frame of this type is that which moves with the solar wind, so that in it $\vec{\mathbf{E}} = 0$ far from the Earth. In the magnetospheric frame (origin at the center of the Earth, axes North-South, noon-midnight, morning-evening), the effect of these boundary conditions is to cause an electric field distribution which must tend towards $\vec{\mathbf{V}}_{rot} \times \vec{\mathbf{B}}_{dipole}$ (the cylindrically symmetric 'corotational' electric field) in the ionosphere, where $\vec{\mathbf{V}}_{rot}$ is the velocity of the terrestrial frame with respect to the magnetospheric frame, and towards $\vec{\mathbf{V}}_s \times \vec{\mathbf{B}}_s$ ('interplanetary' electric field) beyond the magnetopause, where the subscripts refer to the velocity and magnetic field of the solar wind in the magnetospheric frame. Since the atmosphere does not turn exactly with the Earth because of its own motions (tides, etc.) one can expect modulations of the corotational field ('dynamo' field), which are studied in Chapter VII. Similarly, since B_s fluctuates in magnitude and direction with solar activity, it follows that there exists a corresponding modulation of the interplanetary electric field.

The effect of these boundary conditions however does not suffice to account qualitatively for the observed plasma velocity field, or, equivalently, electric field distribution in the magnetosphere because of the discontinuity at the magnetopause. All the evidence points to the existence, in addition to the corotational electric field, of a strong dawn-dusk electric field, parallel to the equatorial plane and perpendicular to the Sun-Earth direction, tending to convect magnetic field tubes and plasma toward the dayside magnetopause (Axford and Hines, 1961). The magnitude of this field called the magnetospheric 'convection' field, corresponds to a total potential

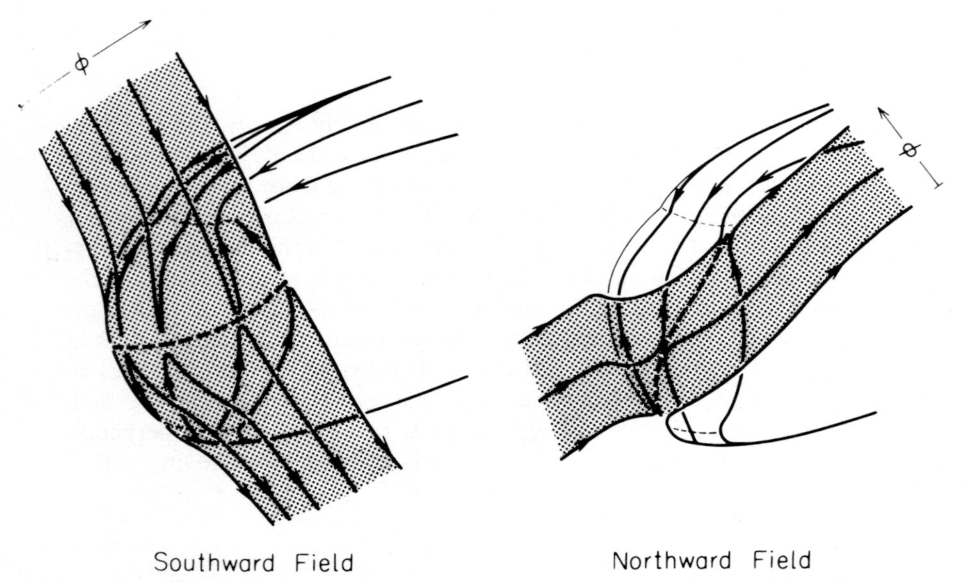

Fig. II.7. Geometry of the dayside reconnection between the magnetospheric and interplanetary magnetic fields. Reconnection occurs along the dashed line, and leads to a dawn-dusk magnetospheric electric field corresponding to the potential difference φ, whether the interplanetary magnetic field has a southward or a northward component perpendicular to the ecliptic.

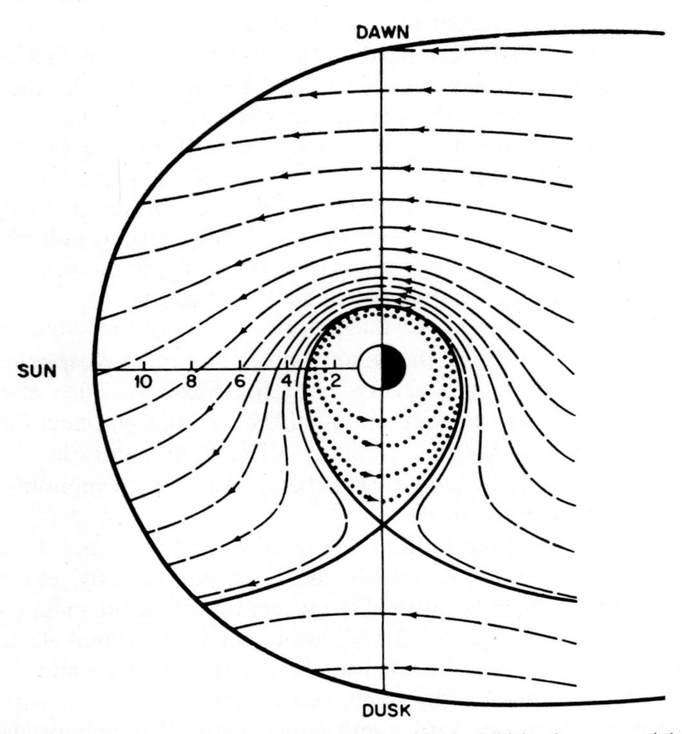

Fig. II.8. Model of equipotentials of the magnetospheric electric field in the equatorial plane obtained when adding to the corotational electric field a constant dawn dusk electric field. Arrows indicate the corresponding ($\vec{\mathbf{E}} \times \vec{\mathbf{B}}$) flow of magnetic flux tubes (perpendicular to the plane of the figure and arching like shown in Fig. II.5) carrying 'frozen' thermal plasma.

drop across the magnetopause of roughly 40 keV. This is an order of magnitude less than the potential, in the magnetospheric frame, across the solar (interplanetary) magnetic field tube 'intercepted' by the magnetosphere. It means then that the interplanetary electric field is 'mapped' inside the magnetosphere in a peculiar way.

In early models of the magnetosphere, the interplanetary magnetic field was practically ignored (Spreiter *et al.*, 1968). No interconnection between terrestrial and interplanetary magnetic field lines occurred and there was no plasma penetration into the magnetosphere. In these models (closed magnetosphere), magnetospheric convection was thought to originate from viscous interaction between the solar wind and the geomagnetic field near the boundary of the magnetosphere (Axford and Hines, 1961). In more recent models (open magnetosphere), the importance of magnetic field reconnection was recognized, at least during dynamically active periods, i.e., during periods of magnetic flux adjustment from the magnetospheric front lobe to the tail: substorms, storms, etc ...

As already mentioned, field reconnection occurs along the magnetic null line, which appears when the interplanetary magnetic field B_s opposes the equatorial geomagnetic field exactly. The short-circuiting of the interplanetary electric field E_s, along this line is not expected to be 100% effective, in contrast to the closed magnetosphere model, in which the magnetopause is an electrical equipotential surface. The net result of the incomplete short circuit is an electric field component, E'_s, parallel to the magnetopause and directed along the reconnection line. Regardless of the orientation of B_s, E_{mp} always has a component from dawn to dusk (as in the closed models), but its magnitude is a maximum when B_s is southward, increasing the reconnection rate (Figure II.7).

The problem of the mapping of the magnetopause potential distribution within the magnetosphere is one of the most complicated of ionospheric and magnetospheric physics and will be discussed in Chapter VII. Figures II.8 and II.9 give a simple-

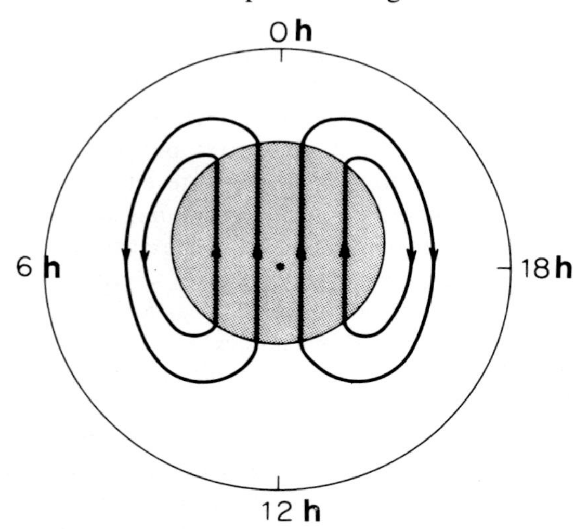

Fig. II.9. Scheme of the electric equipotentials (flow lines of magnetic flux tubes and thermal plasma) over the polar caps. Shaded area is the bundle of high latitude magnetic field lines corresponding to the magnetospheric tail which do not intersect Fig. II.8.

minded qualitative view of a solution where one simply adds a constant dawn-dusk convection field to the corotational field (Brice, 1967).

Because magnetic field lines are themselves equipotentials, the problem is actually two-dimensional. A convenient image of the two-dimensional solution space, where each point is in fact a field line, is usually given by a cross section of the system:

– equatorial cross sections (Figure II.8), where each field line is represented by its apex, have the drawback that they do not intersect polar field lines.

– or hemispherical cross sections at ionospheric heights (Figure II.9), where each field line is represented by its foot in a single hemisphere.

2.3. The Dynamics of Charged Particles

The analysis of the behavior of charged particles in the magnetosphere is strongly dependent on their location and energy, from 'ring current' protons which obey the equations of motions of free particles, to ionospheric ions which must receive an hydrodynamical treatment. The transition from one type of transport to another is in fact more or less continuous, and reflects a progressive variation of the relative importance of large scale fields (gravitational, magnetic, electric) on the one hand, and of collisions on the other hand.

Indeed, it is the continuous character of this transition which integrates all charge transports within the Earth's environment into a single global electrodynamical system extending from the base of the ionosphere rotating with the atmosphere to the magnetospheric bow wave and tail in the streaming solar wind.

Our understanding of this global electrodynamical system will be discussed in Chapter VII. In the following sections we recall the basic features of the motions of the various populations of charged particles.

2.3.1. THE ORBITS OF ENERGETIC PARTICLES

The energy which is needed to transport against the gravitational field of the Earth an H atom from the surface of the planet to infinity is about 10^{-19} J (1 eV): gravity is inefficient to confine protons having energies much larger than 1 eV. Therefore, in the absence of the geomagnetic field, hot electrons and protons streaming across the remote environment of the Earth, which originate either from solar injection or simply from the ionosphere, would not stay around; but as theoretical computations as well as high altitude nuclear explosions have demonstrated, their residence time within the magnetospheric domain can in effect be of the order of years. The geomagnetic cavity thus efficiently confines a swarm of particles whose steady state distribution reflects their orbital statistical mechanics.

For energetic particles, the influence of gravity and partial pressure gradients is negligible. The influence of Coulomb collisions is also very small, except for the relatively low energy (10 eV) photoelectrons. They can thus be treated in first approximation as a collection of non-interactive particles. We shall attempt here a summary description of the dynamics of these particles, stopping short of the adiabatic invariants approach which frames the subject within classical perturbation mechanics (Northrop and Teller, 1960; Roederer, 1970).

The most general form of the equation of motion of a particle of charge $\pm e$ of

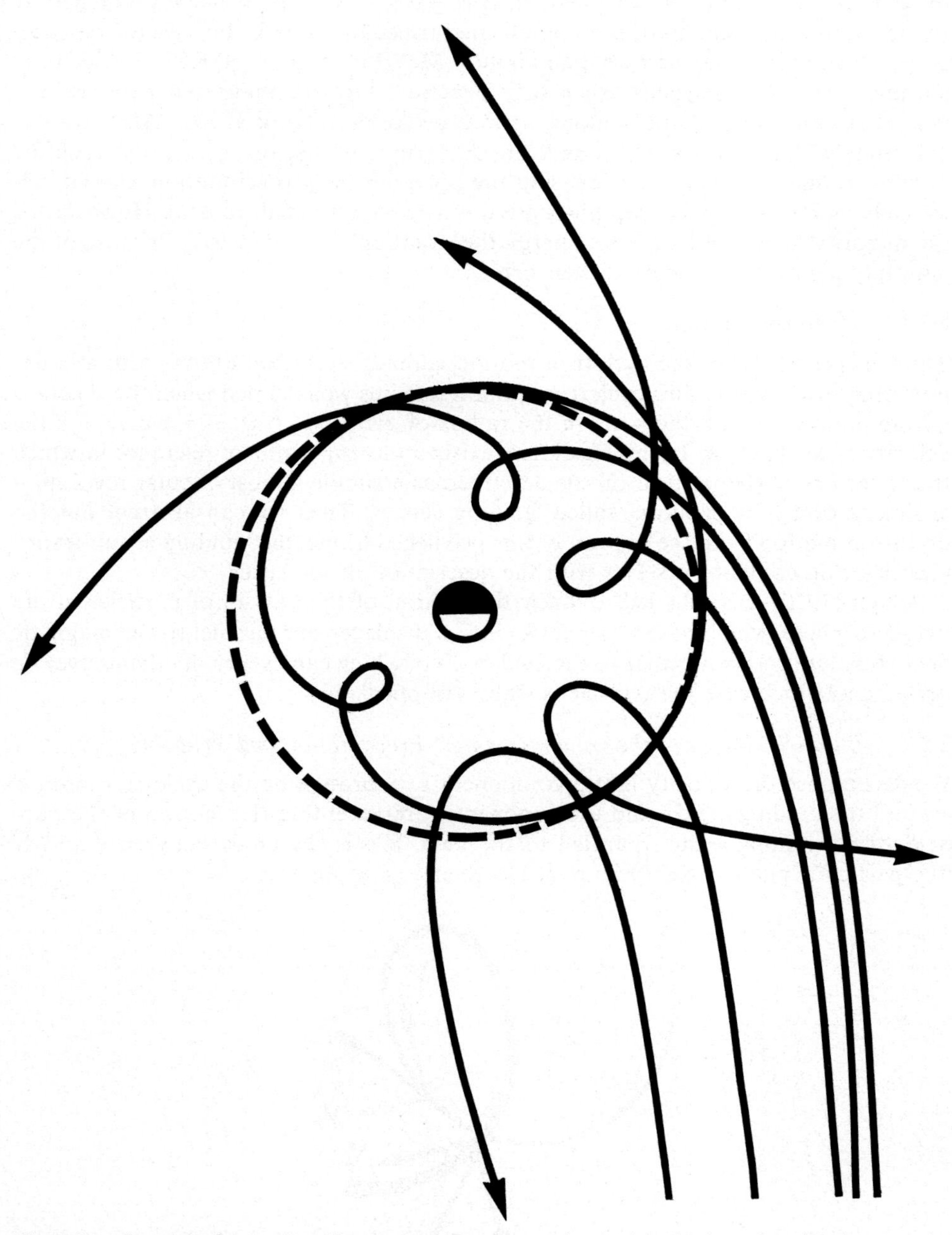

Fig. II.10. Samples of the trajectories of very high energy solar particles in the equatorial plane, to give an idea of the magnetic shielding of the lower latitudes of the Earth.

mass m in a vacuum electromagnetic field is $\pm e(\vec{\mathbf{E}} + \vec{\mathbf{V}} \times \vec{\mathbf{B}}) = m\,\mathrm{d}\vec{\mathbf{V}}/\mathrm{d}t$. Generally the solution of this equation is a complicated trajectory; this is the case for particles of very high energy, say greater than about 1 MeV for protons, 50 KeV for electrons (cosmic rays). Such particles when they penetrate into the magnetosphere see a $\vec{\mathbf{B}}$ field that constantly changes along their trajectories (Figure II.10). Many results, beginning with Störmer's (1917), indicate that eruptive solar protons easily reach the Earth through the polar caps (causing the phenomena of precipitation known collectively as P.C.A. – polar cap absorption – to which we shall return). However for the majority of particles of lesser energy the situation is more orderly because of the 'multiply periodic' character of their orbit.

2.3.1.1. *Cyclotron Motion*

The first periodicity is the cyclotron motion, already described above, with angular frequency $\omega_{\mathrm{B}} = eB/m$. This cyclotron motion remains well defined when the B field is quasihomogeneous on the scale of the radius of gyration $\lambda^{\mathrm{B}} = \omega_{\mathrm{B}} v$, where v is the velocity of the particle. In this case there exists a moving frame of reference in which the trajectory of the particle can be described as a simple almost circular revolution around a center of gyration, called 'guiding center'. Then we can abstract out the cyclotron motion from the motion of this privileged frame, the 'guiding center frame' (G.C.F.), and content ourselves with the description of the latter.

Alfvén (1950) was thus led to study the motion of the G.C.F. of particles in the magnetosphere, which in turn separates into a displacement parallel to the magnetic field and a drift perpendicular to the field, both of which can eventually themselves be periodic in which case the particle is stably 'trapped'.

2.3.1.2. *Parallel Motion of the Guiding Center: Precipitation and Trapping*

We decompose the velocity into a component $v_{\perp}$ representing the cyclotron motion around the guiding center and a component $v_{//}$ representing the motion of the particle, or its guiding center, parallel to the field. If α is the angle between $\vec{\mathbf{v}}$ and $\vec{\mathbf{B}}$ (the particle's 'pitch angle') (Figure II.11), then $v_{\perp}/v = \sin\alpha$, $v_{//}/v = \cos\alpha$.

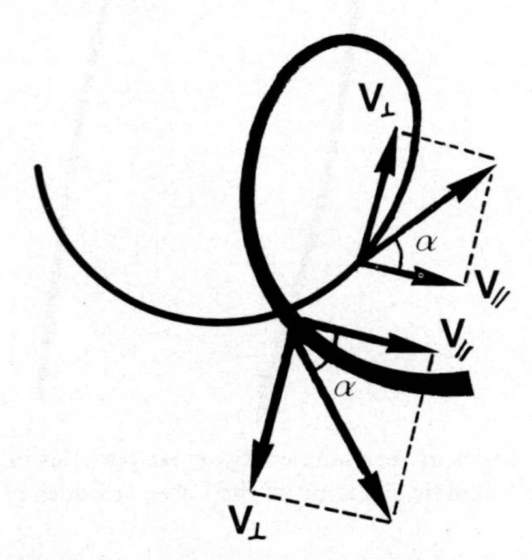

Fig. II.11. The helicoïdal motion of free particles about magnetic field lines to which are referred the parallel and perpendicular directions. α is the 'pitch angle'.

The angular momentum of the particle with respect to its guiding center, $mv\,\lambda^B$, is conserved; since $\lambda^B = mv_\perp/eB$, this implies that $v_\perp{}^2$ varies proportionally to B: if the magnitude of the $\vec{\mathbf{B}}$ field increases along its line of force, followed in the sense determined by the particle's parallel motion, the cyclotron motion speeds up, and vice versa. At the same time, since $\sin\alpha = v_\perp/v$, and v is constant, $\sin^2\alpha$ also varies as B: when B increases the pitch angle increases or, equivalently, the parallel velocity $v_{//} = v\cos\alpha = v(1 - \sin^2\alpha)^{1/2}$ decreases, and vice versa. Alternatively, when the field is not homogeneous, the lines of force are not parallel and, as explained by Figure II.12 which is drawn in a plane containing the guiding center and the magnetic field, the Lorentz-Laplace force tends to push the particles towards the diminishing B region.

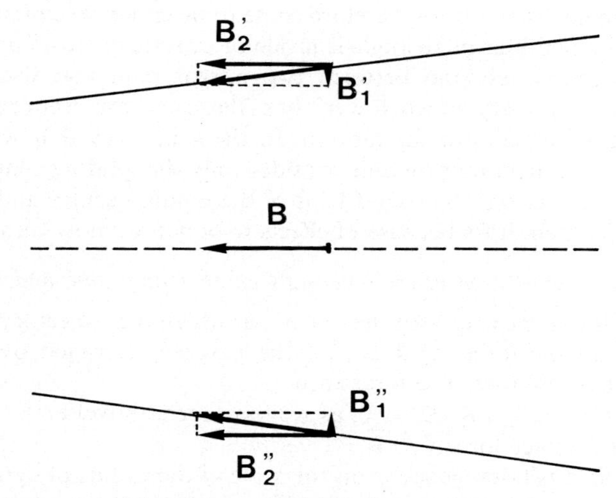

Fig. II.12. When magnetic field lines are not parallel, the Lorentz-Laplace force resulting from the small magnetic components noted 1 acting on the perpendicular velocity is directed along $\overline{B}$ and tends to accelerate the particle towards the region of decreasing B, whichever the sign of the particle's charge.

The lines of force of the magnetospheric field are such that B increases as one approaches the Earth. In their motion along these lines of force, the guiding centers of all particles slow down when moving toward the Earth and accelerate when moving away from it. Let us consider a particle whose pitch angle is α_0 at a distance from the Earth where the field is equal to $\vec{\mathbf{B}}_0$, and is moving towards the Earth along a line of force with parallel velocity initially equal to $v(1 - \sin^2\alpha_0)^{1/2}$. As it moves, it sees an increasing magnetic field and its parallel velocity decreases, vanishing when the field has reached the value $B_0/\sin^2\alpha_0$; at that moment, the pitch angle of the particle is 90°.

Two cases must be considered, depending on the altitude at which this happens. Either the altitude is sufficiently low for the particle to enter into collisions with the molecules of the neutral atmosphere, to which it gives up its energy; it is 'precipitated' or, the altitude is too high for collisions to have occurred. In this case, the guiding center is still subject to the parallel force and starts moving in the other direction: it 'bounces'. We note that the 'mirror point' altitude is uniquely determined by the field geometry, and that it is the same for all particles having the same pitch angle at the same place far from Earth. At that point, those which will be precipitated have a pitch angle less than some maximum value that defines the 'loss cone'. The loss cone angle is greater the nearer one is to Earth, along a given line of force.

It is clear that when injected into 'open' magnetospheric field lines whose roots are in the North and South polar caps, particles of external origin either are precipitated or definitively ejected towards the tail of the magnetosphere, possible after a single bounce if their pitch angle is between the loss cone angle and 90°.

On the other hand, along the 'closed' lines of force of the geomagnetic field, bouncing of particles can occur both at the North and the South end; the particles thus can find themselves 'trapped'. Particles injected in the region of closed field lines (having a certain pitch angle distribution) therefore come to be either precipitated or trapped, depending on whether their pitch angle is inside or outside the loss cone. In the latter case, trapped particles rebound between two mirror points so that another periodicity, of frequency (very much lower than the cyclotron frequency) called the 'bounce frequency', appears in the motion. In the same way that we were able to abstract out the cyclotron motion and consider only the guiding center, we can, for these particles, abstract out the round trips of the guiding center and consider only the 'bounce line' which drifts because of effects to which we now turn our attention.

2.3.1.3. *Perpendicular Motion of the Guiding Center: Convection and Drift*

In the reference frame moving with the G.C.F., in which the trajectory of a particle is nearly circular, all the forces that act on the particle, averaged over a cyclotron revolution, must equilibrate. The forces are:

– the electrical force $\pm eE_{\perp}(E_{//} = 0$, as we have seen above) and
– the Lorentz-Laplace force $\vec{v}_{\perp} \times \vec{B}$.

If the magnetic field is homogeneous on the scale of the radius of gyration (i.e., if we consider particles of energy low enough for $\lambda^{B} \ll B/\nabla B$), the average of the Lorentz-Laplace force is zero, and consequently the mean of $\vec{E}$ is likewise. It follows that if div $\vec{E} = 0$, the $\vec{E}_{\perp}$ field is zero in the G.C.F. In some frame for which $\vec{E}$ is known, the guiding center must therefore have a drift

$$\vec{V}^{0} = -\vec{E} \times \vec{B}/B^{2}$$

in order to cancel $\overleftarrow{E}$ in its frame.

This drift, called zero order, as it does not depend on the energy or charge of the particles, affects protons and electrons in the same way, and thus cannot create currents or space charges. It is in particular the above-mentioned 'convective' velocity of the thermal plasma in Equation (II.5).

If instead the energy of the particles is large enough for the magnetic field not to be altogether homogeneous on the scale of the gyro-radius, that is if λ^{B} is not much smaller than $B/\nabla_{\perp} B$, the average of the Lorentz-Laplace force over a cyclotron revolution does not exactly vanish, as sketched in Figure II.13. It can be shown that there corresponds a supplementary drift of the G.C.F. given (for an irrotational B field) by

$$\vec{V}^{1}_{\perp} = \frac{mv^{2}}{2eB}(1 + \cos^{2}\alpha)\vec{\nabla}_{\perp} B \times \vec{B}$$

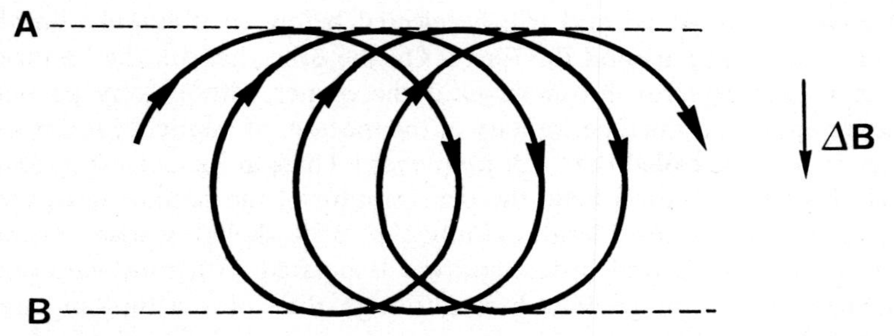

Fig. II.13. A perpendicular gradient ΔB of the magnetic field implies a smaller radius of curvature for the trajectory of charged particles in B that in A, resulting in a drift (toward the right). A particle of opposite sign, rotating the other way around, would drift in the opposite direction.

where α is the pitch angle. This drift, called first order, since it depends on energy and charge and therefore affects electrons and protons differently, can be the cause of transverse currents and polarization electric fields in the magnetospheric cavity.

For particles of high enough energy, the zero order drift becomes negligible with respect to the first order drift. Since the latter drift is perpendicular both to the magnetic field and its perpendicular gradient (orthogonal to the magnetic shells), it carries the bounce lines of the energetic trapped particles around the Earth, along equi L shells (this is actually the exact definition of the L parameter) the electrons eastward, the protons westward. This motion, because of the day-night asymmetry of the magnetospheric field, can 'untrap' the particles whose mirror points are in the zone of unstable trapping indicated on Figure II.14. This zone only contains particles

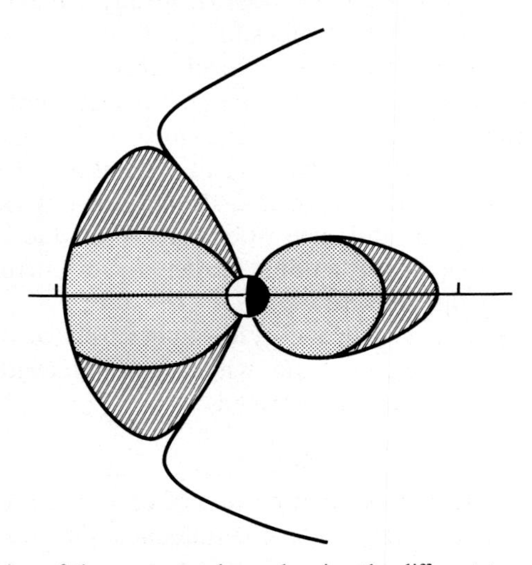

Fig. II.14. Meridian section of the magnetosphere, showing the different trapping zones.
The bounce lines of particles with their mirror points within the stable trapping zone (grey area) can drift around the Earth.
Particles with mirror points within the unstable trapping zone (cross hatched area) cannot complete half a revolution: when injected on the dayside, they are lost to the tail, when injected on the nightside, they are lost to the magnetopause.
On open field lines, no trapping can occur.

that just have been injected and will be ejected before their bounce line has completed a half revolution around the Earth. On the other hand if the bounce line can accomplish a complete revolution around the planet without any particles being ejected, there appears a third periodicity of the motion, of frequency (even lower than the bounce frequency) called the 'drift frequency'. Thus, in the corresponding zone of stable trapping, or 'radiation belt', the drift motion of the bounce lines can be abstracted out, leaving in consideration only the 'drift shells', whose contraction or expansion are due to second order motions associated with time variations of the fields which we shall not go into here although they play a fundamental role in magnetospheric dynamics.

The effect of the perpendicular drifts on the different populations of particles is easy to describe in the plane of the magnetic equator, which is everywhere perpendicular to $\vec{\mathbf{B}}$. Let us consider the particles which have a pitch angle of 90° on this plane. Since B has a minimum there, these particles move only on the plane (they bounce 'in place'). Recapitulating, we can distinguish (Figure II.15):

(a) – 'high energy' particles (proton energy between 1 MeV and 100 keV) for which one can neglect the influence of the electric field. The guiding centers of these particles drift because of the first order effect, conserving their angular momentum (which is proportional to B), thus following the trajectories B = const, electrons moving eastward and protons westward. These motions give rise to the 'ring' current around the Earth, but as long as the charge distribution is symmetric, they cannot cause an electric field.

(b) – 'low energy' particles (proton energy between 100 keV and about 1 keV) which are influenced by the electric field. A parametric adjustment of this field to the observed flux of these particles gives values for the field and the particle trajectories (McIllwain, 1972). It is seen that particles within this energy range can produce a strong morning-evening asymmetry in the magnetospheric charge and ring current distributions. These particles have an energy which varies under the effect of this field, proportionally to the local magnetic field. Therefore, they are 'adiabatically' accelerated while approaching the Earth, and slowed down when moving away from it. Another way of looking at this energy change is as due to a time dependent magnetic field in the frame of the guiding center and a corresponding 'Fermi' or 'betatron' acceleration of the particle.

(c) – 'thermal energy' particles (with energy less than a fraction of a keV) for which the first order drifts become negligible, and whose guiding centers must follow the equipotentials of the electric field as described.

2.3.2. THE CONVECTION OF THERMAL PLASMA

In contrast to the 'free particles' population described in the preceding section, thermal energy particles remain largely under the influence of Coulomb interactions and gravity.

Nevertheless, as we have just seen, above the collision dominated lower ionosphere, thermal plasma is simply drifting perpendicularly to the magnetic field with the velocity $\vec{\mathbf{E}} \times \vec{\mathbf{B}}/B^2$. As far as perpendicular motions are concerned then, once the electric field and magnetic field tubes convection have been described, discussing the convection of the thermal plasma is hardly more than telling the same story with different words. Figures II.8 and II.9 show the flow pattern in the equatorial plane and polar caps.

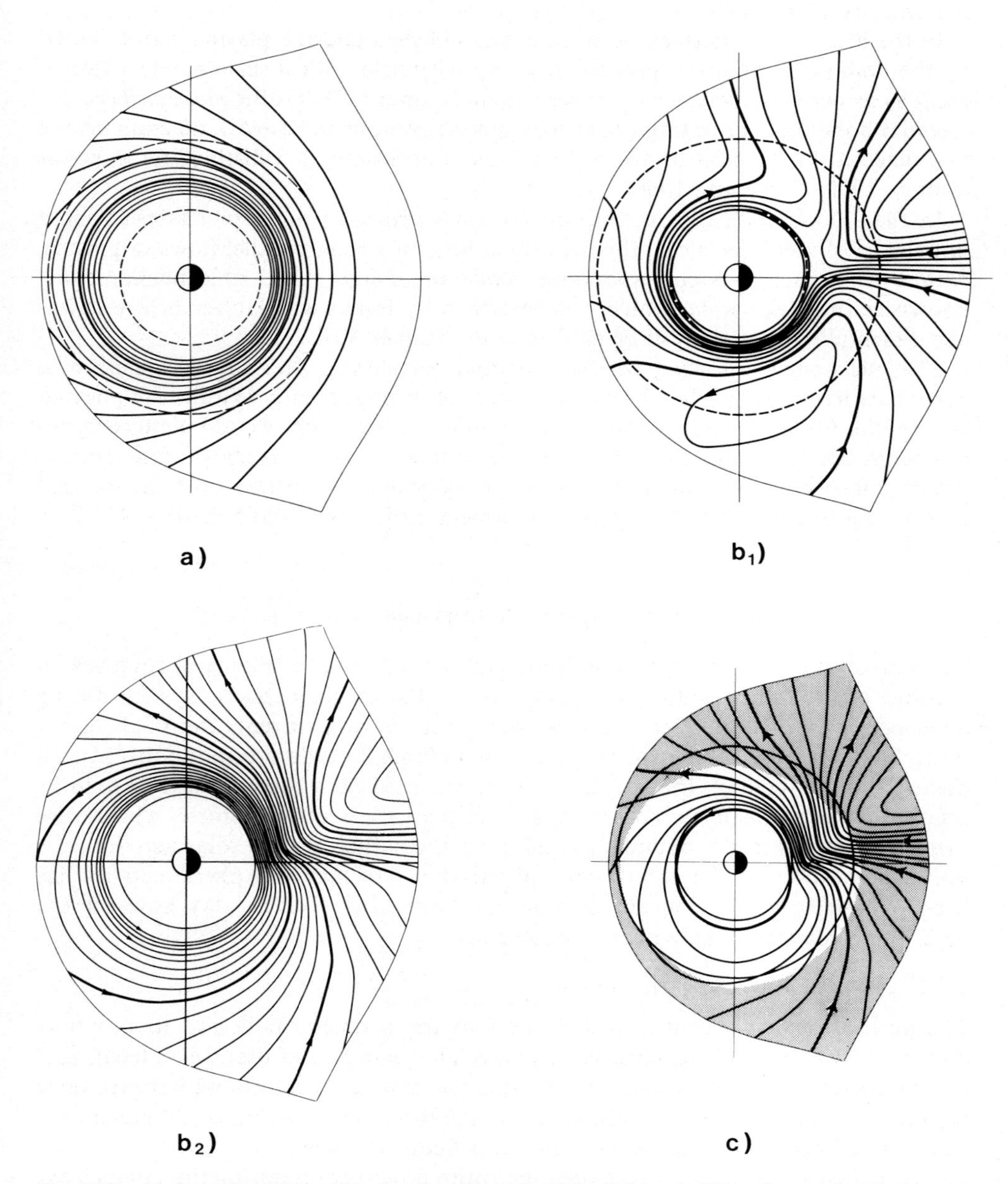

Fig. II.15. Trajectories of the guiding center of particles with pitch angle 90° in the equatorial plane: (a) Particles with sufficient energy (between 100 keV and 1 MeV) following the curves B = const, (b) Protons (b_1) and electrons (b_2) with energy between 1 and 100 keV. (c) Low energy particles following equipotentials of the electric field.

The hydrodynamical situation that results for the thermal plasma within the flux tubes is very different for the corotating (low L) tubes and for the higher L flux tubes which undergo large changes of volume while convecting and eventually open up to the interplanetary medium.

In the first case, a quasi-static regime is established and the plasma is distributed by the ambipolar diffusion process in the gravity field with a scale height twice as great as its parent neutral gas, as described in Chapter I. This is the *plasmasphere*, the toroidal bubble of corotating upper ionospheric protons in hydrostatic equilibrium, first detected by its effect on the propagation of atmospheric 'whistling' radio waves (Storey, 1952; Carpenter, 1966).

In the second case, that is, outside the *plasmapause*, a more or less permanent unbalance of pressures along the field tubes sets up strong parallel flows of plasma, with a net loss and a corresponding depletion of light ions. This phenomenon, known as the '*polar wind*' which was described by Banks and Holzer before it was experimentally observed, will be explained in Chapter VII.

The question of knowing whether thermal particles in the magnetosphere ever behave as free particles has been the subject of a long debate. Within the plasmasphere, the plasma density always remain sufficient to insure the dominance of interactions between charged particles. But Marubashi (1970) has shown that outside the plasmasphere, above an altitude of some thousands kilometers, even the thermal plasma can receive a purely adiabatic treatment, as Lemaire and Scherer (1973) had argued.

2.4. Magnetospheric Activity and Substorms

The description of magnetospheric fields and particles above (Figure II.16) gives an account of the fundamental stationary state of the magnetospheric system during quiet periods (existence of the plasmasphere and the ring current) and of the gross latitudinal structure of ionospheric phenomena (high latitude precipitations, plasma escape and convection). But it must be supplemented by a theory, these days in its infancy, of magnetospheric activity, that is of the variety of phenomena which arise from fluctuations in the boundary conditions, themselves due to solar activity. It is only since 1964 that the great temporal variability of magnetospheric activity has been organized by the concept of magnetic 'substorm'. To this day however the description remains largely phenomenological.

2.4.1. AURORAL AND MAGNETIC STORMS

The auroral oval is eccentric with respect to the magnetic pole and its center is displaced by 3° toward the dark hemisphere. The geometry of the oval is fixed, as a first approximation, with respect to the Sun, and the Earth rotates underneath once per day. The auroral zone is nothing more than the locus of the midnight part of the oval where intense auroral displays are most frequently seen.

The sequence of auroral events over the entire polar region during the events from auroral quiet, through the various active phases, to subsequent calm, has been called after Chapman an auroral substorm which coincides with a magnetic substorm. Akasofu (1964) broke down the phenomenon into an expansion phase followed by a

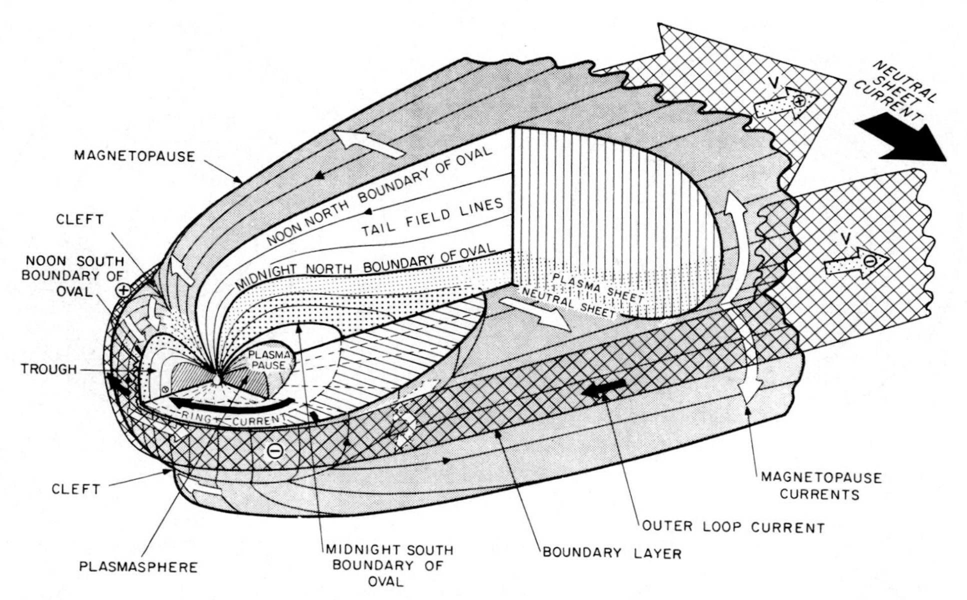

Fig. II.16. A sketch of the quiet magnetospheric system (Heikkila, 1972). It must be kept in mind that the magnetosphere is not a solid object.

recovery phase (Figure II.17). After a quiet phase, during which are observed homogeneous auroral bands, sometimes barely visible, the expansive phase starts with a sudden brightening of one of the quiet arcs a few thousand kilometers in length, approximately centered at the midnight meridian. After a few minutes, the region of min^{-1}. The highest geomagnetic latitude reached by active bands varies with the intensity of the substorm, between 70° and 80° or even beyond. After the formation of the of the substorm, between 70° and 80° or even beyond. After the formation of the bulge resulting from the poleward expansion, folds are formed and move rapidly westward. These propagating folds are called a 'westward travelling surge'. The duration of the expansion phase varies between one and three quarters of an hour. The recovery phase starts when the arcs have reached their maximum poleward location and involves weak loop and surge activity at the poleward border of the auroral excited region, followed by an equatorward drift of auroral forms and the eventual re-establishment of the pre-substorm quiet arc configuration. Each auroral substorm has a lifetime of the order of 1 to 3 h. Several substorms occur at intervals of a few hours or less during moderately disturbed periods. During a fairly disturbed period, a new substorm may start before the second phase of the preceding one is over.

Large-amplitude magnetic perturbations occur in conjunction with auroral substorms. Such magnetic variations, like the negative 'bay' of the horizontal component of the field, have typical amplitudes of several hundred gammas and last for a period of roughly one hour. Early workers believed that the substorm magnetic variations were caused by purely ionospheric current systems. However, the older concept proposed by Birkeland (1908–1913) is closer to the truth: currents run along geomagnetic field lines, westward through the ionosphere (the auroral electrojet),

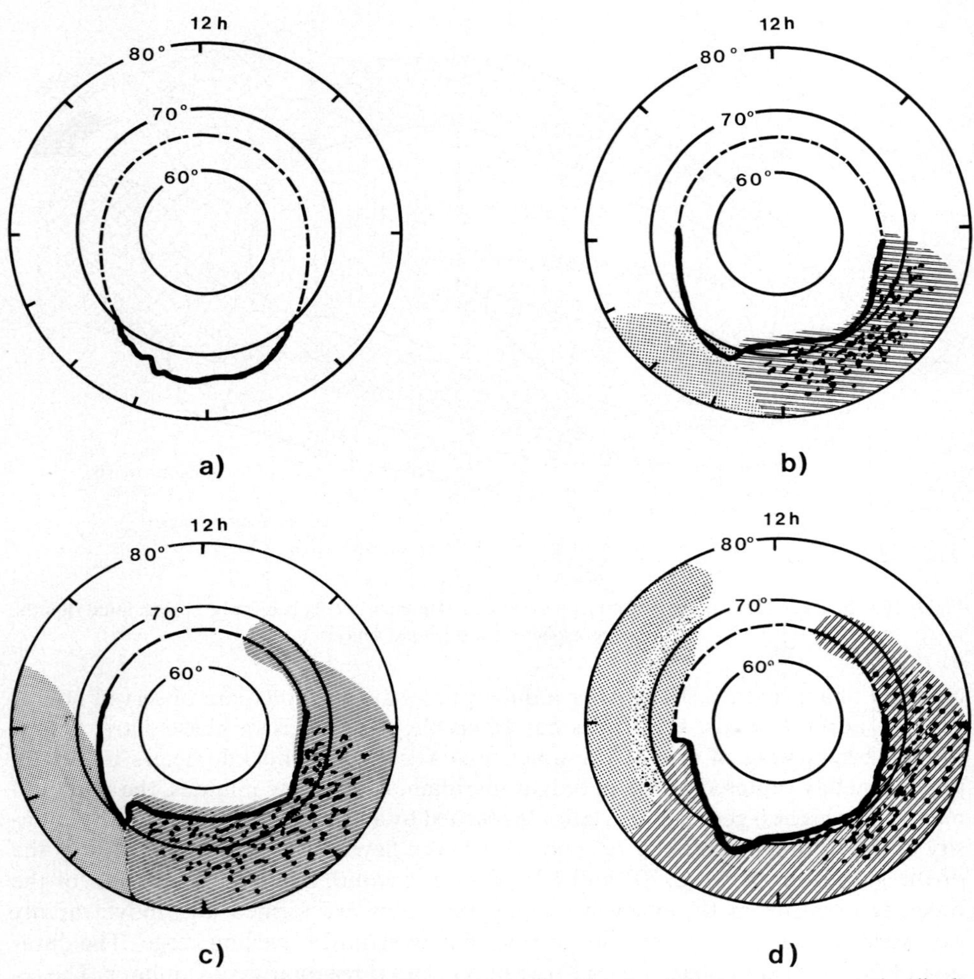

Fig. II.17. Scheme of the development of a substorm in solar (local time) and geomagnetic (dipole latitude) coordinates (Akasofu *et al.*, 1966). (a) quiet situation, (b) beginning of expansion, (c) maximum expansion, and (d) beginning of recovery. Auroral arcs are shown in full black. Departures of the ground magnetic field from its quiet values are shown in dotted (positive) and hatched (negative) areas.

and eastward in the magnetosphere (the ring current), as will be discussed in Chapter VII.

2.4.2. MAGNETOSPHERIC STORMS

It gradually dawned in the last ten years that substorms involve the whole magnetosphere and that auroral substorms are the manifestation in the atmosphere of phenomena of magnetospheric scale. The magnetospheric interpretation of a substorm can be divided into three stages: the growth phase when the energy reservoir in the tail is being filled; the expansion phase when the energy is rapidly released to the inner magnetosphere, and the recovery phase when the currents set up during the expansion phase are decaying. There is some controversy as to whether the growth phase can be unambiguously recognized in ground based data. However, data taken in the magnetosphere show the growth phase quite clearly.

The growth phase is seen as an inward motion or erosion of the sunward magnetosphere, caused by a southward turning of the interplanetary field, according to the process described in Section 2.2.2 of this chapter. The eroded flux is transported by the solar wind to the lobe of the tail. As flux is added to the tail, the magnetic energy begins to increase above the normal value. Consequently the tail current system moves earthward and its strength increases.

The cause of the expansion phase is not yet firmly established. The most popular theory is that the expansion phase begins when the near-Earth tail plasma sheet becomes extremely thin and a so-called X-type (corresponding to the crossing of two lines of force) neutral point forms within the plasma sheet. The cross tail current is disrupted at the neutral point and diverted to the closed field line, magnetic flux being annihilated and plasma energized at the neutral point. Bursts of high energy particles are injected into the inner magnetosphere, with some being precipitated in auroral zones (electrojets), and some trapped as an initially asymmetric addition to the radiation belts (ring current).

The recovery phase begins when the additional energy in the tail has been completely released through field annihilation. This description is highly qualitative and reflects the incomplete state of our knowledge. Quantitative models describing the relationship between auroral and magnetospheric substorms are still beyond us.

2.4.3. IONOSPHERIC AND THERMOSPHERIC EFFECTS

The upper atmosphere exhibits various stormy features, affecting mainly the auroral and polar regions. Strong enhancements of ionization by precipitations in the lower ionosphere cause an heavy absorption of radio waves, in particular of the cosmic radio noise. These absorption events are observed simultaneously at geomagnetically conjugate areas of the auroral zones (Figure III.7). During the early phase of the substorm, absorption is confined to the narrow region where the first sign of the visual aurora is observed. During the expansion phase of the auroral substorm, the absorption extends in all directions. In the morning sector, the absorption region expands both along the auroral oval and the auroral zone; this is the ‘polar blackout’.

In addition to anomalous ionization in the lower ionosphere, the higher polar ionosphere is greatly distorted during geomagnetic storms and presents large deviations from horizontal stratification. The deformation of the ionosphere is seen as an anomalous behavior of the *F* region. Auroral radar echoes are also observed with a lifetime of one to three hours.

Heating of both the neutral and ionized species is also observed, the corresponding perturbation of the atmosphere propagating to mid-latitude regions, and modifying the photochemical ionosphere there. Magnetospheric transient electric fields are observed to penetrate even at mid-latitudes and affect ionospheric dynamics.

Finally, the whole topside plasma distribution system of the plasmasphere and mid-latitude trough between the plasmapause and auroral region is upset, as the enhanced convection ‘peels off’ a chunk of the plasmasphere, with subsequent refilling of the tubes of force.

PART TWO

THE TECHNIQUES OF IONOSPHERIC MEASUREMENTS

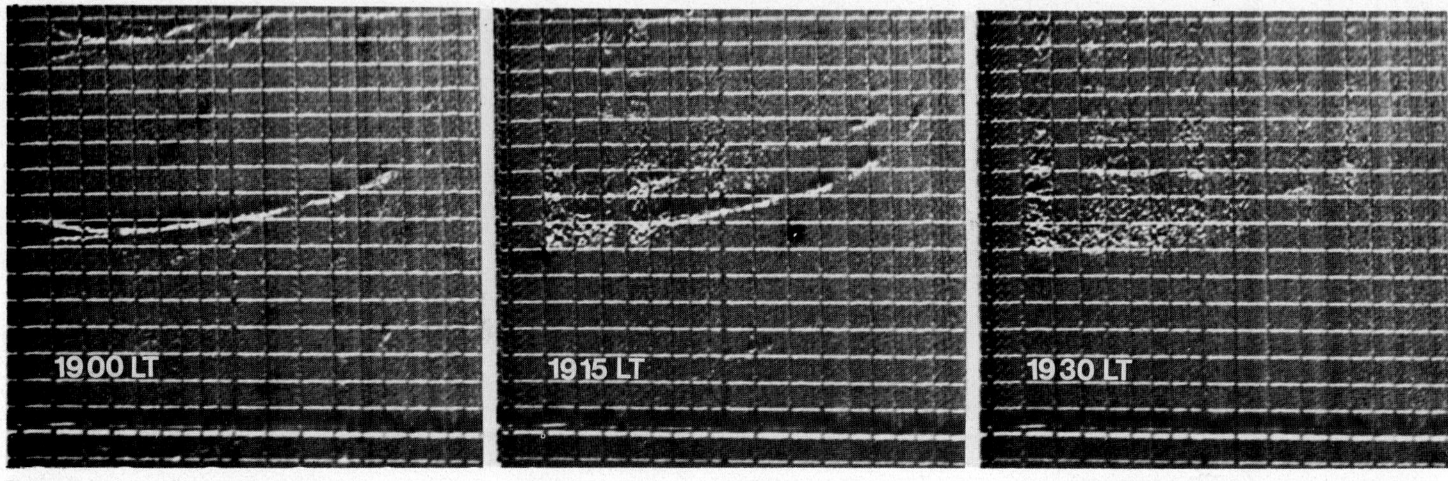

Two generations of ionospherists have labored over 'ionograms', the conventional X band radar spectra of upper atmospheric conditions. There have been at times some one hundred simultaneously operating ionosondes scattered over the globe, that could each produce up to an ionogram a minute. With such countless automatic three-dimensional plots of the radio echoing properties of the ionosphere (frequency in abcissa, range in ordinate, strength in trace intensity), the network yielded a considerable amount of information, so considerable in fact that it was sometimes more complicated to interpret ionogram data than to understand the ionosphere. The art of ionogram reading has been the source of numerous discoveries about upper atmospheric phenomena, and of many of the strange names by which they are known. Here we see for example the development of a 'spread *F*' condition at Sarh (Chad) Central Africa, on April 8, 1971. Needless to say, the *F* region is not spread more at 1900 than at 1930 local time; it is just the ionogram that spreads. (Courtesy of J. M. Faynot, Centre de Recherches en Physique de l'Environnement, Issy les Moulineaux)

CHAPTER III

PROPAGATION OF RADIO WAVES

We summarized in Chapter I the character of the interaction with the atmosphere of that part of the electromagnetic spectrum capable of exciting transitions in individual neutral molecules. Longer wavelengths in the electromagnetic spectrum (microwaves) do not correspond to any strong interaction, whence a large transparent 'window' put to good use by radioastronomers and satellite telecommunication engineers, until the frequency domain capable of exciting collective response of the free electrons in the ionosphere is reached.

Ever since the first long distance radio links were achieved by Marconi at the beginning of the century, the propagation of these longer electromagnetic waves has been the basis from which the essentials of our overall knowledge and practical use of the ionosphere have been derived. The theory of this propagation was developed under the name of magnetoionic theory from Eccles (1912) to Ratcliffe (1959). It has been revived and generalized within the modern framework of plasma physics, which bears to the radio spectrum the same relationship as solid state physics to the optical spectrum. Very many works on these subjects exist. For example the book by Stix (1962) reviews the general theory of waves in plasmas, while those by Budden (1961), Rawer and Suchy (1967) and Davies (1969) are more slanted towards ionospheric applications.

The *propagation* phenomena considered in the present chapter are those in which the ionospheric plasma exhibits the macroscopic scale character of a smooth conducting dielectric 'fluid'. In the following chapter, we consider the *scattering* of electromagnetic waves, due in particular to the fluctuating distribution of the particles constituting the plasma, left out by such a description: the fluid-like ionospheric medium is, as it were, 'boiling'.

After having reviewed some general ideas about plane waves and the propagation of phase and of energy, we examine the propagation of electromagnetic waves in a homogeneous collisionless and cold plasma, the latter qualifier expliciting the fact that we neglect the thermal motion of the particles. The results obtained are next extended to the case of the prevailing one-dimensional (vertical) inhomogeneity, that is to the horizontal stratification of the ionosphere. Phase and group velocity effects on high and low frequency waves, respectively, are then described. Finally, we cover absorption and other phenomena where the role of interparticle collisions becomes preponderant, and which are therefore characteristic of propagation through the lower (D) ionospheric region embedded in the very much denser neutral atmosphere.

3.1. Electromagnetic Waves in Plasmas

3.1.1. PROPAGATION OF PHASE AND ENERGY

3.1.1.1. *Notion of a Plane Wave and Propagation of Phase*

When interested in waves of small amplitude, we can always linearize, with respect to the wave impressed perturbation, the equations that describe the medium. Since Maxwell's equations are themselves linear, we finally obtain linear differential equations with constant coefficients (if the medium is homogeneous). These admit solutions which depend on the space variable $\vec{\mathbf{r}}$ and time t by a factor $\exp(i\,\vec{\mathbf{k}}\,.\,\vec{\mathbf{r}} - i\omega t)$, provided that ω and $\vec{\mathbf{k}}$ satisfy a relation of the form $F(\vec{\mathbf{k}},\omega) = 0$ which depends only on the characteristics of the medium and is called the dispersion relation. The said solutions represent plane waves, the surfaces of equal phase being planes normal to the vector $\vec{\mathbf{k}}$, whose equations are $\vec{\mathbf{k}}\,.\,\vec{\mathbf{r}} - \omega t = \text{const}$. As a function of time, the wave front characterized by a constant phase moves parallel to $\vec{\mathbf{k}}$ with a velocity $v_\phi = \omega/k$, which is the phase velocity. The index of refraction N is defined by $N = c/v_\phi = kc/\omega$, where c is the velocity of electromagnetic waves in a vacuum. The quantity ω is called the angular frequency and characterizes the time variations of the wave. The wave vector $\vec{\mathbf{k}}$ describes the spatial variations of the wave and the wavelength λ is related to $\vec{\mathbf{k}}$ by the relation $\lambda = 2\pi/k$.

If ω is constrained to be real, there may exist solutions such that $\vec{\mathbf{k}}$ and N are also real; this represents propagation without attenuation. On the other hand, $\vec{\mathbf{k}}$ and N might be purely imaginary and the corresponding wave is then evanescent, the field decreasing exponentially with distance. When $\vec{\mathbf{k}}$ and N have both a real part and an imaginary part, the propagation is accompanied by attenuation (or growth).

3.1.1.2. *Notion of Group Velocity*

The plane waves we have written about, fill all of space if ω and $\vec{\mathbf{k}}$ are real. Their propagation is not associated with any transport of energy; this implies that their phase velocity v_ϕ can very well be greater than c. An actual physical phenomenon on the other hand, bounded in time and space, can be represented by a superposition of monochromatic waves $A(\vec{\mathbf{k}})\, e^{i(\vec{\mathbf{k}}\cdot\vec{\mathbf{r}} - \omega t)}$, the values of ω and $\vec{\mathbf{k}}$ satisfying the dispersion relation and $A(\vec{\mathbf{k}})$ being such that they remain close to central values ω_0 and $\vec{\mathbf{k}}_0$. At the center of such a wave packet, the various components interfere constructively, while outside the central region they interfere destructively. As the phase of each component evolves with its own velocity; the region in which the interference is constructive moves, and this corresponds to a propagation of the energy associated with the motion of the wave packet, at a velocity known as the group velocity. To be precise, let the field have the form:

$$E(\vec{\mathbf{r}},t) = \int A(\vec{\mathbf{k}})\, e^{i(\vec{\mathbf{k}}\cdot\vec{\mathbf{r}} - \omega(\vec{\mathbf{k}})t)}\, \overrightarrow{\mathbf{dk}}$$

where $A(\vec{\mathbf{k}})$ differs from zero only when $\vec{\mathbf{k}}$ is near $\vec{\mathbf{k}}_0$. The monochromatic waves interfere non-destructively when the phase is stationary i.e., when

$$\partial(\vec{\mathbf{k}}\cdot\vec{\mathbf{r}} - \omega(\vec{\mathbf{k}})t)/\partial\vec{\mathbf{k}} = 0$$

or

$$\vec{\mathbf{r}} - \partial\omega/\partial\vec{\mathbf{k}}\, t = 0.$$

Therefore, the group velocity $\vec{\mathbf{v}}_g$ is $\partial\omega/\partial\vec{\mathbf{k}}$, its components along the x, y and z axes being respectively $\partial\omega/\partial k_x$, $\partial\omega/\partial k_y$, $\partial\omega/\partial k_z$. It is different from the phase velocity v_ϕ, unless the magnitude of the latter is independent on the magnitude of $\vec{\mathbf{k}}$ (non-dispersive medium).

The preceding analysis applies if ω and $\vec{\mathbf{k}}$ are real. It remains valid if the imaginary parts of ω or $\vec{\mathbf{k}}$ are small compared with the respective real parts. Otherwise, the group velocity $\partial\omega/\partial\vec{\mathbf{k}}$ can no longer be identified with the velocity of propagation of energy. Hines (1951) presents a general discussion of the relations between these two velocities. In the cases which we shall have to consider, both velocities indeed are identical.

3.1.2. PROPAGATION IN A COLD, HOMOGENEOUS, COLLISIONLESS PLASMA

Before turning to the properties of waves in a magnetoplasma, let us briefly study the case of a plasma without magnetic field in order to bring out the essential physics of the problem.

3.1.2.1. *Plasma without Magnetic Field*

Under the influence of an electromagnetic wave, charged particles are put into motion, and become oscillators which, in turn, radiate in all directions. Thus, at a point in the medium illuminated by an incident wave, the total field is a superposition of the field of the incident wave and the field radiated by the particles acted upon by this total field; there results a wave that propagates at a velocity different from that of the incident wave.

Under the influence of an electric field $\vec{\mathbf{E}}e^{i(\vec{\mathbf{k}}\cdot\vec{\mathbf{r}}-\omega t)}$, a charge acquires a velocity of the form $\vec{\mathbf{v}}\,\mathrm{e}^{i(\vec{\mathbf{k}}\cdot\vec{\mathbf{r}}-\omega t)}$, with

$$-\,i\omega\vec{\mathbf{v}} = e\vec{\mathbf{E}}m \tag{III.1}$$

where m is the mass of the particle: this expresses the proportionality of acceleration to force. When the motion of the ions is neglected with respect to that of the electrons, which are several thousand times lighter, and therefore much more mobile, there results a current density:

$$\vec{\mathbf{j}} = ne\vec{\mathbf{v}} = \frac{+\,ie^2n}{m\omega}\,\vec{\mathbf{E}} \tag{III.2}$$

where n is the concentration of the electrons (and m is now the electron mass). Maxwell's equations

$$\nabla \times \vec{\mathbf{E}} = -\,\mu_0\partial\vec{\mathbf{H}}/\partial t$$
$$\nabla \times \vec{\mathbf{H}} = \varepsilon_0\partial\vec{\mathbf{E}}/\partial t + \vec{\mathbf{j}}$$

become, for waves of the form $\mathrm{e}^{i(\vec{\mathbf{k}}\cdot\vec{\mathbf{r}}-\omega t)}$ (this factor, which multiplies all the terms, is omitted):

$$\vec{\mathbf{k}} \times \vec{\mathbf{E}} = \omega\mu_0\vec{\mathbf{H}}$$
$$\vec{\mathbf{k}} \times \vec{\mathbf{H}} = -\varepsilon_0\omega\vec{\mathbf{E}} - i\vec{\mathbf{j}}.$$

These two equations combine into

$$\vec{\mathbf{k}} \times (\vec{\mathbf{k}} \times \vec{\mathbf{E}}) = \omega\mu_0(-\varepsilon_0\omega\vec{\mathbf{E}} - i\vec{\mathbf{j}})$$

so that when $\vec{\mathbf{j}}$ is replaced by its value found from Equation (III.2), we obtain:

$$\vec{\mathbf{k}} \times (\vec{\mathbf{k}} \times \vec{\mathbf{E}}) = -\,\omega^2\mu_0\varepsilon_0\vec{\mathbf{E}} + \mu_0\frac{ne^2}{m}\vec{\mathbf{E}} \tag{III.3}$$
$$= \mu_0\varepsilon_0\vec{\mathbf{E}}\left[\frac{ne^2}{m\varepsilon_0} - \omega^2\right].$$

Equation (III.3) is a very simple version of a dispersion relation $F(\vec{\mathbf{k}}, \omega) = 0$. If the $\vec{\mathbf{E}}$ field is parallel to $\vec{\mathbf{k}}$ (longitudinal polarization), the left hand side is zero, and Equation (III.3) reduces to:

$$\omega^2 = \frac{ne^2}{m\varepsilon_0}.$$

The only possible longitudinal oscillations thus take place with angular frequency ω_p, where $\omega^2{}_p = ne^2/m\varepsilon_0$, and is independent of $\vec{\mathbf{k}}$. We recognize the angular frequency ω_p and the corresponding frequency $f_p = \omega_p/2\pi$ as the *plasma frequency* which, we recall, is of the order of several MHz in the ionosphere. Longitudinal oscillations in a plasma do not involve propagation of energy, since $\vec{\mathbf{v}}_g = \partial\omega/\partial\vec{\mathbf{k}}$ is zero.

We now consider a wave whose electric field $\vec{\mathbf{E}}$ is perpendicular to $\vec{\mathbf{k}}$ (transverse propagation). Because of the general identify:

$$\vec{\mathbf{k}} \times (\vec{\mathbf{k}} \times \vec{\mathbf{E}}) = -k^2\vec{\mathbf{E}}_{\perp}.$$

Equation III.3 is equivalent to:

$$-k^2 = \mu_0\varepsilon_0({\omega^2}_{\mathrm{p}} - \omega^2)$$

and since $\varepsilon_0\mu_0c^2 = 1$, and the index of refraction N is defined as Kc/ω,

$$N^2 = 1 - {\omega^2}_{\mathrm{p}}/\omega^2.$$

In ionospheric notation ${\omega^2}_{\mathrm{p}}/\omega^2$ is usually called X, and the above equation is written:

$$N^2 = 1 - X. \tag{III.4}$$

Thus the index of refraction of a plasma for transverse electromagnetic waves depends on the frequency, approaching unity for frequencies much higher than the plasma frequency. Waves of frequency less than the plasma frequency have a purely imaginary index of refraction; therefore they are evanescent and cannot propagate.

This simple example of a cold, unmagnetized plasma shows how the calculation of the current induced in the plasma by an electric field, together with Maxwell's equations, allows us to determine the dispersion relation governing the parameters ω and $\vec{\mathbf{k}}$ of waves that can propagate in the medium. We also see how the direction of the electric field with respect to the wave normal, i.e. what is usually called the *polarization* of the wave, has to obey certain conditions for wave propagation to be allowed.

A more refined treatment of the problem consists mostly in replacing Equation (III.2) by a more rigorous equation, i.e. in a more precise determination of the conductivity tensor relating the current to the field. The following features may have to be taken into account: the presence of ions, the magnetic field, and thermal motions of the particles. To take account of the last effect, one can either introduce a pressure term using physical assumptions or one can study the evolution of the particle distribution function by means of a kinetic equation such as the Vlasov equation. We only mention this problem as a reminder, referring the reader to specialized works for more details, since the description in terms of a cold plasma is sufficient for this chapter.

3.1.2.2. *Magnetoplasma*

We again neglect the presence of ions, which play practically no role if the wave frequencies considered are much greater than the ion gyrofrequency. If we also neglect collisions and thermal motion, then it is sufficient to replace Equation (III.1) by an equation which takes account of the presence of the terrestrial magnetic field $\vec{\mathbf{B}}$:

$$-i\omega m\vec{\mathbf{V}} = e(\vec{\mathbf{E}} + \vec{\mathbf{V}} \times \vec{\mathbf{B}}).$$

The calculation is performed in a way quite analogous to that for plasmas without magnetic field, and we need only conclude as to the principal characteristics of the waves that can propagate in a cold magnetoplasma.

The presence of the magnetic field introducing an anisotropy in the electrons' motion, the angle θ between the vectors $\vec{\mathbf{k}}$ and $\vec{\mathbf{B}}$, is clearly a fundamental parameter.

The electronic gyrofrequencies ω_B or $f_B = \omega_B/2\pi$ now enter into the expression for the index of refraction (f_B is of the order of 1 MHz in the ionosphere). In ionospheric notation we usually use the reduced variables $Y = \omega_B/\omega$, $Y_L = Y\cos\theta$ and $Y_T = Y\sin\theta$, the parameter X, as before, being $\omega^2{}_p/\omega^2$. The index of refraction is then shown to be given by the Appleton-Hartree equation:

$$N^2 = 1 - \frac{X}{1 - \dfrac{Y_T{}^2}{2(1-X)} \pm \sqrt{Y_L{}^2 + \dfrac{Y_T{}^4}{4(1-X)^2}}}. \tag{III.5}$$

Since there is a $\pm$ sign before the square root in the denominator, two modes of propagation exist simultaneously for each frequency; this is a phenomenon similar to the optical birefringence of certain crystals.

Figure III.1 shows the variation of N^2 with X, Y being held constant. The behavior of the curves is different in the cases $Y > 1$ and $Y < 1$. The angle θ enters as a parameter that defines a family of curves. Negative values of N^2 evidently correspond to evanescent waves, and only positive values of N^2 can be associated with waves that really propagate.

In the case where $\vec{\mathbf{k}}$ is parallel to $\vec{\mathbf{B}}$ ($\theta = 0$), Equation (III.5) reduces to

$$N^2 = 1 - X/(1 \pm Y).$$

The two modes corresponding to the double sign are represented on Figure III.1. Both waves are circularly polarized, one to the right (R), the other to the left (L).

If $\vec{\mathbf{k}}$ is perpendicular to $\vec{\mathbf{B}}$ ($\theta = \pi/2$), the two values of the index are

$$N^2 = 1 - X$$

and

$$N^2 = 1 - \frac{X}{1 - \dfrac{Y^2}{1-X}}.$$

The first of the two corresponding waves is linearly polarized along the magnetic field (hence transversely); its index does not depend on the magnetic field; it is called the 'ordinary' mode (O). The second wave or 'extraordinary' mode (X) is 'mixed', with polarization neither purely longitudinal nor purely transverse; its electric field vector describes an ellipse in a plane perpendicular to $\vec{\mathbf{B}}$; this ellipse degenerates into a circle when N goes to zero, into a straight line when N approaches infinity.

For oblique propagation, the curves for the refractive index occupy a position intermediate between the $\theta = 0$ and $\theta = \pi/2$ cases. The various types of propagation can be classified according to the zone on the figure in which the curve $N^2(X)$ is found:

– Zone 1 is bounded by the modes at $\theta = 0$ and $\theta = \pi/2$ labelled with the letters L and O. We call the corresponding mode *ordinary* (O) an extension of the case $\theta = \pi/2$. The index vanishes for $X = 1$, or $f_o = f_p$.

– Zone 2 is bounded by the R and X modes, and the corresponding mode is called extraordinary (X). The index vanishes for a frequency f_x corresponding to $X = 1 \pm$

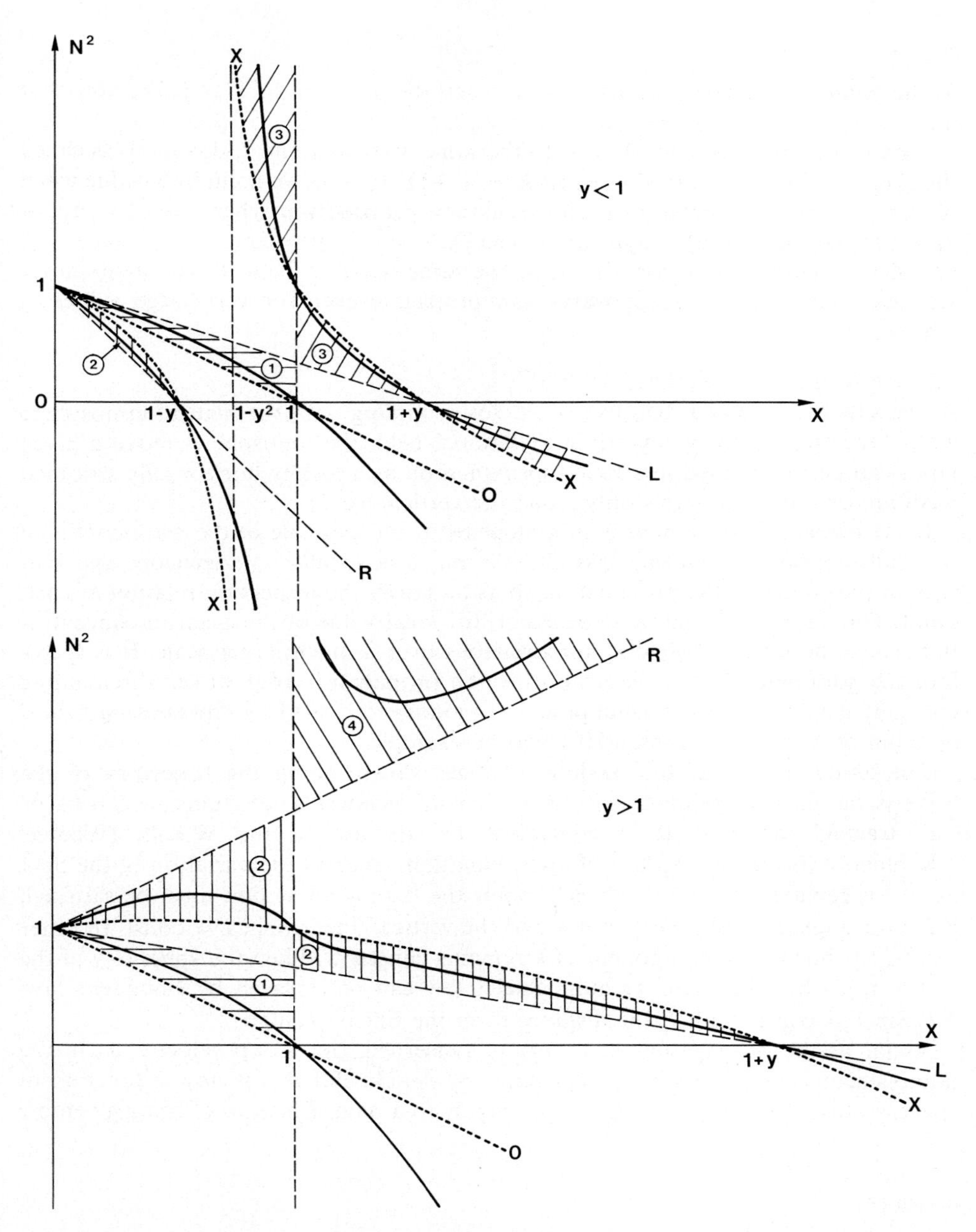

Fig. III.1. The double valued index of refraction of a magnetoplasma as a function of the magnetoionic parameters X (square of the ratio of the electron plasma frequency to the wave frequency) and Y (ratio of the electron gyrofrequency to the wave frequency). The dispersion curves shown give N^2 for Y constant and X variable, as in the ionosphere the electron gyrofrequency varies much less than the electron plasma frequency. Dashed curves are for waves along the magnetic field, dotted curves for waves perpendicular to the magnetic field. They bound the different zones identified in the text corresponding to the various propagation modes for oblique waves (full curves): zone 1 ('ordinary' mode) is hatched horizontally, zone 2 ('extraordinary' mode) is hatched vertically, zone 3 ('Z' mode) and zone 4 ('whistler' mode) are hatched obliquely. Negative values of N^2 correspond to evanescent waves.

Y, the minus sign corresponding to $Y < 1$ and the plus sign to $Y > 1$. We note that if $f_p \gg f_B$, then, very nearly, $|f_X - f_p| \simeq f_B/2$.

– Zone 3 corresponds, for $Y < 1$, to the same wave as Zone 2; this mode is called the Z mode. The index vanishes when $X = 1 + Y$. It takes on an infinite value when $X = (1 - Y^2)/(1 - Y^2 \cos^2\theta)$, or, for transverse propagation, when $X = 1 - Y^2$, or $f_T = (f^2{}_p + f^2{}_B)^{1/2}$ (upper hybrid frequency).

– Zone 4 corresponds, for $Y > 1$, to the same wave as Zone 1. The index never vanishes and non-evanescent waves can propagate even for very large values of X ($f \ll f_p$).

3.1.3. PROPAGATION IN A STRATIFIED MEDIUM

As we saw in Chapter I, because of its close coupling with the neutral atmosphere within the spherically symmetric gravitational field, the ionosphere above a given point can be represented in a first approximation as a locally horizontally stratified medium, i.e., one that varies only along the vertical axis.

If the medium is only slightly inhomogeneous on the scale of the wavelengths of the fields considered, one can take the plasma to be locally homogeneous and k to vary in the course of the propagation so as to satisfy the dispersion relation at each point. The dispersion relation depends on the local value of the electron concentration, since the magnetic field is homogeneous over a much larger scale. It is therefore the parameter X that varies during the propagation of a wave, Y remaining constant. This explains the usual practice in ionospheric work of representing N as a function of X, of which Figure III.1 was an example.

The assumption that the medium changes slowly along the trajectory of the wave is the basis of geometrical optics (limit of Maxwell's equations as $\lambda \to 0$), or 'ray tracing' methods. It is equivalent to the use of the W.K.B. (Wentzel – Kramers – Brillouin) method of approximation, wherein the variation of the field as $e^{i\vec{k}\cdot\vec{r}}$ is replaced by $e^{i\int \vec{k}\cdot d\vec{r}}$. At each point the laws of refraction must be satisfied; if I is the angle between the vector $\vec{\mathbf{k}}$ and the vertical, then $N \sin I = \text{const.}$ In other words, the horizontal component of $\vec{\mathbf{k}}$ remains constant. When the sphericity of the Earth needs be taken into account, Snell's law can be replaced by Bouguer's law: $NR \operatorname{Sin} I = \text{const.}$, with R the distance from the Earth's center.

As far as the propagation of energy is concerned, the group velocity, as in the homogeneous case, is given at each point by $\partial\omega/\partial\vec{\mathbf{k}}$. But this is now a function of space, so that the time for energy to propagate to a point $\vec{\mathbf{r}}$ instead of being given by:

$$t = \vec{\mathbf{r}} \cdot \partial\vec{\mathbf{k}}/\partial\omega$$

becomes

$$t = \int_0^{\vec{r}} \frac{\partial\vec{\mathbf{k}}}{\partial\omega} \cdot d\vec{\mathbf{r}}.$$

3.2. The Reflection of Radio Waves by the Ionosphere

3.2.1. TOTAL REFLECTION

Let us consider a plane wave propagating into the ionosphere. For a plane stratified ionosphere, at each point, $N \sin I = \text{const.}$ must be satisfied. Outside the ionosphere, propagation takes place in a medium with index near the vacuum value, $N = 1$. If I_0 designates the angle between $\vec{\mathbf{k}}$ and the vertical entering of the ionosphere, then N must satisfy

$$N \sin I = \sin I_0.$$

This can only be true if N remains greater than $\sin I_0$. If the electron concentration, and therefore X, reach values (Figure III.1) for which N is too small, the equality cannot be satisfied; there will occur a phenomenon of total reflection similar to the optical phenomenon at the surface of separation between two media of different indices.

Total reflection can take place even for vertical incidence, $I_0 = 0$, when the index of refraction of the medium vanishes. The wave becomes evanescent and cannot propagate past the level above which the index is imaginary, meaning that all the energy of the incident wave is reflected back.

The analysis we have made is only approximate, since we have implicitly taken the group velocity of the waves, $\vec{\mathbf{v}}_g$, to be in the same direction as the phase velocity. The relation $\vec{\mathbf{v}}_g = \partial\omega/\partial\vec{\mathbf{k}}$ shows that this property, which holds for an isotropic medium, is no longer true for an anisotropic medium. Nonetheless, if the index does not vary too rapidly with the angle θ, group velocity and phase velocity diverge only slightly and the above analysis remains qualitatively correct.

For waves of low frequency (e.g. less than 100 kHz), the wavelength (greater than 3 km) is no longer small compared with the scale of variation of the electron concentration. Then it is necessary to rely on a complete treatment, such as is described, for example, in Budden (1961) under the name 'full wave method'. We shall not go into this method any further, because it is not essential for the types of ionospheric measurements which will be presented below.

3.2.2. VERTICAL SOUNDING

Vertical sounding has long been the most prevalent method for studying the ionosphere, and through it we have acquired the bulk of our knowledge of the structure and behavior of the electron concentration in the Earth's atmosphere: profile as a function of altitude; diurnal, seasonal and 11 yr variations; variations with geographical and magnetic latitudes, etc.

3.2.2.1. *Ionosondes*

We have seen that waves of suitable frequency are reflected by the ionosphere. If the frequency f is increased, the reflection can take place for larger and larger values of f_p, and thus of electron concentration. If the electron concentration increases monotonically with distance from the radar, the reflection thus occurs further and further away, and the echo takes more time to return.

A mechanical model of variable frequency vertical sounding, 'may be constructed by rolling a marble under gravity up a hill for which the vertical profile is the ionospheric ionization density profile. The slowing down of the marble as it rolls up the hill corresponds to the reduction in group velocity of a wave packet as it enters the ionosphere. The fact that slow marbles cannot roll over the hill, while fast ones can, corresponds to the fact that waves of sufficiently low frequency are reflected from the ionosphere whereas waves of sufficiently high frequency penetrate it. Delay time for the marble as a function of initial velocity and angle of incidence reproduces the corresponding functions for a wave packet as a function of frequency and angle of incidence on the ionosphere' (Booker, 1974).

Actually, the transmitted wave can be considered as the sum of two waves having the polarization of each of the two modes that can propagate. Since each model has its own height of reflection, two echoes per transmitted frequency can be received.

The time delays are usually measured in terms of virtual distance by the relation $h = ct/2$, c being the velocity of electromagnetic waves in a vacuum. The quantity h would be equal to the range of reflection if the propagation occurred at speed c, which is not true. Thus there is a problem of how to convert virtual to real range, to which we shall return.

In some modern ionosondes, the use of pulses is replaced by a more sophisticated modulation of the transmitted wave, such as a linear increase of frequency versus time ('chirp' radar). The signal to noise and the protection against interference signals are improved.

3.2.2.2. *Ionograms*

Ionosonde results are usually displayed as frequency-range traces called 'ionograms'. Because there are several modes of propagation, ionograms always show several traces. Usually, ionograms measured from the ground are oriented with their axis of virtual heights pointing upward, while the same axis points downward on ionograms taken from a satellite looking downward.

A model of an ionogram taken from below is given in Figure III.2. Essentially, we see traces O and X, corresponding to the O and X modes of Figure III.1. The Z mode is normally not observed, since the wave of corresponding polarization has been reflected at the height where $X = 1 - Y$ (see Figure III.1). However, it can appear at high latitude, as a result of coupling between modes under conditions close to longitudinal propagation. In general, ionograms have more complex structures: multiple traces due to multiple reflections between the ionosphere and the ground; diffuse traces due to departures from horizontal stratification; *Es* traces corresponding to thin layers from which the waves are totally or partially reflected, absorption phenomena etc.

A model of an ionogram taken from above, obtained on-board satellite, is represented in Figure III.2. We again find the O and X modes, as in the ground probe. A significant difference is the systematic presence of the Z trace. This results from the fact that the emission takes place inside the plasma, so that the value of X at the transmitter is no longer zero, but, for suitable frequencies, has the value necessary for the Z trace to appear, independently of any coupling phenomenon. Another consequence of transmitting inside the plasma is the appearance of 'spikes', characteristic of resonance phenomena, which will be explained further on.

The problem of the reconstruction of the electron concentration profile $n(h)$ starting from the observation of the apparent ranges $h'(f)$, is to invert the equation that gives $h'(f)$: $h'(f) = (c/2)\int_0^{\vec{r}} \frac{\partial \vec{\mathbf{k}}}{\partial \omega} \cdot d\vec{\mathbf{r}}$. It is easy to see that as long as the ionization increases

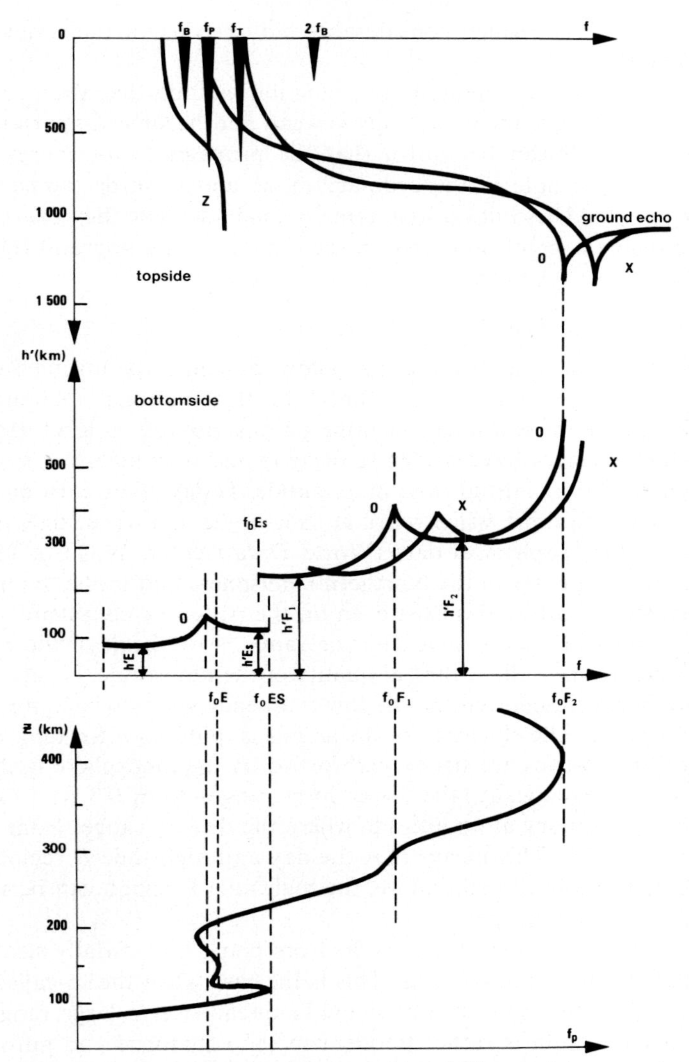

Fig. III.2. Typical topside (upper panel) and bottom side (middle panel) ionogram traces. The lower panel gives the corresponding height profile of the plasma frequency. The conventional notation for the various characteristic frequencies and virtual heights is indicated. Note the twin ordinary (O) and extra-ordinary (X) traces on both ionograms, and the Z trace and resonance spikes on the topside ionogram.

uniformly with distance from the ionosonde, the problem has a step-by-step solution: we let the frequency f increase by a quantity δf when exploring a new stratum between ranges of reflection $h_r(f + \delta f)$ and $h_r(f)$, while, considering as an example soundings taken from below, all the altitudes less than or equal to $z_r(f)$ have been explored in the course of earlier probings at frequencies less than f.

On the other hand, the reconstruction is impossible if there are 'valleys' of ionization, i.e. relative minima in the concentration, as sometimes happens between the E and F layers. Luckily, other techniques have proved that these depressions are in general not pronounced, which considerably reduces their actual impact on ionograms interpretation.

The most generally used methods of reduction assume that the propagation of energy is vertical, because the anisotropy is weak and becomes important only near the reflection point. It can be shown that the propagation of energy is diverted towards the magnetic pole in the ordinary mode and towards the equator in the extraordinary mode. These deviations remain small, so that they can be neglected without introducing a significant error in the reduction of ionograms (Doupnik and Schmerling, 1965).

3.2.2.3. *Bottomside Sounding*

The relatively low cost of ionosondes has allowed them to be installed quite extensively over the surface of the globe. Since the 'International Geophysical Year' (I.G.Y., 1957–58), the international scientific unions, notably U.R.S.I. (International Union for Radio Science), have striven to develop and then maintain a full coverage of the ionosphere both in latitude and in longitude. Today, there exist more than one hundred permanent stations, which work at fixed hours and whose data are collected in world centers for geophysical data (World Data Center, W.D.C.). The coverage achieved is evidently denser in the Northern hemisphere and in the temperate zones, although permanent ionosondes exist even on the Antarctic continent.

The upper limit of the accessible altitude range is the peak of the *F* layer. The lower limit depends upon the technical quality of the ionosonde, i.e. upon the signal to noise ratio which is achieved for the low frequency waves which are reflected by the lower ionosphere. The efficiency of the aerials is quite poor for long wavelengths and moreover these waves are strongly absorbed by the ionosphere itself, as will be described later on. Practically, the lower limit ranges from 0.3 to 1 MHz, which places the lower boundary at an altitude where the density ranges from 10^9 to 10^{10} m^{-3} (10^3 to 10^4 cm^{-3}). This means that the day and nighttime *D* region cannot be explored with ionosondes. Furthermore, the nighttime *E* region can be studied with the best quality ionosondes only.

Occasionally, when the ionosphere is far from being horizontally stratified, it becomes very difficult to use ionograms. This is the case when the so-called 'spread *F*' phenomenon is present: the return of signal is spread over a large range of virtual height. This phenomenon is rather frequent in the equatorial and auroral regions. The situation is even worse during the 'lacunae', in the polar zone, when the echo is no longer observed.

When dense sporadic *E* layers are present, the overlying regions may be occulted (blanketed), a typical midlatitude phenomenon, equatorial and polar sporadic *E* being always of the diffuse (transparent) type.

Without going through the process of reducing ionograms in true heights, however, very valuable results may be obtained from direct reading of the so-called critical frequencies corresponding to the maximum electron concentration of the layers (f_0F_2). An illustration is given by Figure III.3, where the existence of a winter increase of the peak in the *F* region is clearly apparent as well as its dependence on the solar cycle. It must be realized that such a figure results from the recording and analysis of many thousand ionograms.

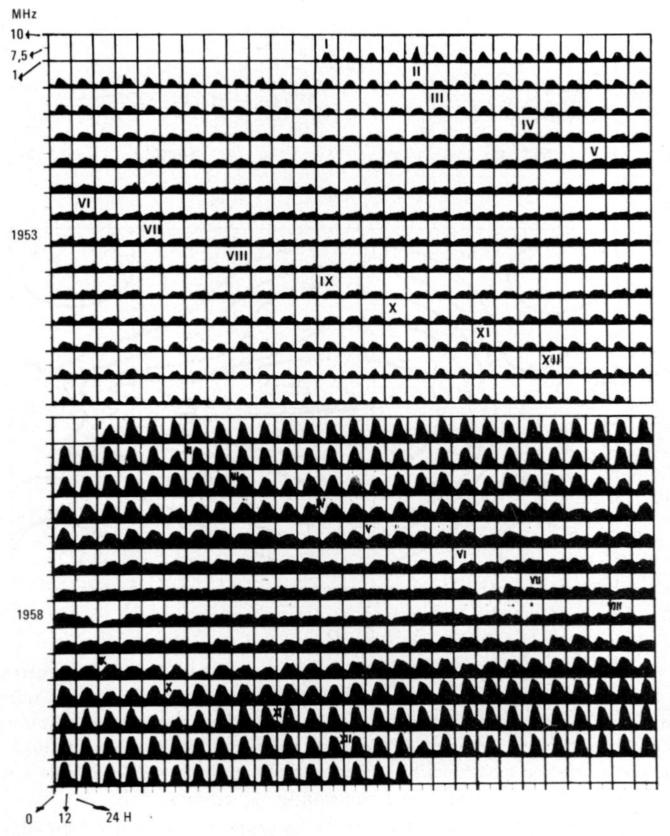

Fig. III.3. Variation as a function of local time of the characteristic frequency f_0F_2 (proportional to the square root of the peak ionospheric electron concentration) observed at Lindau (52°N) for each day of years 1953 (minimum solar activity) and 1958 (maximum solar activity). Each 'line' of 27 days corresponds to a solar rotation; the beginning of each month is indicated with a roman numeral. The strong day time maximum observed during winter months, especially marked in 1958, is the so-called 'winter anomaly' of the *F* region.

3.2.2.4. *Topside Sounding*

The Canadian-American satellite ALOUETTE 1, launched in 1962 into a circular orbit at 1000 km, carried the first ionosonde that routinely operated from above. It was followed by other topside sounders, EXPLORER 20, ALOUETTE 2, ISIS 1 and ISIS 2, which assured and continue to assure a permanent coverage of the upper ionosphere.

The electron concentration is thus explored from the altitude of the satellite (maximum 3000 km for the existing topside sounders) down to the peak of the *F* layer. Consistency between bottomside and topside sounding has been checked by comparing the altitude of the *F* layer peak and the corresponding concentration deduced from the two techniques. Comparison has also been performed with other techniques, such as incoherent scattering, which permits (Chapter IV) to observe both the

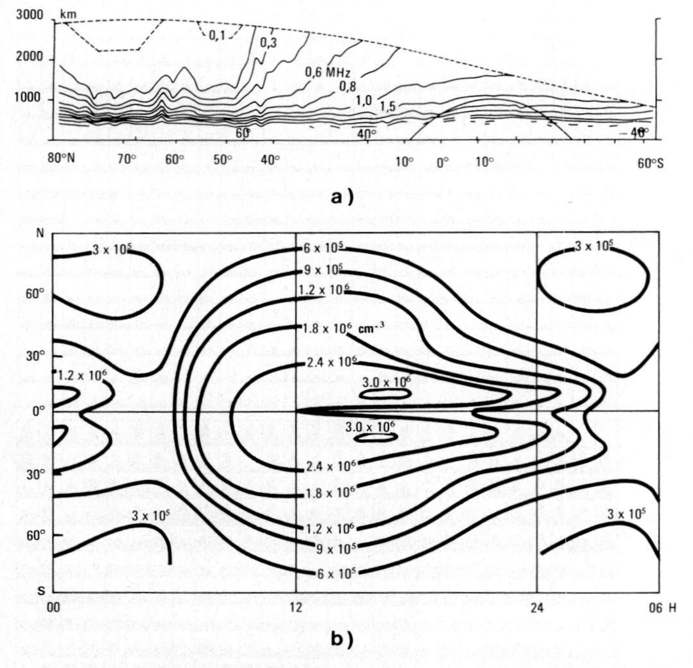

Fig. III.4. (a) Contours of constant electron concentration (labelled with the corresponding plasma frequency) as a function of altitude and latitude (figures above the horizontal axis refer to geomagnetic latitudes, figures below to geographic latitudes), observed by the topside sounder onboard ALOUETTE 2 in polar orbit. Note the 'trough' between 45° and 65° degree of geomagnetic latitude. (b) Contours of constant electron concentration as a function of geomagnetic latitude and local time at the *F* layer peak, deduced from the network of ground based ionosondes at equinox. The day time crests on either side of the geomagnetic equator is the ionospheric 'equatorial anomaly'.

bottom and topside parts of the ionosphere. Generally speaking, the electron concentration values are consistent within a few percent (Fleury and Taieb, 1971). The major error quoted by Jackson (1969) is a systematically lower value of the altitude of the topside profiles. The error, however, is only a few percent of the total propagation path, which means an error of 20 km maximum on the altitude of the peak.

Thus, ground-based and space observations have a complementary aspect for studying the *F* layer: ground observations watch the time behavior at one place, while topside soundings are scanning along the orbit, which is usually explored in less than 2 hr. This good geographical coverage led to a better understanding of the morphology of the equatorial anomaly and to the discovery of a 'trough' in the *F* region between medium and auroral latitudes, as shown by the latitudinal maps in Figure III.4.

3.3. Phase Effects with High Frequency Waves

If a wave of frequency much higher than the electron gyrofrequency f_B and the plasma frequency f_p propagates in the ionosphere, its index of refraction remains too close to unity for strong refraction to be produced. Experimentation is still possible by using transmission between a space vehicle and the ground, or reflection of waves from a distant object such as the Moon. In a large frequency range, the index is sufficiently different from unity for the change in phase along the trajectory to be significantly different from what it would be in a vacuum. This difference depends on the electron concentration along the wave path, and therefore can be used to measure it.

As a first approximation, the two modes have the same refractive index for high frequencies and the same phase change. However, the index of the two modes which result from the presence of the Earth magnetic field is actually slightly different and a Faraday effect corresponds to the difference in phase propagation of the two modes.

3.3.1. DOPPLER EFFECT

Let us consider a plane wave propagating between a fixed point A and a moving point M. The phase change is written:

$$\varphi = -2\pi f \int_{A}^{M} \mathrm{d}s/v_{\phi}$$

$$= -\frac{2\pi f}{c} \int_{A}^{M} N \, \mathrm{d}s$$

where f is the frequency, v_{ϕ} the phase velocity and N the index of the mode considered.

The phase φ varies with time, which means that the frequency of the signal received at M is no longer f, but $f + \Delta f$, where $\Delta f = (1/2\pi)\, \mathrm{d}\varphi/\partial t$ is given by:

$$\Delta f = -\frac{f}{c}\frac{\mathrm{d}}{\mathrm{d}t}\left[\int_{A}^{M} N \mathrm{d}s\right]$$

$$\Delta f = -\frac{f}{c}\left[N_{\mathrm{M}} V + \int_{A}^{M} \frac{\mathrm{d}N}{\mathrm{d}t} \mathrm{d}s\right]$$

where N_{M} designates the value of the index at point M, V the projection of the velocity of M on the direction of phase propagation and $\mathrm{d}N/\mathrm{d}t$ the time derivative of the index integrated along the wave path. The Doppler effect therefore depends both on the value of the index at point M and the time derivative of the index all along the trajectory.

For frequencies, much higher than f_{p} and f_{B}, X and Y are much less than 1, and from Equation (III.5), the two modes have, to second order in X and Y, the same index

$$N^2 = 1 - X \tag{III.6}$$

or, since X is small,

$$N = 1 - X/2.$$

The deviation of the index from unity is thus proportional to the electron concentration n. The difference between the observed Doppler effect and that which would have existed in a vacuum, Δf_v (Δf_v corresponds to $N = 1$)

$$\Delta f - \Delta f_v = -\frac{f}{c}\left[(N_M - 1)V + \int_A^M \frac{dN}{dt}\, ds\right]$$

is therefore composed of two terms, one of which is proportional to the electron concentration at M and the other to the time derivative of the concentration along the trajectory of the wave.

3.3.2. OCCULTATION OBSERVATIONS

Let us briefly mention that a similar method has been used to detect the ionospheres of other planets. Just before or just after occultation of an interplanetary craft by a planet, the wave transmitting the radio telemetry propagates through the ionosphere of the planet. Observed Doppler shift is recorded as a function of the distance of closest approach of the ray. This frequency shift differs from the free space Doppler shift calculated from the spacecraft trajectory. The observed difference results from the superposition of two effects of the planetary ionosphere: first the index of refraction differs from unity along the ray path and second this path increases due to refraction. The refractivity depends (Equation (III.6)) upon the electron concentration

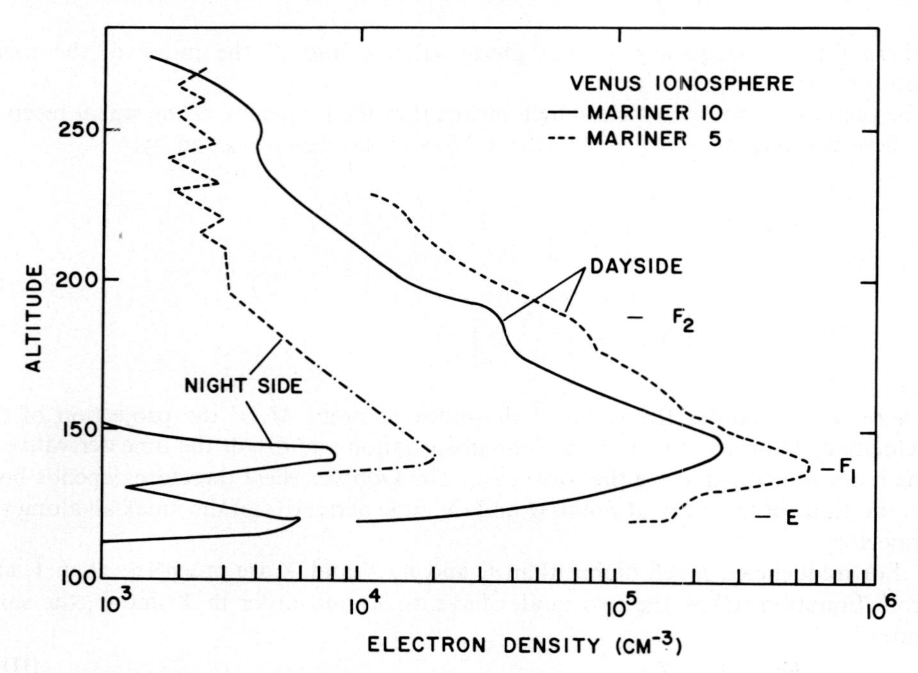

Fig. III.5. Dayside and nightside electron concentration profiles obtained from the Mariners occultation experiments. A tentative labelling of ionospheric regions similar to those of the Earth is indicated on the right hand side (Howard *et al.*, 1974).

$n(e)$. An altitude profile of $n(e)$ can be deduced from the measurements, through a conventional inversion process, under some simplifying assumptions such as spherical symmetry (see Figure III.5). Telemetry transmitters or tracking beacons are used for such observations which do not require any special onboard device; they have frequencies in the range 100 to 1000 MHz. When two harmonically related frequencies are transmitted (49.8 MHz and 423.3 MHz for the MARINER 5 spacecraft), the measurement sensitivity is greatly improved (Fjeldbo *et al.*, 1965).

When the point of closest approach of the ray is close enough to the planet surface, the dispersivity of the neutral atmosphere causes a positive Doppler shift, while the ionospheric one is negative. Information on the neutral atmosphere can thus also be obtained.

3.3.3. FARADAY EFFECT

We have seen that if $X \ll 1$ and $Y \ll 1$, then to second order in X and Y, $N^2 = 1 - X$. In a development of Equation (III.6) to higher order, we find, if θ is not close to $\pi/2$:

$$N^2 = 1 - \frac{X}{1 \pm Y_L}$$

or

$$N^2 = 1 - X \pm XY_L.$$

This expression remains true as long as $Y_T^4/4 \ll Y_L^2$, i.e. as long as the deviation of θ from $\pi/2$, expressed in radians, remains larger than Y, which is almost always true if Y is small enough.

If two modes with indices N_1 and N_2 propagate along the same trajectory AM, the difference between the phase shifts undergone by the two waves will be:

$$\varphi_1 - \varphi_2 = \frac{2\pi f}{c} \int_A^M (N_1 - N_2) \mathrm{d}s.$$

Here $N_1{}^2 - N_2{}^2 = 2X\,Y_L$, and $N_1 - N_2$ is very small compared with N_1 and N_2, which are of order 1, so that

$$N_1 - N_2 = XY_L = \frac{f_p{}^2 f_B \cos\theta}{f^3} = \frac{e^2 n}{4\pi^2 m\varepsilon_0} \frac{f_B}{f^3} \cos\theta = \frac{K f_B}{f^3} n \cos\theta$$

where $K = e^2/4\pi^2 m\varepsilon_0$.

Then finally, we have

$$\varphi_1 - \varphi_2 = \frac{2\pi}{c} \frac{K}{f^2} \int_A^M n f_B \cos\theta \, \mathrm{d}s.$$

To the extent that $f_B = eB/2\pi m$ varies only slightly along the wave trajectory, we can take it out of the integral, arriving at a formula of the type:

$$\varphi_1 - \varphi_2 = H_2 \frac{B \cos\theta}{f^2} \int_A^M n \mathrm{d}s.$$

The difference $\varphi_1 - \varphi_2$ is therefore proportional to the integral of the electron concentration along the trajectory, the factor of proportionality decreasing with frequency as $1/f^2$.

The polarizations of the two principal waves are circular or quasi-circular; let us suppose that the emitted wave is linearly polarized. We can decompose it into the sum of two waves of opposite circular polarization that propagate with the relative phase shift, $\varphi_1 - \varphi_2$, which we have determined. The resultant wave, which is obtained by taking their sum, is linearly polarized, but its plane of polarization undergoes a rotation. This is customarily referred to as the Faraday effect.

In all rigor, the two waves, having different indices, have different group trajectories, and it would be necessary to take account of the real trajectory of each wave when calculating its phase shift. This could be done by using models of the ionosphere and iterative processes, adjusting the parameters of the model in order to retrieve the observed values of the Faraday effect. Fortunately, such refinements are only needed on exceptional occasions, when the propagation is nearly horizontal. Usually it is possible, with the measurement of the Faraday effect, possibly accompanied by that of the Doppler effect, to determine, for a small cost, values of the 'total content' (of electrons) along the trajectory in the ionosphere, $\int n\, ds$, by receiving on the ground with very simple receivers the signals from radio beacons aboard satellites. The use of several frequencies eliminates certain experimental difficulties, such

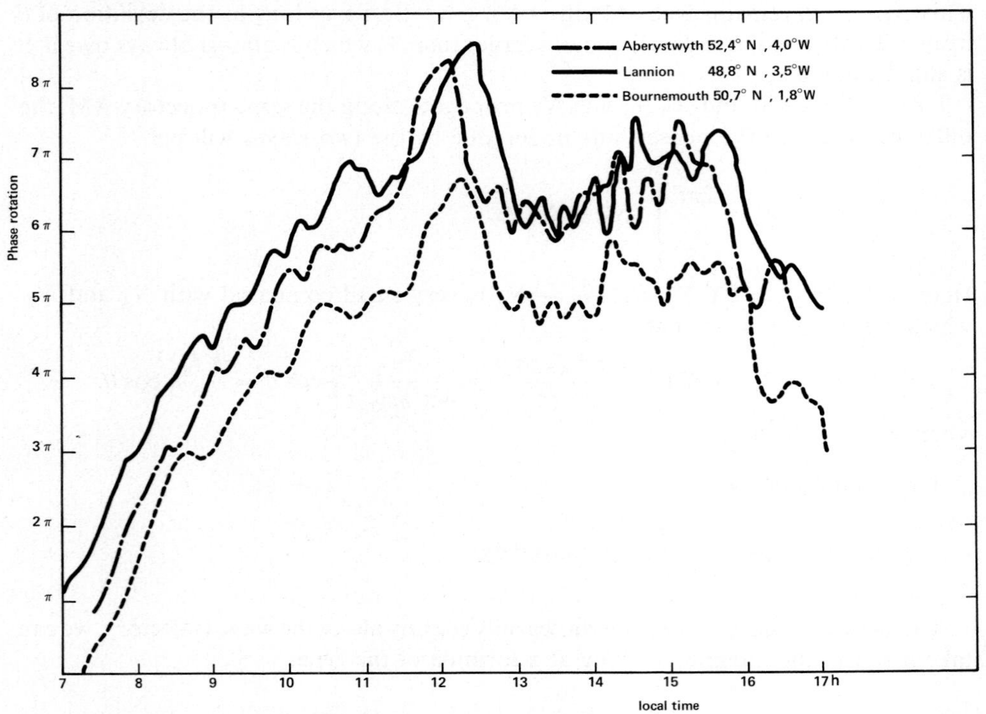

Fig. III.6. Faraday phase rotation of the signal received at three different stations from a beacon (frequency 136 MHz) on a geostationary satellite (INTELSAT II F2), showing the propagation of so called 'Travelling Ionospheric Disturbances' (courtesy F. Bertin).

as uncertainty in the real value of the phase, which can only be known up to a round number of complete rotations, or the proper spinning of space vehicles, which can bring about an apparent rotation of the polarization independently of all ionospheric effects.

Geostationary satellites are specially suitable for Faraday measurements, for they are always in view of the same point and permit a continuous coverage. The measurement in time of the total content evidently gives, above all, information on the variations of the F_2 region of maximum electron concentration (Bowhill, 1958). In this respect, Faraday rotation studies are particularly useful to observe the oscillations and disturbances of the F region and their horizontal propagation (see Figure III.6), by using a network of receivers permanently recording the signal transmitted by a geostationary satellite.

Otherwise the study of the Doppler and Faraday effects in a wave propagating between a rocket and the ground has been used to determine the profiles of electron concentration in the D and night E regions of the ionosphere. This method is probably the most accurate one to measure the very low values of electron density there (Mechtly *et al.*, 1967; Derblom and Ladell, 1973).

3.4. Group Effects with Lower Frequency Waves

Some of the waves sustained by a plasma propagate with a group velocity which is very different from the free space velocity and from the phase velocity. Peculiar phenomena are thus obtained, such as transformation of a short pulse in a long duration signal resulting from the propagation of energy along paths on which the medium imposes severe constraints. Atmospheric whistlers and resonance spikes in topside ionograms belong to this category of phenomena.

3.4.1. ATMOSPHERIC WHISTLERS

The study of atmospheric whistler modes was used, before the advent of the space era, to estimate the electron concentration well above the peak of the ionosphere.

Atmospheric whistlers are natural electro-magnetic waves which, applied after amplification to the coil of a loud speaker, are heard as a pure note whose frequency decreases with time (hence their name). The correct interpretation of this phenomenon, proposed by Storey (1952), is the following: lightning flashes of short duration are accompanied by the emission of electromagnetic energy contained in a wide frequency band. This energy propagates in the mode which we have designated by number 4 in Figure III.1, and it is the propagation, which depends on frequency that develops the signal. We have already noted that mode 4 involves frequencies much smaller than the electron gyrofrequency and the plasma frequency, such as acoustic frequencies (of the order of kHz) which are of interest to us here. One can show that for this mode the group vector is contained within a cone with a half angle of 20° whose axis is parallel to the terrestrial magnetic field. The energy is thus *guided* along the lines of force of the field. On the other hand, the group velocity depends on the frequency; more precisely, if the frequency is much smaller than the electron gyrofrequency, the time of propagation τ is given by the expression:

$$\tau = f^{-1/2}\left[\frac{1}{2c}\int\limits_{S}\frac{f_p \mathrm{d}s}{(f_B \cos\theta)^{1/2}}\right]. \tag{III.7}$$

This equation shows that signals emitted simultaneously during the lightning flash are received at the magnetic conjugate point after a time the longer the smaller their frequency, thus creating a whistling sound. The change of τ with f increases with distance traveled, and this permits us to distinguish various types of whistlers. In the most simple case a signal from a storm propagates along the lines of force from one hemisphere to the other, where the phenomenon is observed; this is a one hop whistler. The energy can also be reflected and observed in the hemisphere from which it came: two hop whistlers. Successive reflections thus can result in a large number of bounces, the products $\tau f^{1/2}$, or dispersion, of the various whistlers observed at a given point being in the ratios 2, 4, 6 ... or 1, 3, 5 ... depending upon whether the original flash was in the hemisphere of observation or the opposite one.

The interpretation of whistlers has also brought about an important contribution to the knowledge of the magnetosphere, through the discovery and the study of the plasmapause (Carpenter, 1966). For this, one uses frequencies closer to the electron gyrofrequency, for which Equation (III.7) is no longer valid, and must be replaced by an expression which has a minimum at $f = f_H \cos\theta/4$. The curve $T(f)$ then has a dip, and it is the coordinates of this dip which are used in order to determine the electron concentration as a function of altitude in the equatorial plane.

3.4.2. Plasma Resonances

We have previously mentioned the existence of 'resonance spikes' in topside ionograms. These spikes correspond to an identifiable signal which lasts several tens of milliseconds after transmission of the sounding pulse only some 100 μs long. They occur when the transmitted frequency is equal to one of the characteristic frequencies of the plasma in the immediate neighbourhood of the satellite. Such signals received at the plasma frequency, the upper hybrid frequency, the electron gyrofrequency and its multiples, are explained in terms of waves having a group velocity small enough to be comparable to the spacecraft velocity. These waves are electrostatic; they are correctly described only when the thermal motion of the electrons is taken into account, in the framework of a hot plasma theory (as opposed to the cold plasma theory to which corresponds the Appleton-Hartree formula).

An oversimplified picture describes the resonance signals as waves travelling with the same velocity as the satellite which receives them during a long time. A more accurate theory involves so-called oblique echoes. Let us explain the basic phenomenon for the case of the resonance at the plasma frequency f_p. A small variation of a few per thousand in electron concentration is sufficient to change drastically the direction of the group velocity of an electrostatic wave having a frequency close to f_p. A gradient of electron concentration permanently exists in the ionosphere and the energy of a wave packet can be refracted back to the satellite from which it has been transmitted. This path is extremely frequency-dependent, and only two waves having slightly different frequencies (the difference is typically 1 kHz for usual ionospheric conditions where f_p is a few MHz) satisfy after a different path the rendezvous condition with the spacecraft (McAfee, 1968, 1969). The frequency of each wave varies by a few kHz as a function of time delay after pulse transmission. The beating of the two waves explains the fringe pattern of some modulated spikes. The variation of the beating frequency versus time delay depends upon electron temperature; this property has been used to measure it (Warnock *et al.*, 1970; Feldstein and Graff, 1972).

Similar phenomena occur for the other characteristic frequencies of the plasma. The gradient of the Earth's magnetic field supplements and replaces the electron concentration gradient. Interested readers will find in Muldrew (1972) a good review on these problems.

Resonance signals have been used to measure very low values of electron density in the polar topside ionosphere (Hagg, 1967; Timleck and Nelms, 1969). They have also been used successfully as a diagnostic tool for the dilute magnetospheric plasma, on spacecrafts GEOS and ISEE.

3.5. The Effects of Collisions

Collisions between electrons and neutral particles play only a secondary role in propagation phenomena in E and F regions and the upper ionosphere; they cause a slight damping of the waves particularly near the reflection points where $N \to 0$ (deviative absorption) as well as coupling between certain modes, but do not affect in an essential way the diagnostic methods that we have described. In the D region, on the other hand, collisions are a dominant process and play a fundamental role in all the methods of experimental study using radiofrequency waves. After introducing collisions into the equations of propagation, we shall discuss absorption, partial reflections, and intermodulation techniques.

3.5.1. INTRODUCTION OF COLLISIONS INTO THE EQUATIONS OF PROPAGATION

The most simple way of taking account of collisions in a cold plasma consists in adding to the right hand side of Equation (III.1), a collision term of the form $-m\nu\vec{v}$, where ν is an effective collision frequency, independent of the velocity. The Appleton-Hartree equation then takes the following form, where $U = 1 - i\nu/\omega$:

$$N^2 = 1 - \frac{X}{U - \dfrac{Y_T^{\,2}}{2(U - X)} \pm \sqrt{Y_L^{\,2} + \dfrac{Y_T^{\,4}}{4(U - X)^2}}}.$$

However laboratory experiments by Phelps and Pack (1959) have established that the frictional force actually depends upon the velocity v of the electron in collisions with N_2 gas, which is the major constituent of the D region. The cold plasma approximation is thus no longer valid and the velocity distribution function of the electrons has to be taken into account. Using Phelps and Pack's results (ν is proportional to v^2), Sen and Wyller (1960) have generalized the Appleton-Hartree theory and their formula is used in most of the experimental studies which will be described below. As soon as the collision frequency is less than one tenth of the wave frequency f, the Appleton-Hartree formula is recovered, provided the collision frequency ν is properly defined. Nevertheless Sen and Wyller's corrections are essential when ν is of the same order of magnitude as f.

3.5.2. ABSORPTION MEASUREMENTS

The theoretical studies quoted in Paragraph 5.1 have established that the absorption coefficient is a linearly increasing function of both the electron density $n(e)$ and the collision frequency ν, provided ν is much smaller than the wave frequency f. When ν is comparable to f or greater, the absorption coefficient is still an increasing function of $n(e)$, but as a function of ν, it presents a maximum when ν is about equal to $\omega = 2\pi f$. It is proportional to n_e/ν, if $\nu \gg \omega$.

Provided the collision frequency profile is known either from independent measurements or from an atmospheric model, absorption measurements can be used to determine the electron concentration in the *D* region.

3.5.2.1. *The Attenuation of Waves Reflected by the E and F Regions*

Study of the amplitudes of waves with frequency equal to several MHz reflected off the *E* or *F* regions gives a measure of absorption in the *D* region. One can use either vertical reflection of pulses (a method called A.1) or oblique reflection of continuous waves (called A.3). An essential difficulty comes from the fact that the absorption does not take place exclusively in the *D* region. In particular, when the wave is reflected by the *E* layer, collisions play a role in producing absorption in the neighborhood of the point of reflection (deviative absorption) which can be comparable to the non-deviative absorption in the *D* region (non-deviative means that the refractive index is nearly unity).

These methods measure the change in absorption, i.e. in electron concentration with time at a given point, within the region where $\nu \sim \omega$. They have led to discovery of the winter anomaly in the *D* region: on many winter days the absorption at low and medium altitudes is abnormally high (Appleton and Piggott, 1948, 1954; Beynon and Davies, 1954; Thomas, 1961). This phenomenon has been correlated only with increases in the temperature of the stratosphere far below (Bossolasco and Elena, 1963; Shapley and Beynon, 1965).

3.5.2.2. *'Riometers'*

'Riometers' (Relative Ionospheric Opacity meters) are low noise receivers that measure the level of cosmic radiofrequency noise received at ground level after having traversed the ionosphere (this is also called the A.2 method). Of course the level received depends on the diurnal variation in the input from galactic radio sources, but, by comparison of observations made at the same sideral time, one obtains the relative day-to-day variation of absorption. Several frequencies can be used (usually from 5 to 30 MHz) to find the variations in electron concentration at different altitudes in the lower ionosphere (Parthasarathy, *et al.*, 1963).

Riometers are used mostly to detect the arrival of high-energy particles in the polar and auroral regions in connection with magnetospheric phenomena (see Figure III.7). In particular, they can be used to study the effect of solar eruptions of very high energy protons, which show up as a particularly intense long duration absorption in the polar caps (called P.C.A., for polar cap absorption), and which were discovered through their effect on the propagation of long waves.

CHAPTER IV

SCATTERING OF RADIO WAVES

The use of the word 'scattering' or 'scatter' is rather loose, and the distinction between the propagation of waves as discussed in the preceding chapter, and in this one, is somewhat artificial. What we have been concerned about so far is the effect on wave propagation of the fundamentally one-dimensional, plane stratified structure of the ionospheric plasma (except for magnetic field direction); furthermore we have supposed that its characteristics are homogeneous and slowly varying on the scale of the probing electromagnetic wavelength and frequency. We now turn to propagation phenomena and techniques which result from the more or less ubiquitous and permanent departure from such simple conditions in the ionosphere, that is to say, to the influence on wave propagation of the existence of ever changing irregular blobs of excess or deficiency in ionization density. In general, as we shall see, a new dimension is brought in by scattering techniques as they allow the observation of ionospheric *motions*.

Of course, scattering of radio waves is always basically due to the wave induced electron oscillations radiating a secondary electromagnetic field. Nevertheless a useful distinction can be made depending on whether the electric field on each electron must or need not include the secondary field radiated by all the other electrons. In the first case, which is just an extension of the discussion in the last chapter, we talk of 'quasi-specular reflection'. In the second case the scattering of waves already scattered is neglected, a simplification known as 'Born's approximation'. An enlightening discussion of the conceptual problems involved can be found in Booker's 1974 article 'Fifty Years of the Ionosphere – Electromagnetic Theory'.

4.1. Quasi-Specular Reflexion from Ionospheric Irregularities

Even though surfaces of constant plasma concentration are roughly horizontal, they can depart considerably from idealized smooth plane stratification. As Booker remarks, 'There is almost certainly a considerable tendency to overlook this in order to arrive at simple theories of wave propagation. For example, the coupling between upgoing and downcoming waves in a reflecting stratum was outlined in Chapter III on the basis of a wave solution in the stratum, assumed to be plane stratified. However, such a stratum is automatically a place where the derivative of the relevant refraction parameters with respect to ionization density is high. A reflecting stratum is therefore a location that is unusually sensitive to any irregularities in ionization density that may exist.'

4.1.1. FRESNEL AND RAYLEIGH CRITERIA

Thus, when considering actual three-dimensional ionospheric reflection experiments, the questions arise: What region of the reflecting surface is involved in the process? To what extent will roughness of the reflecting surface affect reflection? Both questions have a well-known answer in the field of classical optics.

4.1.1.1. *Location and Size of the Reflecting Region*

To answer the first question, let us consider (Figure IV.1) a smooth surface illuminated by a source at the point A. The locus of all points in the surface, from which

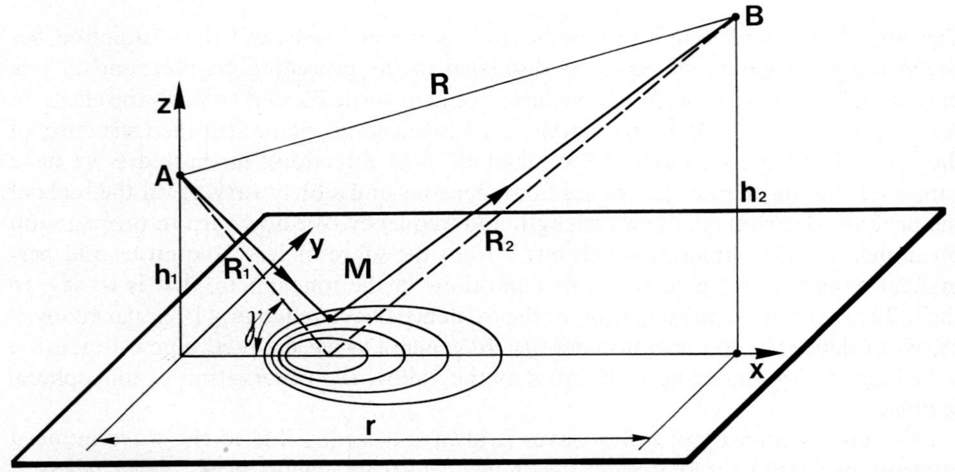

Fig. IV.1. Fresnel zones are the intercepts by a smooth reflecting surface of the nested ellipsoids with transmitter and receiver as foci, corresponding to successive constant phase path loci of increment Π. The contributions of adjacent zones do not quite cancel, and the main part of the reflected radiation originates in the first Fresnel zones centered on the geometrical reflection point (cf. text).

the secondary radiation arrives at B with a constant phase difference δ is given by the condition $AM + MB = C_{st}$ which is the equation of ellipsoids of revolution with foci at A and B. Hence the desired loci in a plane are ellipses determined by the intersection of that plane with the ellipsoids. If δ now is increased in steps of $\lambda/2$, starting from the value δ which corresponds to the geometrical point of reflection, one obtains the so-called Fresnel zones; the averaged phase of the radiation from each zone bounded by two neighbouring ellipses differs from the adjacent zone by π, so that the elementary secondary waves radiated by successive zones interfere destructively. It is usually claimed that the first Fresnel zone makes the most important contribution to the total field received at B: the contribution of each zone is found by integrating the amplitude of excitation over the zone. Since successive zones are in phase opposition, the contribution of adjacent zones will tend to cancel, but not quite, as the amplitude of excitation decreases slowly from zone to zone, leaving the radiation of the area of approximately half the first Fresnel zone as the final total.

If the transmitter A and the receiver B are located at the same point, the Fresnel ellipsoids are spheres and the field reflected by a plane, or a line is then obtained by the conventional construction, which uses the Cornu spiral and is described in most textbooks of optics.

4.1.1.2. *Specular and Rough Scattering*

To answer the second question, Rayleigh considered two rays (Figure IV.2) incident on a surface with irregularities of height h at a grazing angle γ. The path difference between the two rays is:

$$\Delta_2 = 2h \sin \gamma$$

and hence the phase difference is:

$$\Delta\varphi = \frac{2\pi}{\lambda} \Delta_2 = \frac{4\pi h}{\lambda} \sin \gamma.$$

If this phase difference is small, the rays are almost in phase as they are in the case

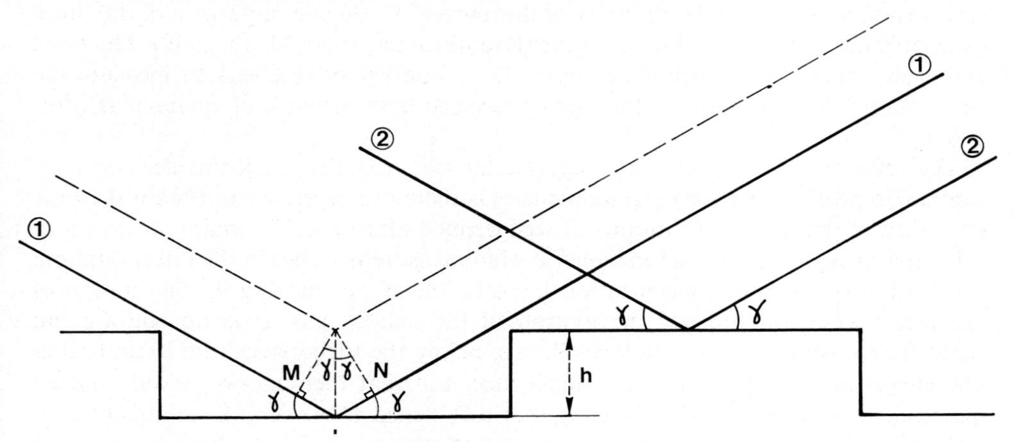

Fig. IV.2. The Rayleigh criterion expresses the fact that an irregularity on a reflecting surface intervenes significantly when it introduces a path difference greater than $\Pi/2$. (Cf. text.)

of a perfectly smooth surface. If the phase difference increases up to $\Delta\varphi = \pi$, the two rays will be in phase opposition and cancel. If there is no energy flow in this direction, then it must have been redistributed in other directions and the surface scatters. A limit between the smooth reflecting surface and the rough scattering surface is given by the Rayleigh criterion, which is obtained by arbitrarily choosing the value $\Delta\varphi = \pi/2$ (half way between the two cases):

$$h = \frac{\lambda}{8 \sin \gamma}.$$

The surface will thus tend to be effectively smooth only under two conditions: $h/\lambda \to 0$ (the irregularity is small compared to wavelength) or $\gamma \to 0$ (grazing incidence).

This implies that scattering surfaces are rough or smooth depending on the probing wavelengths and the geometry. In particular, it can be seen that the Fresnel approach can be crude and incorrect, as the different Fresnel zones are being illuminated with different angles of incidence. In most actual cases, the reflected field is composed in various proportions of a specular component, and of a rough scatter, as just looking at any surface (paper for instance) under some light source demonstrates (cf. Beckmann and Spizzichino, 1963).

4.1.2. 'Fading' on Ionospheric Radio Links

When the ionosphere is illuminated by a wide beam radio transmitter in the H.F. range, the field at ground level, resulting from reflection on the E region always exhibits a fluctuating character with a wide spectrum of fades and enhancements (a fact that usually goes unnoticed with frequency swept ionosondes, but not on daytime A.M. broadcasting for instance). Because the ionospheric irregularities responsible for these fluctuations are stable enough, and have a horizontal bulk motion the fading pattern on the ground moves, with twice the velocity of this motion, as can be easily checked from elementary laws of reflection in geometrical optics. Reception at several different places permits study of this motion through comparison of the times of occurrence of similar fading patterns (Krautkramer, 1950; Mitra, 1949). The most elaborate experimental techniques use a large number of receivers to measure the correlation functions between the signals received by a network of antennae (Fedor, 1967).

The reflection height varies throughout the day, together with the electron concentration profile. Using several frequencies is therefore necessary to obtain the time evolution of the observed motions at well defined altitudes. The major weakness of this experimental method of ionospheric wind measurement lies in the interpretation: the motion of electron concentration irregularities is not necessarily the motion of the ionization itself nor yet the motion of the neutral gas. It is not certain but perfectly plausible that, at the lower levels, below the turbopause, the irregularities are caused mainly by neutral gas turbulence, and that their motion should indeed match the neutral gas winds advecting that turbulence. However, this argument does not apply about the level of turbulence cut-off; the irregularities there are induced mainly by gravity waves, the phase velocity of which is completely different from the wind velocity. Hines (1968) has shown that the interference pattern imposed on the atmosphere by waves of certain type is simply transported as if by the wind and in that case the fading pattern would still reveal the wind velocity. The dominance of such waves over all the other waves is yet to be established, by either theoretical or observational means, before interpreting ionospheric drift observations for the upper *E* region in terms of neutral gas winds.

On 'forward' paths between distant transmitters and receivers, scattering can propagate waves to much higher frequencies because of the grazing incidence (Bailey *et al.*, 1952). Field strength and direction of arrival measurements on such V.H.F. links show wild fluctuations, except when blanketing sporadic *E* layers appear, in which case the reflection can sometimes be for tens of minutes as perfect as on a silver mirror; Giraud (1966) has shown that such smooth layers of enhanced electron concentration are slightly tilted from the horizontal, a property in support of the 'wind shear' theory of their formation (Chapter VII).

4.1.3. METEOR SCATTER

Small meteorites that penetrate the atmosphere continuously create in the lower thermosphere numerous ionized trails (cf. Chapter VI) with any direction and a lifetime against diffusion of the order of a second. Those trails are able to reflect waves in the V.H.F. range (30 to 300 MHz), as evidenced by the large number of echoes observed on forward scatter links (Vogan and Campbell, 1957).

According to their mass and velocity, meteorites can create electron trail concentration covering many orders of magnitude. The scattering of radio waves from meteoric trails can take place in two different ways depending whether the trail plasma frequency is smaller or greater than the radiofrequency used. In the case of 'underdense' trails, we have partial reflection from an inhomogeneity; in the case of 'overdense' trails, there is total reflection as if from a metallic cylinder. In both cases, the reflection is strongly aspect dependent, as only the first Fresnel zone significantly contributes to the reflected field. This means that the only observed trails are those which happen to form tangentially to the Fresnel ellipsoids within the aerial lobes (in the case of a monostatic radar, perpendicular to the beam). This strong aspect dependence is an asset for the astronomical study of faint meteors, as it permits reconstruction of their velocity vector, that is of the meteors' orbits (McKinley, 1961).

The main geophysical use of meteor radars is with the study of the Doppler shift of the echo received after the trail has formed and before it diffuses into the environment. This allows the determination of its displacement, which at these low altitudes, where the plasma is mostly collision dominated, follows the bulk motion of the neutral atmosphere (of the order of several tens of m s^{-1}). The localization of the small echoing main Fresnel zone of each trail can be made rather precise with range and angle of arrival measurements (Revah, 1969). Because of the numerous data, this is a very powerful method for the study of the vertical wind profile, and the monitoring of its time variation. It has contributed fundamentally to the experimental study of the motions of the neutral atmosphere in the region where trails form (between about 80 and 100 km): turbulence, gravity waves, tides, planetary waves (Spizzichino and Revah, 1968).

4.1.4. THE SPREADING OF IONOGRAM TRACES

Swept frequency ionosondes usually produce clean range traces, corresponding to returns from the main Fresnel zone at the reflecting layer. For instance, at 3 MHz and 100 km altitude, the first Fresnel zone is only 5 km in diameter, and the range spreading over such a small area is negligible.

When strong departures from horizontal stratification of the electron concentration arise, however, ionograms exhibit 'spread' traces.

Klemperer (1963) has shown that 'spread *F*' echoes have a strong specular component; they would correspond to the onset of quasi-specular reflection over several ranges simultaneously. Such a criterion can be met by guided propagation along ducts of ionospheric irregularities aligned along the geomagnetic shells. But the scattering theory, as well as the theory of formation of spread *F* (cf. Chapter VII), is still somewhat confused.

The long wavelength regime responsible for spread *F* echoes is one aspect of ionospheric instabilities. Other types of diffuse traces on ionograms, as well as other ground based and *in situ* measurements have shown that the spectrum of ionospheric irregularities extends down to very small dimensions and gradients, as discussed in the next section.

4.2. Scattering from Volume Plasma Fluctuations

The fact that the ionosphere is the seat of a more or less constant small scale unrest is evidenced by the 'radiostar scintillation' phenomenon. Radioastronomers working in the V.H.F. and U.H.F. bands observe a rapid fluctuation in the strength and phase of pin point cosmic radio sources, particularly in the higher latitudes. There is no correlation between the fading at pairs of receivers separated by more than a few km, and this identifies the scintillation as an ionospheric effect (Booker, 1958).

A similar effect occurs with radio signals from artificial satellites; scintillation even occurs in the GHz band at the equator (Skinner *et al.*, 1971).

Such plasma fluctuations are able to scatter a minute fraction of the radio waves which, well above the plasma frequency, propagate as a first approximation as if in a vacuum. In the following, we describe the scattering process involved by assuming each fluctuation in electron concentration occurs on a scale which is small compared to the probing wavelength; the current induced in each irregularity has then a well defined phase, and the total scattered field is the sum of the fields due to each scatterer, all assumed to be identical, and neglecting multiple scattering (Born's approximation).

4.2.1. THE FOURIER SPECTRUM OF FLUCTUATIONS AND BRAGG'S SCATTERING FORMULA

Consider a volume containing N scatterers, occupying positions M_p ($p = 1 \ldots N$). In the case of the ionosphere, the dimensions of the volume are usually small compared with the distances OT and OR that separate some origin O, taken in the interior of the scattering volume, from the transmitter T and the receiver R. We can therefore take all the segments TM_p to be parallel to each other, and all the segments M_pR to be parallel to each other. All the incident waves thus have the same vector $\vec{\mathbf{k}}_1$ and all the received waves the same $\vec{k}_2$. Under these conditions, the field scattered by the scatterer of index p is given at the point of reception by:

$$E_p = \frac{E_o}{TM_p} e^{i\vec{k}_1 \cdot \vec{TM}_p} K \frac{1}{M_pR} e^{i\vec{k}_2 \cdot \vec{M_pR}}$$

where E_o is the incident field at unit distance from the transmitter T and K is a coefficient identical for all scatterers. The total scattered field is

$$E_T = \sum_p E_p$$

$$= \sum_p \frac{E_o}{TM_p} \frac{1}{M_pR} K\, e^{i(\vec{k}_1 \cdot \vec{TM}_p + \vec{k}_2 \cdot \vec{M_pR})}.$$

Again using the small size of the scattering volume, we set $TM_p = TO$ and $M_pR = OR$ in those terms that represent attenuation with distance; we can thus write

$$E_t = \frac{E_oK}{TO \cdot OR} \sum_p e^{i(\vec{k}_1 \cdot \vec{TM}_p + \vec{k}_2 \cdot \vec{M_pR})}.$$

From $\vec{TM}_p = \vec{TO} + \vec{OM}_p$ and $\vec{M_pR} = \vec{M_pO} + \vec{OR}$ there results

$$\vec{k}_1 \cdot \vec{TM}_p + \vec{k}_2 \cdot \vec{M_pR} = \vec{k}_1 \cdot \vec{TO} + \vec{k}_2 \cdot \vec{OR} + (\vec{k}_1 - \vec{k}_2) \cdot \vec{OM}_p,$$

so that

$$E_t = \frac{E_oK}{TO \cdot OR} e^{i(\vec{k}_1 \cdot \vec{TO} + \vec{k}_2\, \vec{OR})} \sum_p e^{i(\vec{k}_1 - \vec{k}_2) \cdot \vec{OM}_p}.$$

We see that aside from a factor depending only on the experimental conditions, the total scattered field is equal to $\sum e^{i(\vec{k}_1 - \vec{k}_2) \cdot \vec{OM}_p}$. This last quantity is nothing but the spatial Fourier component of the scatterer density at the characteristic wave number $\vec{k} = \vec{k}_1 - \vec{k}_2$. This can be shown very simply by the application of the definition of Fourier transform, with the scatterer density being written in the form:

$$n = \sum_p \delta(\vec{OM} - \vec{OM}_p)$$

where the Dirac δ functions express the discrete nature of the scatterers that are found at the various points M_p.

The field received at each instant from volume distributed fluctuations is therefore simply proportional to the spatial Fourier component of the scatterer's density $n(\vec{k}, t)$, where the wave vector is the difference between those of the incident and scattered waves (Figure IV.3).

An alternative derivation of the scattering properties of a random medium is to consider the medium as a continuum characterized by its dielectric permitivity (Booker, 1956). This permitivity ε may be expressed as the sum of the average permitivity $\langle\varepsilon\rangle$ plus a fluctuating part $\Delta\varepsilon$, which is usually assumed to be much smaller than $\langle\varepsilon\rangle$. A similar calculation (Yeh and Liu, 1972) shows that the received field is proportional to the spatial Fourier component of $\Delta\varepsilon$, where the wave vector is again the difference between those of the incident and scattered waves.

When a transmitter illuminates a random distribution of scatterers, the energy

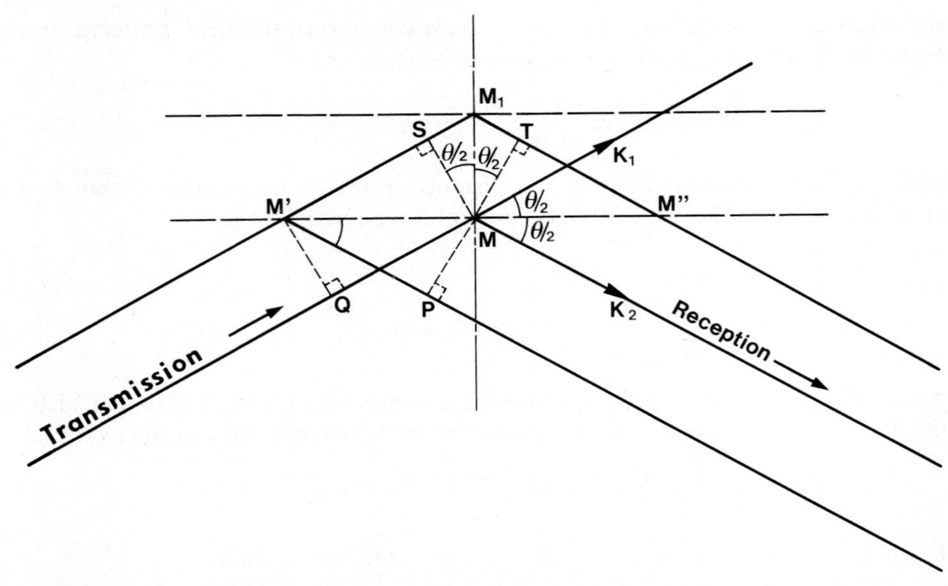

Fig. IV.3. The Bragg condition states that volume distributed scatterers add their contribution when they lie on equiphase loci. This means that a scattering experiment is a 'space spectrometer', selecting a particular spatial Fourier component of the scatterers' distribution (cf. text).

scattered in a direction θ is thus due only to that particular component of the Fourier spectrum of irregularities with wavelength $\Lambda = 2\pi/|\vec{\mathbf{k}}| = \frac{\lambda}{2}\sin(\theta/2)$. This means that the probing wavelength and geometry 'selects' that part of the irregular structure which satisfies Bragg's reflection condition for crystal lattices (Figure IV.3).

4.2.2. SCATTERING FROM FIELD ALIGNED IRREGULARITIES

A particular case of volume scattering occurs in the ionosphere with the existence of field aligned irregularities in the regions where strong currents flow (equatorial and auroral electrojets, cf. Chapter VII). In that case, Bragg's condition is felt as 'magnetic aspect sensitivity'; for a monostatic radar, this means that the beam must be nearly perpendicular to the magnetic field lines for scattering to be observed.

Such equatorial and auroral backscatter are detected with ionosondes as peculiar traces ('equatorial sporadic *E*', 'slant sporadic *E*' etc . . .) They also have been studied with specialized equipment, like 'auroral radars', at higher frequencies.

Auroral scattering also known as 'radio aurora', is detected in the V.H.F. (30 to 300 MHz) and U.H.F. (300 to 3000 MHz) bands, between 90 and 150 km altitude. The average efficiency of the scattering falls off sharply with frequency (approximately as f^{-5} according to Chesnut *et al.*, 1968). The echoes exhibit a Doppler shift roughly proportional to the probing frequency, indicating scatterer velocities of several hundred ms^{-1}, sometimes as high as 1 or 2 kms^{-1}. The velocity depends on echo intensity, largest shifts corresponding to strongest intensities. A smooth changeover from westerly to easterly drift around magnetic midnight fits well with the motion of visual auroras (Egeland, 1971).

4.2.3. SCATTERING FROM RANDOM THERMAL FLUCTUATIONS

The most general case of volume scattering corresponds simply to the random distribution of the smallest elementary scatterers, that is, each electron taken individually, with its 'Thomson' scattering cross-section.

$$\sigma = (\mu_0 \varepsilon^2/4\pi m_e)^2 \sin^{-2} \psi,$$

(only 8×10^{-30} m^2 for the most favorable polarization angle ψ).

Fabry (1928) showed that if the electrons are really randomly distributed, that is if no interaction exists between the particles constituting the plasma, the Fourier component $n(k, t)$ has a non-vanishing mean square value $\langle |n(k, t)|^2 \rangle$ proportional to the electron concentration, and the Doppler spectrum of the return is a MHz wide Gaussian corresponding to the electron thermal speeds, for a scattered wave in the G Hz range.

Actually, as shown by Fejer (1960), Salpeter (1961), Dougherty and Farley (1960), Buneman (1961), Hagfors (1961), Farley *et al.*, (1961), Rosenbluth and Rostoker (1962), the random fluctuations of plasma particles exhibit a large degree of correlation because of Coulomb interactions, and the determination of $n(k, t)$ and its power spectrum requires a statistical treatment of the Vlasov and Maxwell equations.

We borrow from Hagfors the following very simplified picture of the fundamental processes at work: the ions and the electrons of a fully ionized plasma where two-body collisions are unimportant to first order travel as if completely free and non-interacting. The first order correction to this motion can be obtained by considering an individual particle to still be traveling along an unperturbed path, but to be surrounded by a cloud slightly depleted in particles of like sign and by a cloud slightly enriched in particles of the opposite sign. In a plasma consisting of electrons and one type of positive ions, the electron concentration fluctuation can be thought of as consisting of three contributions. One contribution is caused by the concentration fluctuation associated with the fact that the electron is a point particle in space. A second contribution is one associated with the cloud surrounding the electron of an electron concentration depletion. Note that when considered on a reasonably large scale, as compared with the cloud size, the two contributions to the electron concentration fluctuation mentioned so far counteract each other. A third contribution is associated with the electron cloud surrounding an ion. There will be no intrinsic counterpart to counteract this contribution, and this will be the important part of the fluctuation in electron concentration at least in the low frequency region of the fluctuation spectrum.

The size of the clouds 'dressing' the particles is typically the Debye length λ_D, and it is understandable that the power spectrum of $n(k, t)$ should be very different whether the wavelength Λ of the fluctuations selected by the Bragg scattering condition (depending on the probing wavelength and the geometry) is much smaller or much greater than λ_D (depending on the plasma concentration and temperature). In the first case ($\Lambda \ll \lambda_D$), the scattering is by uncorrelated electron fluctuations as considered by Fabry, and the power spectrum of the return reflects the thermal velocity distribution of these electrons. In the second case ($\Lambda \gg \lambda_D$), the scattering is by collective electron fluctuations organized by plasma waves; the power spectrum narrows down

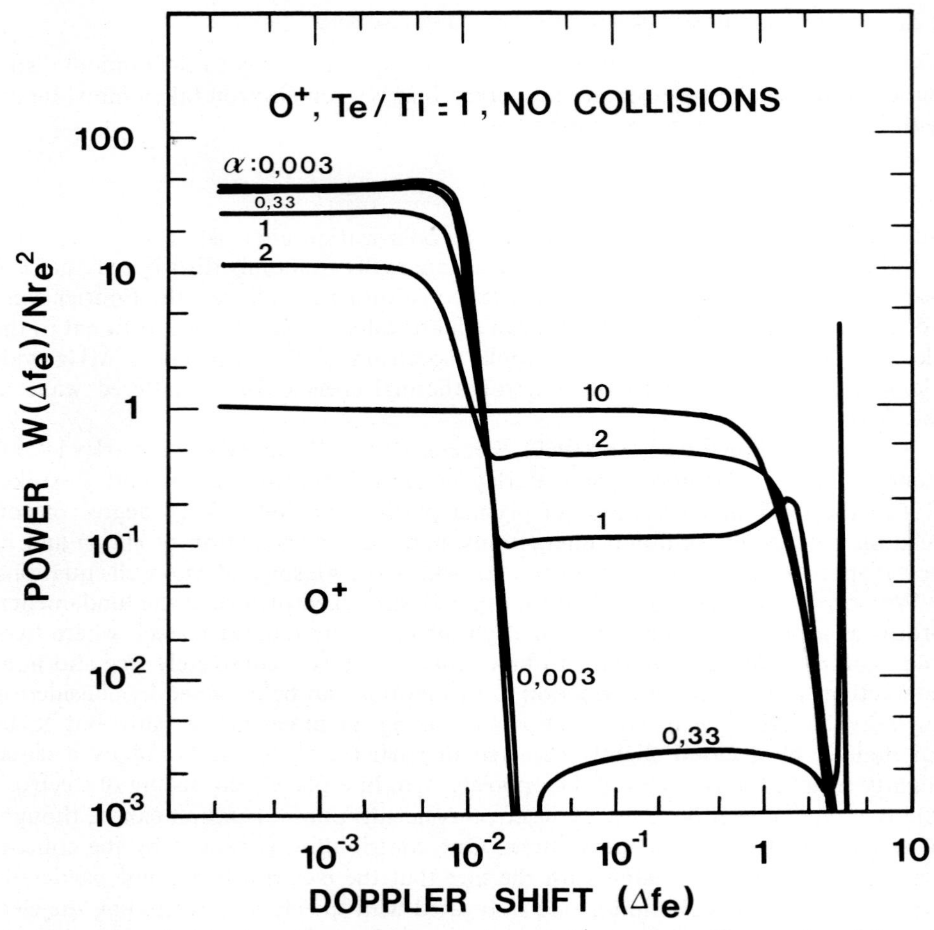

Fig. IV.4. Theoretical incoherent scatter spectra of a thermal O^+ plasma for different values of the ratio $\alpha = 2\pi\ \lambda_D/\Lambda$ of the plasma Debye length λ_D to the wavelength Λ of the Fourier component selected by the Bragg scattering condition. The indicated Doppler shift is symmetrical on either side of the transmitted frequency when the plasma has no body motion.

to a central 'ion line' roughly corresponding to the Doppler effect tied in with the propagation of ion acoustic waves, and to a residual electronic component consisting of two 'plasma lines' up and downshifted by the plasma frequency, corresponding to plasma oscillations (Figure IV.4).

In all cases, when the plasma is in body motion, the whole spectrum is shifted with respect to the transmitted frequency by an amount corresponding to the body Doppler along the scattering direction.

4.3. Implementation of Incoherent Scatter Sounding

The use of incoherent scattering by thermal fluctuations of the ionospheric plasma as a sounding method was initiated in 1958 by Gordon and Bowles (1958, 1961) with a high power V.H.F. radar system designed and used for meteor scatter research. The technique has revealed to be much more potent than anticipated and developed into a major mean for studying the Earth's environment despite the drawback of its cost: because of the small value of the Thomson scattering cross section, powerful transmitters, large antennas and sophisticated signal processing are required.

This explains why there have been few specially designed incoherent scatter facilities actually budgeted (four in all).

On the other hand, the richness of the incoherent scatter technique lies in the fact that a large number of plasma parameters intervene into, and can be deduced from, the shape of the scattered field's power spectrum. Thus it provides a simultaneous measurement of many quantities within the observed scattering volume. Furthermore, this scattering volume can be made to scan rapidly the whole vertical extent of the ionosphere.

4.3.1. THE OPERATION OF INCOHERENT SCATTER RADARS

Since all the electrons in the ionosphere encompassed within the lobes of the transmitting and receiving antennas can participate in the phenomenon of incoherent scattering, the first experimental problem is to reduce the scattering region to such dimensions that the various plasma parameters within it are nearly constant. This constraint mostly involves the vertical dimension, ionospheric gradients always being strongest in that direction.

4.3.1.1. *Modulation by Single Pulsing*

A very simple idea consists in applying the classical method of pulse radar, transmitting and receiving at the same point with a vertically directed antenna. The electromagnetic wave under consideration propagates with the speed of light *in vacuo*, since its frequency is well above ionospheric plasma frequencies. Therefore, the wave transmitted at time 0 in the form of a brief pulse travels a distance ct in time t, and the scattered signal received at time t has been scattered at a distance $ct/2$. If the pulse has finite duration τ, small compared with t, the response at the instant t will therefore come from a slab of ionosphere of thickness $c\tau/2$, situated at altitude $ct/2$. Thus if $\tau = 100\ \mu s$, the slice being studied is 15 km thick.

We then run into the following difficulty: a pulse of duration $\tau = 100\ \mu s$ has a power spectrum whose width $1/\tau = 10$ kHz is comparable to the width of the ionic part of the incoherent spectrum itself. The observed spectrum is then, in fact, the convolution of this parasitic spectrum with the useful spectrum, so that its interpretation is fraught with supplementary errors. The width of the pulse spectrum cannot be reduced without increasing pulse length, but this would then bring in inadmissible values of altitude resolution. An almost satisfactory compromise can be found for the upper ionosphere. However at low altitude below about 200 km there is no satisfactory solution, for the double reason that the useful ionic spectrum is narrower, and the altitude resolution needed must be better since the vertical gradients of the various parameters are steeper; the geophysical reason for these two phenomena is the same: decrease in plasma temperatures and increase of the mean ion mass.

In conclusion, the method of single pulsing has certain advantages of simplicity: only one antenna and a simple modulation system are necessary. On the other hand its performance is limited, especially at low altitude, by the problem of convolution of the useful spectrum with the spectrum of the emitted pulse. Besides this, there exists a similar problem related with the receiver gating, but this does not qualitatively affect the simplified discussion that we have presented.

4.3.1.2. *Modulation by Double Pulsing*

As before, emission and reception are vertical and at the same point. Let us suppose that at time 0, a pulse 1 with a certain polarization is emitted, followed at instant τ by pulse 2 with orthogonal polarization, so that the scattered signal coming from pulse 1, which we call $x_1(t)$, can be separated from that coming from pulse 2, $x_2(t)$. The signals $x_1(t)$ and $x_2(t + \tau)$ come from the same altitude $z = ct/2$ (Figure IV.5).

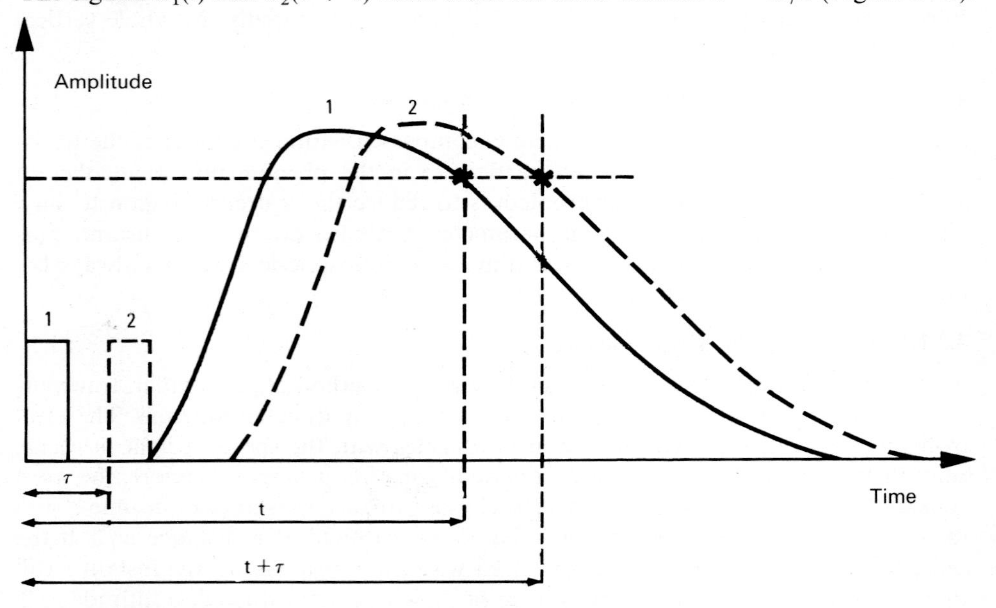

Fig. IV.5. Amplitude as a function of time of the incoherent scatter ionospheric echo from two successive pulses 1 (full line) and 2 (dashed line). The amplitude maximum corresponds to the peak electron concentration in the *F* layer.

But signal x_2 has been scattered τ seconds later than x_1; more precisely $x_1(t)$ is proportional to $n(k, t/2)$ at altitude z and $x_2(t)$ is proportional to $n(k, t/2 + \tau)$ at the same height z. The quantity $\langle x_1(t)\, x_2(t + \tau)\rangle$, the average being taken over many pairs of pulses, is therefore proportional to $\langle n(k, t/2)\, n(k, t/2 + \tau)\rangle$, which is nothing but the autocorrelation function, with delay time τ, of $n(k, t)$. Letting τ vary systematically enables us to know the autocorrelation function, from which the power spectrum can be deduced by a simple Fourier transform, according to the Wiener-Khintchin theorem. We are thus completely freed from the problem of spectral convolution mentioned in the preceding paragraph and we can improve the resolution in altitude, choosing pulse lengths as small as we wish without any difficulty except weakening of the received signal.

Now we shall see that the use of two orthogonal polarizations is not essential. Let us imagine the two pulses to be emitted with the same polarization. The signal received at time t will be $x(t) = x_1(t) + x_2(t)$, without it being possible to separate x_1 and x_2.

If we form the product $x(t)\,x(t + \tau)$ and average as before over many pulses, we obtain:

$$\langle x(t)\,x(t + \tau)\rangle = \langle [x_1(t) + x_2(t)]\,[x_1(t + \tau) + x_2(t + \tau)]\rangle$$

or

$$\langle x(t)\,x(t + \tau)\rangle = \langle x_1(t)\,x_2(t + \tau)\rangle + \langle x_1(t)\,x_1(t + \tau)\rangle + \langle x_2(t)\,x_2(t + \tau)\rangle + \langle x_2(t)\,x_1(t + \tau)\rangle.$$

The first term in the above equation is the autocorrelation function that we found previously. As for the other three terms, they are zero, since they contain the average values of spatial Fourier components of the electron density taken over different volumes (one can easily convince oneself of this by looking at Figure IV.5), which are not correlated. These extra terms contribute only to statistical errors for a finite time of observation, and this contribution is only important if the peak signal to noise ratio approaches one.

Therefore, the method of double pulsing, at the price of complicating the system of modulation, enables us to resolve elegantly the problem of convolution with the spectrum of the emitted pulse. We must mention, however, that certain practical difficulties arise from the fact that one cannot emit and receive at the same time and consequently, at certain altitudes, it is impossible to measure the autocorrelation function for certain useful delay times. Multiple pulsing is an obvious extension of the double pulsing, which permits one to obtain the autocorrelation function for several time delays simultaneously. Phase and frequency coding within the pulse (Gray and Farley, 1973; Evans, 1974) permit to achieve an extremely good height resolution at no expense to the signal/noise ratio.

4.3.1.3. *Multiple-antenna systems*

If two distant antennae are sufficiently directional for the main lobes of their radiation patterns to have a width of the order of one degree, the intersection of these lobes defines a volume of satisfactory size. A continuous wave can be transmitted from one and the scattering received with the other, eliminating all spurious modulation effects on the received frequency spectrum. It is necessary that one of the antennae be movable so that the altitude of the volume studied can be varied. Figure IV.6 gives an example of an experiment having this general scheme: the transmitting antenna is fixed and the receiving antenna movable, allowing reception of scattered signals from any altitude.

Multiple antenna systems, technically more complex, allow very good results and are particularly well adapted to the study of the lower ionosphere. At high altitude, on the other hand, pulsed systems seem to be more attractive, thanks to the large peak power that can be employed in order to improve the signal to noise ratio. Although steerable monostatic pulsed radars can have access to plasma motion observation by pointing in different directions, the simultaneous measurement of the three components of the plasma body velocity at one ionospheric location make necessary in any case the use of at least three receiving stations.

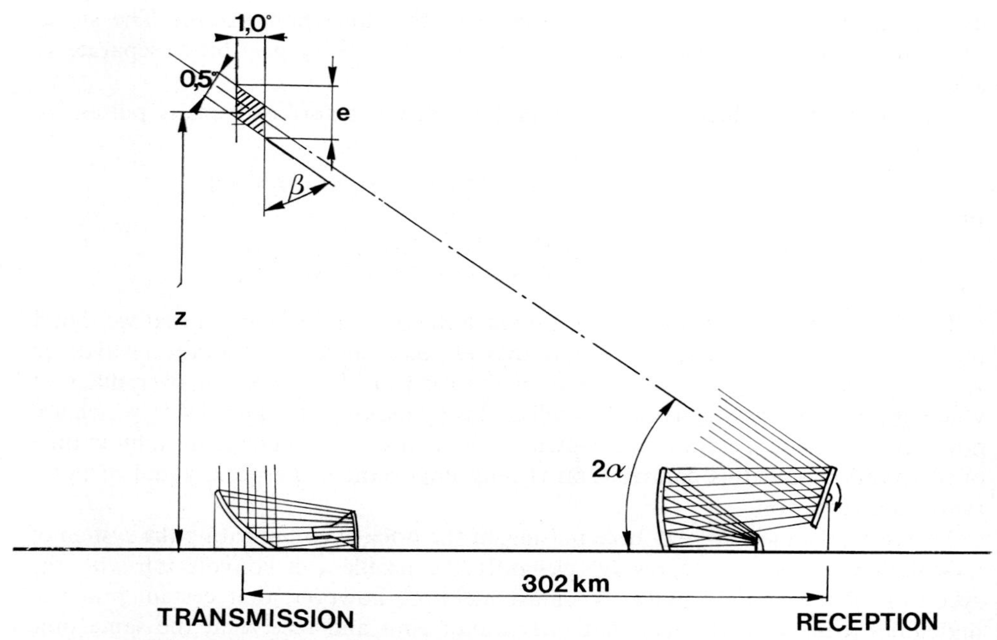

Fig. IV.6. Sketch of the double beam method used in the first version of the French incoherent scatter experiment. The variable tilt α of the plane reflector of the receiving antenna allows one to explore an altitude range z between 100 and 500 km. The vertical extent e of the scattering volume varies at the same time from 3 to 30 km, while the bisector of the scattering angle β, along which body Doppler is measured, keeps approximately parallel to the magnetic field.

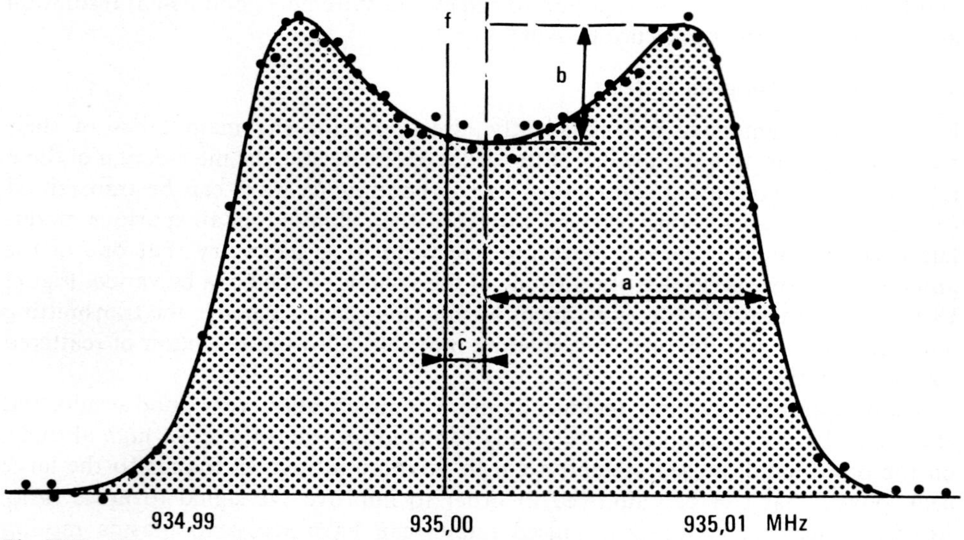

Fig. IV.7. Best fit of a theoretical incoherent scatter spectrum to experimental data points. The parameters being fitted are – the width a, which depends on the electron temperature and mean ion mass; – the shoulder height b, which is mainly a function of the ratio of electron to ion temperature; and – the shift c from the transmitted frequency (here exaggerated for the benefit of clarity).

Actually, the choice of geometry and frequency depends on a large number of factors, and the existing incoherent scatter facilities (Table IV.1) represent a wide choice of compromises.

If it is the primary aim of the experiment to determine plasma velocity components with good time and space resolution using a multi-static system, one must strive to keep the beam of the antennae narrow, and a high frequency around 1 GHz is automatically dictated to keep the size of the antennae within reasonable dimensions. If on the other hand one wants to observe the scalar plasma parameters with good time resolution over a wide height interval, one should choose a pulsed monostatic system operating in the vicinity of 200 MHz, strive to make the peak to average transmitter power as high as possible, and use coded pulses to increase height resolution. This explains why the high latitude European incoherent scatter facility (EISCAT), which was designed without *a priori* constraints to fulfill both objectives, is a dual system.

4.3.2. THE INTERPRETATION OF INCOHERENT SCATTER SPECTRA

The raw data from an incoherent scatter experiment are power spectra (or autocorrelation functions) of the wave received from well defined scattering volumes, integrated over a measurement period that varies from a few seconds to several minutes. Extracting the useful data consists in fitting theoretical spectra, in the sense of 'least squares', to these experimental spectra (Petit, 1968). The agreement between theory and experiment is excellent, as shown in Figure IV.7 which is an example taken from tens of thousands of others.

The errors of measurement are essentially of a statistical type, so that they would tend to zero if one could allow the time during which the received signal is integrated before recording its power spectrum to approach infinity. The residual fluctuations are in most cases due to receiver noise, very often more powerful than the received signal, which always remains small because of the small Thomson scattering cross section. Under such conditions, the error in the measured parameters varies as the noise to signal ratio. However, the signal received is itself of a random nature. Thus when one succeeds in attaining a signal to noise ratio that exceeds unity, the signal fluctuations play an important part, and they make illusory the search for a larger peak signal to noise ratio.

4.3.2.1. *Electron Concentration*

To a first approximation, the electron concentration enters the expression for the ionic spectrum only as a factor of proportionality, upon which, indeed, the total received power depends. However, certain technical characteristics of the experiment, such as the transmitted power and antenna gain, also have a direct influence on the power received. Therefore, the error in measurement of the electron density is not mainly statistical, but comes from imprecision in the determination of these technical characteristics, some of which can change with time. Since these variations are especially important over long times, it is often useful to normalize the observed vertical profile, that is to multiply the values obtained at all altitudes by the same factor, in such a way that the maximum concentration is equal to that which can easily be deduced from a classical ionosonde, even a very simple one. A relative precision of

TABLE IV.1
Incoherent Scatter Facilities

Location (date operational)	Geographical Coordinates	Geomagnetic Latitude	Frequency (MHz)	Peakpower (kW)	Type of Antenna and Steerability
Jicamarca (Peru) (1961)	11°95′ S 76°87′ W	2° N (L = 1.04)	50	4,000	290 m × 290 m array with limited beam steering about vertical (cf. Plate on page 84)
Arecibo (Porto-Rico) (1963)	18°30′ N 66°75′ W	30° N (L = 1.43)	430 40	2,000 2,500	300 m fixed spherical reflector with limited beam steering about vertical
Saint-Santin (France) (1965)	44°60′ N 2°19′ E	47° N (L = 1.76)	935	140 (CW)	20 m × 100 m reflector with fixed (vertical) beam and 3 remote receiving sites, with 40 m × 200 m reflector at Nançay and 25 m dishes at Mende and Monpazier (cf. Plate on page 224)
Millstone Hill (Mass., U.S.A.) (1962)	42°60′ N 71°50′ W	53° N (L = 3.12)	440 1295	3,000 4,000	68 m fixed (vertical beam) dish 25 m steerable dish
Chatanika (Alaska, U.S.A.) (1972)	65°60′ N 147°27′ W	65° N (L = 5.51)	1300	4,000	26 m steerable dish (cf. Plate on page vi)
Tromsöe (Norway) (1979)	69°35′ N 19°10′ E	66° N (L = 6.1)	224 933	6,000 2,500	5,000 m² reflector with meridional beam steering. 32 m steerable dish and 2 remote receiving sites with 32 m steerable dishes at Kiruna (Sweden) and Sodankyla (Finland)

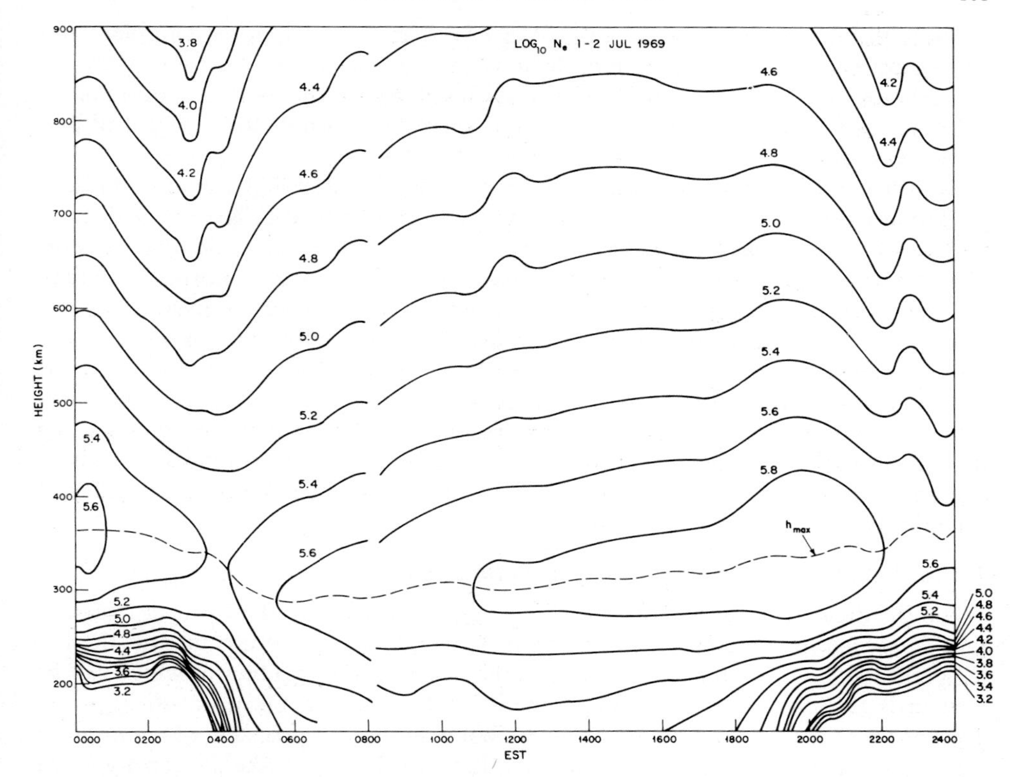

Fig. IV.8. Height versus local time electron concentration contours measured at Millstone Hill from the total returned power.

the order of 1% can thus be obtained (although the error in absolute value can reach 10%, depending on the care with which the power calibration is carried out). Figure IV.8 gives an example of contours of electron density measured by the Millstone Hill facility.

The incoherent scatter data by themselves however can yield under certain conditions a precise determination of the electron concentration profile, either using the plasma lines on either side of the ion line, whose position yields with good precision the plasma frequency within the scattering volume, or using the Faraday effect, when polarization measurements are available as well as a model magnetic field.

4.3.2.2. *Body Motion of the Plasma*

If the plasma is stationary with respect to transmitter and receiver, the incoherent scatter spectrum is symmetric with respect to the transmitted frequency. However, if plasma has a bulk motion, a Doppler effect shifts the entire spectrum in frequency. Furthermore, if the electrons have a body motion, different from the mean velocity of the ions (current), it can be shown that the spectrum becomes asymmetric, but this effect is appreciable only if the relative velocity of the electrons reaches values of the order of their thermal velocity. From this it results that, for velocities small compared with the thermal velocity of the electrons there is only a global Doppler effect corresponding to the bulk motion of the ions and not that of the electrons. Again the preponderance of the ions corresponds to the fact that it is the thermal motion of the ions that causes the fluctuations in electron density responsible for this central part of the spectrum.

Such frequency shifts of the entire spectrum (Carru *et al.*, 1967) are now routinely used to measure the bulk motion of ions (winds and electric fields, Figure IV.9) difficult to obtain by other measurement techniques. The precision that can be obtained is of the order of several meters per second. Here, the only cause of error is normally statistical, the problem of transmitter's frequency stability having been resolved in a satisfactory way by the techniques currently in use. Of course, with only one emitter and one receiver, only one component of the vector mean velocity of the ions (in the direction of the vector $\vec{\mathbf{k}}$) can be measured at one ionospheric location, and three receiving stations must be used in order to reconstruct the vector in space. Assuming that the ion velocity is the same over large horizontal scales, this difficulty has been overcome with monostatic systems by tilting the beam to get $\vec{\mathbf{k}}$ vectors at different orientations.

4.3.2.3. *Electron Temperature, Ion Temperature, Ion Composition*

When the ionic composition is known, one can easily measure the electron temperature T_e, which determines the extension in frequency of the spectrum, and the ion temperature T_i, which, for fixed T_e, affects the height of the two bumps which can be seen in Figure IV.7 on either side of the central frequency. Only statistical errors need be taken into account, and under the best conditions a precision of the order of 1% can be achieved. We note in passing that the ratio T_e/T_i has, like the density N, an influence over the total power received and that the measurement of the total power alone gives a composite parameter practically equal to $N\ (1 + T_e/T_i)$ from which either N or T_e/T_i can be extracted if the other parameter is independently determined.

Actually, the hypothesis of known ionic composition applies well below 120 km, where only molecular ions exist, O_2^+ of mass 32 and NO^+ of mass 30, masses close enough that they need not be distinguished from each other. Similarly, between 250 and 500 km of altitude, it is known that one species, the O^+ ion of mass 16, largely dominates. On the other hand, in the transition region between the molecular ions and the O^+ ions, the ionic composition cannot be considered as known. It is the same at very high altitudes, where the proportion of H^+ ions becomes important.

When the ionic composition is unknown, it is harder to deduce the electron and ion temperatures from a given spectrum. In fact, a change in the ionic composition affects the extent of the spectrum and the height of the shoulders, so that it becomes very difficult to distinguish the effect of this new parameter from that of the temperatures. For a mixture of molecular ions and O^+, Figure IV.10 shows, as a function of the parameter p, the relative concentration of O^+ ions, the errors in the various parameters deduced from the same spectrum, in the case where p is assumed known *a priori* and in the case where it is assumed unknown. We see that under these conditions the errors have greatly increased, to the point where they sometimes destroy the significance of the measurement. When this happens, it is impossible to deduce N, T_e, T_i and p from the observed spectrum if an outside source of information about

Because of the sharp decrease in electron density below 90 to 95 km, the received power does not permit measurements to be performed in the lower regions. Therefore ion-neutral collisions can be studied directly by incoherent scattering only in a thin slice of the ionosphere. In this region, the electron and ion temperatures are equal and the mean ion mass is known. It is therefore very easy to deduce from the shape of the spectrum the value of the ion-neutral collision frequency, and, from this, to study the behavior of the atmospheric density at the altitudes considered. At this altitude, N_2 is the major constituent and the measurements can be interpreted in terms of N_2 density (Figure IV.12).

4.3.2.5. *Effect of the Magnetic Field*

The terrestrial magnetic field plays a role in the phenomenon of incoherent scattering only if the vector $\vec{\mathbf{k}}$ is perpendicular, within a few degrees, to the magnetic field. If this is the case, theory predicts that the 'ionic' spectrum will show oscillations at the ion gyrofrequency, superposed on the usual power spectrum. It had been hoped that this phenomenon could be used to study the ionic composition of the ionosphere. The experimental attempts in this direction were thwarted, the theoretical explanation of this failure being that even a very low ion-ion Coulomb collision frequency suffices to make this modulation disappear. The phenomenon has actually been observed in the dilute plasma of the upper ionosphere, at altitudes of about 1500 km. However, the only ion found there is H^+, so that the measurement has little aeronomic importance.

If the vector $\vec{\mathbf{k}}$ is exactly perpendicular to the magnetic field, a contraction of the spectrum also occurs, all the scattered energy being concentrated in a narrow band. This phenomenon has been used profitably in order to measure more easily the Doppler effect corresponding to the bulk motion of the ions. Also, Woodman (1971) showed that by clever use of this spectral contraction the direction of the terrestrial magnetic can be measured with an accuracy of the order of a minute of arc (Figure IV.13).

4.3.2.6. *Lines at the Plasma Frequency*

The plasma oscillations of the electron gas result in the appearance in the spectrum of two lines situated on either side of the emitted frequency, at a distance from the latter equal to the plasma frequency. Theoretically, their width is very small, and in most cases, the observed width is determined by the variation of the plasma frequency within the scattering volume (Figure IV.14). Therefore, this is an extremely precise means for measuring the plasma frequency, i.e. the electron density, simply through the position in frequency of these lines, as mentioned above.

Under thermal equilibrium conditions, the intensity of these plasma lines is much smaller than the one of the ionic part of the spectrum layer by a factor $k^2 \lambda_p{}^2$ which is, in actual experiments, much smaller than unity (and has to be smaller than 1 for the plasma lines to be present). However, their level is enhanced by a high energy tail of photoelectrons that permanently exists when solar radiation ionizes the atmosphere. Fast electrons traveling through a plasma leave a wake of plasma waves (Pines and Bohm, 1952) and feed energy into these waves in proportion to $f(v\varphi)$, where f is the one-dimensional electron velocity distribution and $v\varphi$ is the phase velocity ($v\varphi =$

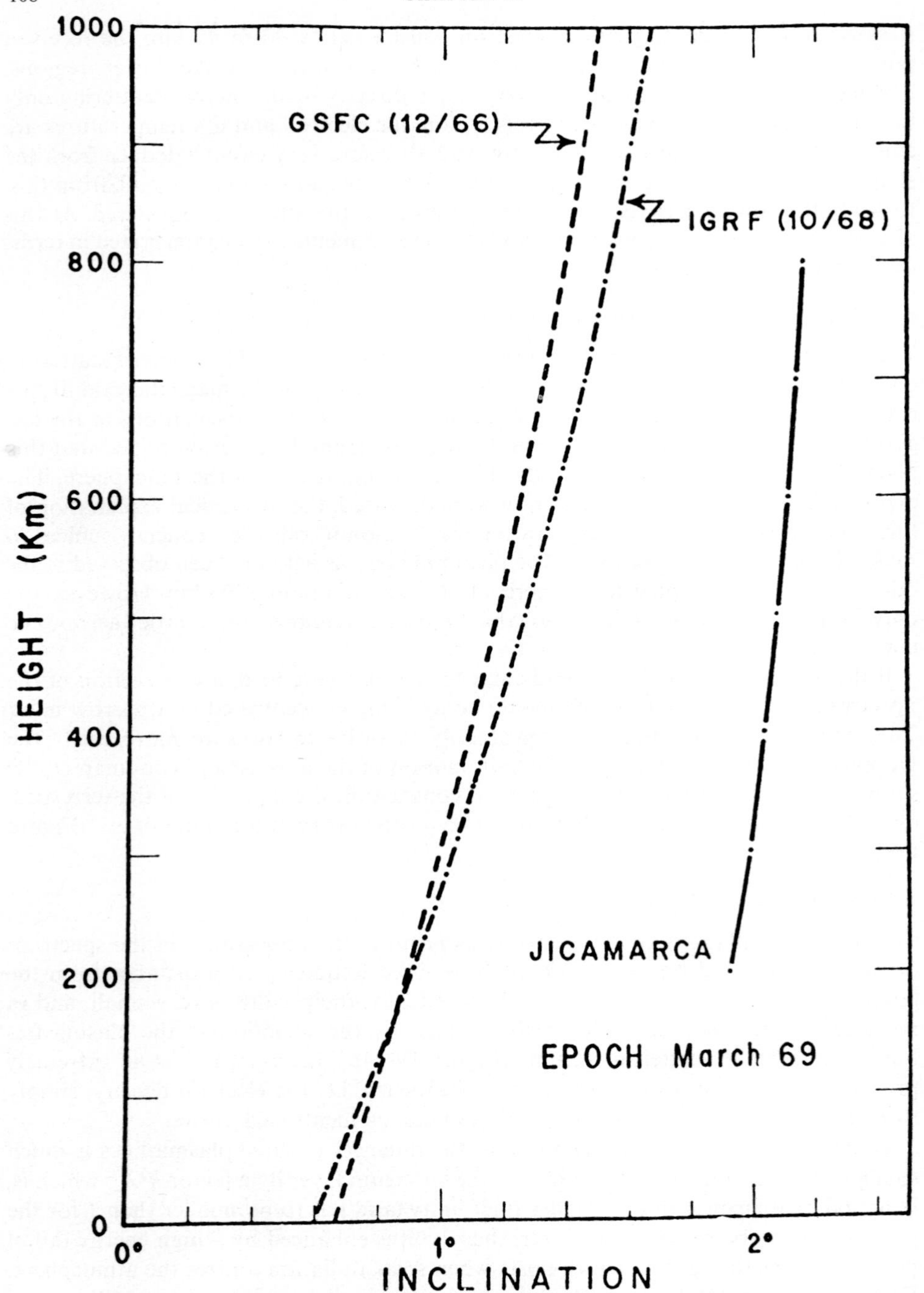

Fig. IV.13. Dip angle of the geomagnetic field above Jicamarca. The interrupted solid line is the value experimentally determined with the incoherent scatter facility; the other curves are from geomagnetic field models.

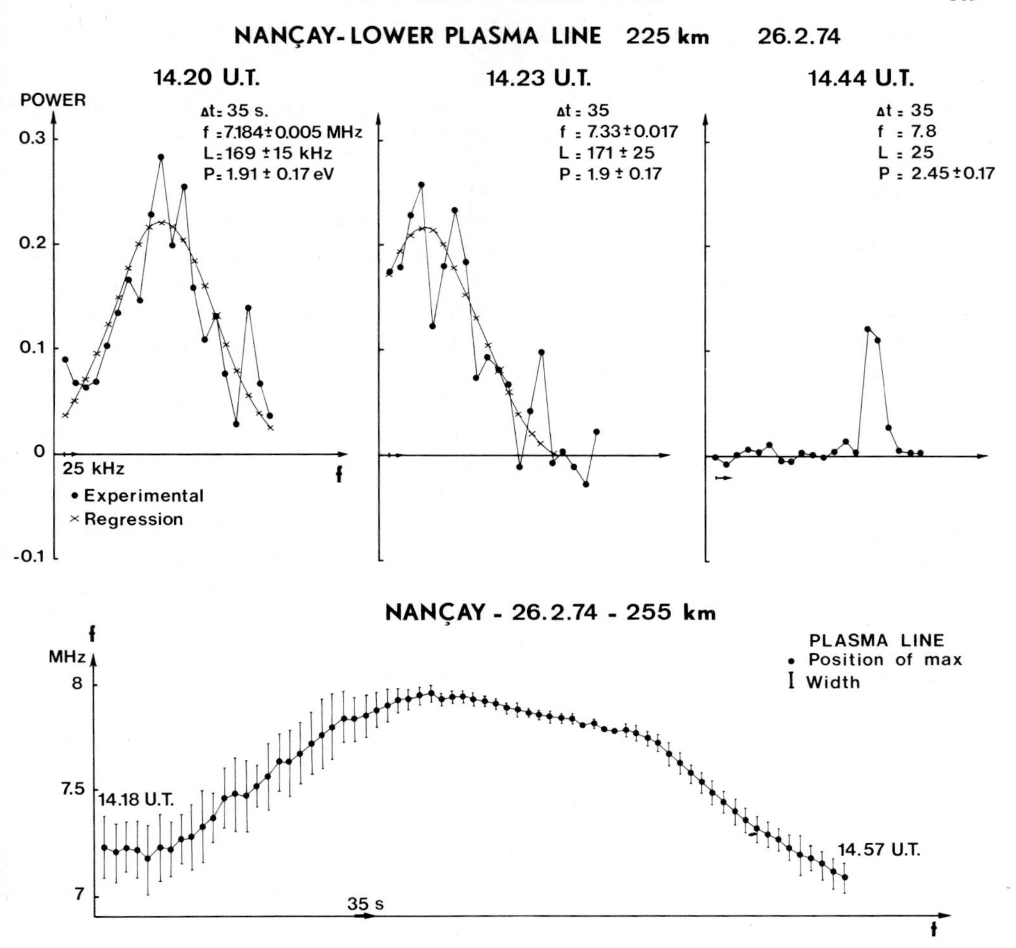

Fig. IV.14. Observations of the plasma line at Nançay. The upper panel shows three typical spectra of the up-shifted plasma line; the third one is very narrow because the scattering volume was located exactly at the F peak. The lower panel represents a 40 minutes variation of the plasma line shift, that is, of the plasma frequency at an altitude of 255 km.

$\omega_p/|k|$) of the plasma wave. This wave is damped by Landau damping (Stix, 1962) which is proportional to $(\partial f/\partial u)_{u = v_\phi}$. The resulting steady state level of plasma lines is then proportional to $-\left[\frac{\partial}{\partial u} \ln f(u)\right]^{-1}_{u = v_\phi}$. At energies where the photoelectrons dominate the Maxwellian electrons, f is much flatter than a Maxwell distribution and the intensity of plasma waves is correspondingly much greater. In applications to the ionosphere, we must include into the discussion the effects of electron ion collisions and the geomagnetic field. Plasma waves are excited and damped by collisions and as a result the dependence of their level on f is more complex (Perkins and Saltpeter, 1965) than indicated above: in particular when the phase velocity is much greater than the thermal velocity of the electrons, there are usually so few photoelectrons with velocities in the neighbourhood of v_ϕ that the collision effect dominates and the presence of photoelectrons has no influence on the level of plasma waves even though $\left[\frac{\partial}{\partial u} \ln f(u)\right]_{u = v\phi}$ is very different from the value given by a Maxwellian distribution.

Yngvesson and Perkins (1968) have investigated, both theoretically and experimentally, the effect of the geomagnetic field. When the angle θ between the propagation vector $\vec{\mathbf{k}}$ and the magnetic field is intermediate between 0° and 90°, the Landau damping can be orders of magnitude greater than the field free expression and depends strongly on θ. As a result, the lowest plasma frequency for which a non-thermal enhancement of the plasma line is observed, depends on the angle θ.

Measurements of the plasma line have been interpreted in terms of one-dimensional velocity distribution and flux of photoelectrons, under simplifying assumptions (Yngvesson and Perkins, 1968; Cicerone and Bowhill, 1970; Wickwar, 1971). One major difficulty is that a different phase velocity (linked to plasma frequency by the relation $v_\phi = \omega_p/k$) is measured at different altitudes in the ionosphere. To reconstruct the velocity distribution, one must assume that this distribution is independent of altitude, which is only valid for altitudes and photoelectron energies large enough. Another difficulty lies with the fact that the velocity distribution is given by an integral equation, the solution of which requires an initial value for a particular value of energy. Here again, an assumption has to be made, usually based on the theoretical estimations of the photoelectron flux.

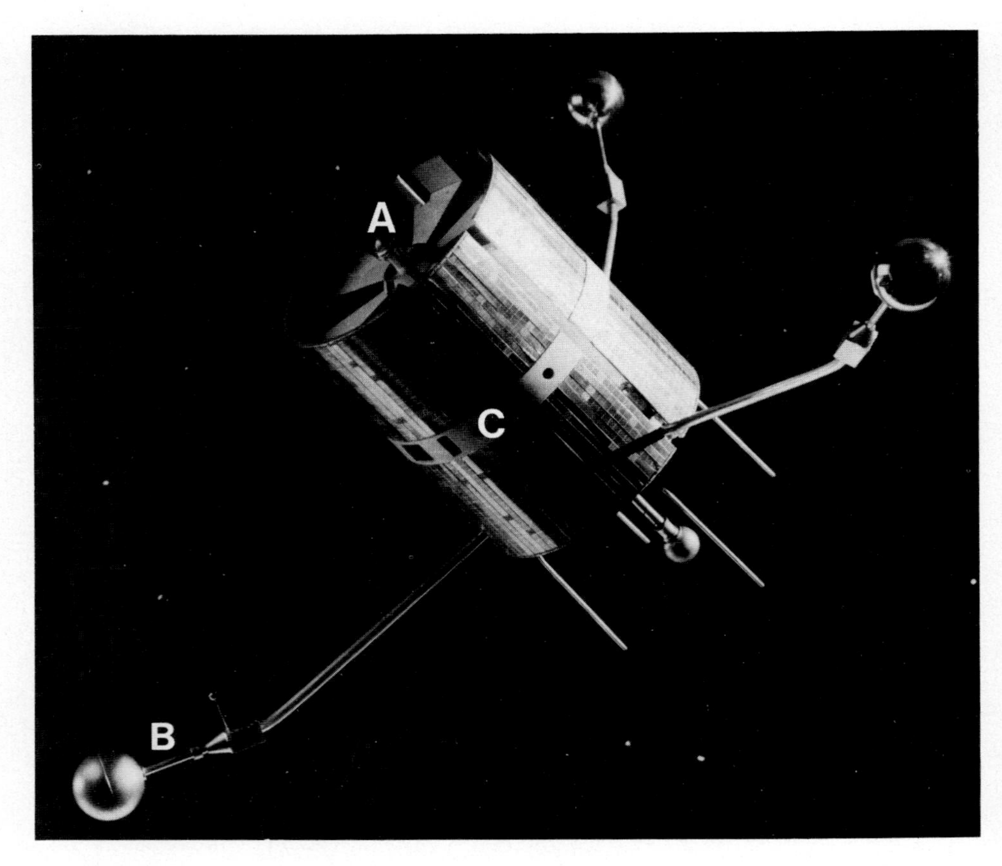

Launched on November 22, 1972, the ESRO 4 satellite is a good example of the generation of small scientific payloads which investigated upper atmospheric conditions when mastery of *in situ* measurements was reached. This satellite weighted 113 kg, was spin-stabilized, and remained 510 days aloft on an elliptical polar orbit. The perigee made five complete meridional revolutions about the Earth, lowering from an initial altitude of 245 km to 115 km an hour before desintegration on April 15, 1974, allowing measurements to be taken close to the source region for many atmospheric and ionospheric processes. The satellite, equipped with onboard data storage capacity permitting continuous coverage, carried three instruments of particular aeronomic relevance: A. a neutral gas analyzer provided by the University of Bonn, consisting of an electron impact ion source, a radiofrequency mass-analyzer and a current multiplier detector. B. thermal plasma probes mounted on a boom, provided by University College, London, consisting of a spherical gridded Langmuir probe enabling in principle mass identification, number density and temperature determination for the major positive ion species, and a small Langmuir probe for measuring electron density, electron temperature and spacecraft potential. C. an auroral particles spectrometer provided by the Kiruna Geophysical Observatory, consisting of various analyzers and detectors to measure the energy spectrum and pitch angle distribution of precipitating electrons and protons in the range 0.2 to 150 keV. The data acquired shed light in particular on the response of the upper atmosphere to geomagnetic disturbances (cf. Von Zahn, 1975). (Courtesy of the European Space Agency)

CHAPTER V

SAMPLING WITH SPACE-BORNE PROBES

The possibilities offered by spacecraft extended considerably the domain of investigation open to methods using:

(a) Radio techniques (topside sounding; local excitation of the characteristic frequencies of the plasma; bistatic Doppler and Faraday methods; reception of the natural radiofrequency noise emitted by the plasma in the very low frequency range), and (b) Optical techniques (spectrophotometric observation of the auroras, of the natural airglow: red line of OI at 6300Å, lines of CaII at 4215Å, of NII at 3934Å, of MgII at 2800Å, of HeII at 304Å; or of artificial tracers, neutral (sodium, trimethylaluminium) or ionized (cesium, barium) injected by rocket probes or satellites).

But above all, it gave experimentalists the opportunity to apply to the ionospheric medium the techniques of direct measurement used to study plasmas in the laboratory: d.c. plasma probes, particle energy analyzers, and mass spectrometers.

5.1. The Langmuir Characteristic

Direct plasma measurements, first used by pioneers in the physics of ionized media (J. J. Thomson, I. Langmuir) in the first half of the XXth century, consist in extracting from within the plasma, with the aid of electrodes or grids followed by a collector, a sample of particles, more or less selected by charge, energy, or mass. The extraction and the selection of particles is done through the application of potentials to the system of electrodes or grids, and the flux of collected particles is transduced to an electronic signal by means of appropriate detectors and amplifiers. As the applied potentials are varied, there corresponds a change in the flux of sampled particles and therefore in the signal, from which the desired plasma parameters can be, in principle, deduced through solution of the Poisson and Boltzmann equations of the system determining the transfer function of the apparatus. In practice a simplified theory is often used; then corrections are introduced.

The simplest system consists of one electrode, – e.g. a simple metal wire whose end is inserted into the plasma, connected to one terminal of a battery while the other terminal is connected to a reference 'ground' having a known difference of potential U_0 with respect to the plasma. The current collected by the probe depends on the potential V of the probe with respect to the plasma, which is related to the potential U applied across the terminals of the battery by

$$V = U + U_0.$$

When $V = 0$ (probe at the plasma potential), the electrons and the ions arrive at the electrode with their thermal velocities. If the velocity distributions are Maxwellian (isotropic) it can easily be shown that the collected currents are

$$J_{e,i} = \pm\, en\, v_{e,i}\, A/4,$$

where A is the surface area of the probe, $n = n_e = n_i$ is the concentration of the unperturbed plasma (assumed neutral) and $v_{e,i} = (8\, kT_{e,i}/\pi m_{e,i})^{1/2}$ (k is Boltzmann's constant, T the temperature and m the mass).

Since, because of the large mass difference between the ions and the electrons, $v_e \gg v_i$, the electron current J_e is much larger than the ion current J_i; therefore, the total collected current $J_e + J_i$ is negative.

If now, from zero, V increases to positive values, the ions will begin to be repelled by the probe and the component J_i will decrease, the total collected current becoming slightly more negative. When V becomes large compared with the mean energy of the ambient ions, no ion will reach the probe. There will have formed around the probe a region, called a 'sheath', which is void of ions, and in which the electrons are accelerated (Figure V.1).

If, conversely, V takes on negative values, the electrons will be more and more repelled and the ions attracted. The total current, less and less negative, will pass through zero at a value of V called the 'floating potential', at which the residual electron current just balances the ion current; this is the equilibrium potential to which the probe would spontaneously adjust itself if it were disconnected from the battery. For values of V more negative than this, the collected current will become increasingly ionic, i.e., positive. When no more electrons can reach the probe, the current will be purely ionic, and this time the sheath will be totally void of electrons.

In the two cases where the current (called 'saturation current') is purely electronic or purely ionic, it is still proportional to $en\, v_{e,i}\, A/4$, but the collecting surface A is no longer the area of the probe, it is that of the sheath. The thickness of the latter, the 'stand off' distance of repelled charged particles, is given very roughly by the screening distance, or distance beyond which the plasma naturally recovers its neutrality. This is of the scale of the Debye length, which we have encountered several times before. But the screening distance, and the surface area of the sheath, depend in a complicated way upon the geometry of the system, and increase with the

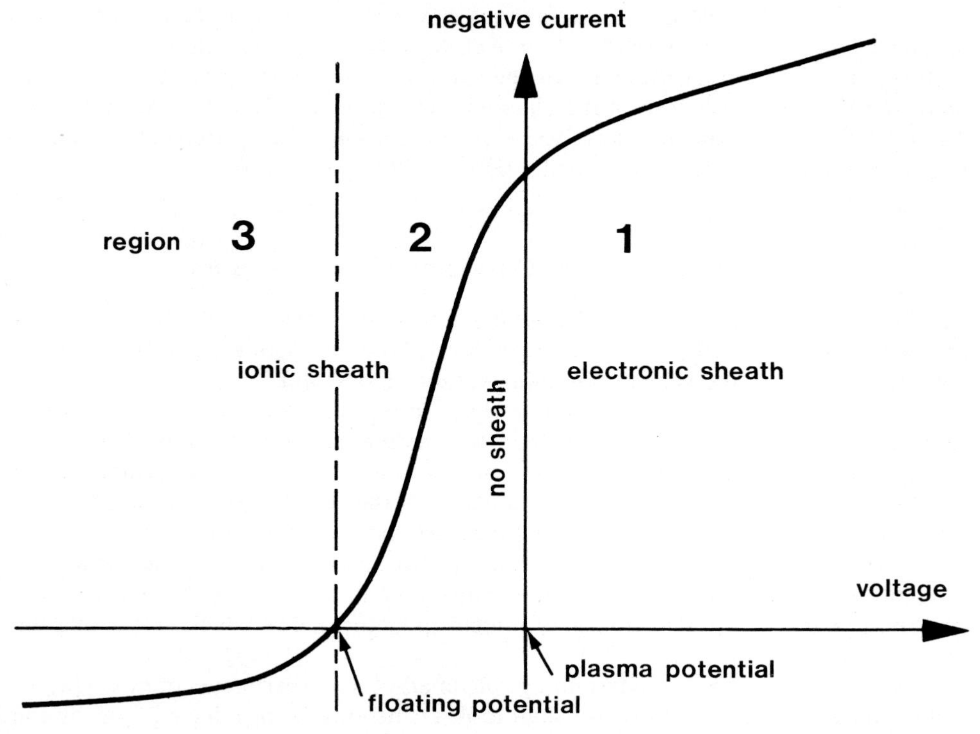

Fig. V.1. The current-voltage characteristic of an electrode in a thermal plasma. The scale of the linear voltage abcissa is given by the difference between the floating and plasma potentials, which is of the order of a volt.

potential on the probe. This effect explains the slow increase in saturation current with the absolute value of the applied potential, and the differences in properties among plane, cylindrical and spherical electrodes.

It should be noted that throughout the ionosphere the Debye length is much smaller than the size of a space vehicle, and at most of the order of the size of commonly used probes.

Between the positive and negative values of V that correspond to the electron and ion saturation currents, the collected current varies abruptly with V. We shall see below that this variation, which represents in effect a scan of the distribution in kinetic energy of the electrons, leads to the determination of their temperature.

If instead of a single collecting electrode (Langmuir probe), the probe is a grid, or a plate pierced by an orifice, the 'current–voltage characteristic' given in Figure V.1 applies to the flux of particles passing through it. This is the case of the part of various types of ion energy analyzers and mass spectrometers.

The above analysis is valid if the trajectories of ions and electrons, in the zone electrically perturbed by the potential applied to the probe, are determined only by the electric field within the sheath. The presence of a magnetic field can have an influence if the radius of gyration of the particles under consideration is not large compared with the Debye length. The data of Chapter I show that in the ionosphere there is no fear of this happening to the ions, but that, on the other hand, in the case of the electrons, one can expect to observe effects due to the orientation of the probe with respect to the magnetic field. In the same way interparticle collisions will intervene if the mean free path of the charged particles is less than the Debye length; Table I.1 shows that this is always so below 80 km. Direct measurements in the *D*-region therefore obey very special rules (Hoult, 1965; Sonin, 1967).

5.2. Specific Problems Connected with Space Vehicles

The fundamental constraints for the application of these methods to the ionosphere are tied in with two irreducible characteristics of the interaction between the ionospheric plasma and the vehicle (sounding rocket or satellite):

– The vehicle almost always, except perhaps in the case of a rocket at the apex of its trajectory, has a velocity greater than the thermal speed of the ions (Table I.1). Consequently, the velocity distribution of the ions in the reference frame of the vehicle is strongly anisotropic, and the 'attitude' of the vehicle, and wake effects, play an essential role in the collection of the particles, especially the ions.

– Even though the plasma is electrically neutral, it is difficult to generate a fixed reference potential. The vehicle is not 'grounded', and acquires, with respect to the plasma, a charge which depends on a number of factors, all of which vary along the trajectory or the orbit.

Since these effects have a determining role in all of the experiments of this type, we shall examine them first. Next, we shall look at the specific problems posed by the various systems of analyzers that are used to track back the ambient particles' concentration, energy or mass from the collected currents.

5.2.1. THE REFERENCE POTENTIAL

The equilibrium potential reached by a body (the rocket or satellite in this case) immersed in a plasma, an unbounded source of charges of two signs (the ionosphere in this case), is determined by the condition that the net current which it receives must be zero. We have seen that in the case of an isotropic Maxwellian velocity distribution of the particles, this 'floating' potential is slightly negative with respect to the plasma potential so that it repels most of the electrons, equilibrating the incident electron and ion currents. For a vehicle of a size large compared with the Debye length and fixed with respect to the plasma, it can be shown (Whipple, 1965) that this potential is, in principle, equal to $kT_e/e \ln (T_i m_e/T_e m_i)^{1/2}$. Depending on the average mass of the positive ions, this lies between 3 kT_e/e and 6 kT_e/e, i.e. in the neighborhood of -1 Volt.

5.2.1.1. *Effects of Changes in Attitude*

Actually, this potential is changed in absolute value by the anisotropic character of the ion velocities, due to the supersonic velocity, v_s, of the vehicle in the medium, which acts to increase the ion current by a factor proportional to the ionic Mach number v_s/v_i.

In addition the photoelectric current emitted by the vehicle under the action of solar photons as well as the incidence of energetic particles can drastically alter the potential from that expected in thermal conditions. (The combination of these two effects has been reported to produce electrical breakdowns of satellites under storm conditions.) The photoelectric current, which has the same sign as, and is comparable in magnitude to, the ion current in the E and F regions, can, when the ambient charged particle concentration is low (D region or protonosphere) become preponderant and cancel or turn positive the potential of the vehicle with respect to the plasma.

Furthermore, especially in the case of a long vehicle, the effect of the terrestrial magnetic field on the anisotropy of the electrons can also act to reduce the electron current.

The capacity of the plasma-vehicle system is of the order of that of a condenser whose plates have an area equal to that of the surface of the vehicle and are separated by about a Debye length. The time constant for reaching equilibrium can be shown to be:

$$\tau = e^{-1}(m/n_e)^{1/2} \simeq 10^{-4} n_e^{-1/2} \text{ s},$$

which is negligible compared with the characteristic time for variations in attitude, usually of the order of a second. It follows that the potential of the vehicle continually adjusts itself to the changes in equilibrium conditions caused by its gyroscopic motions, since in general its properties are not symmetric with respect to its velocity vector, the flux of incident photons, or the lines of force of the geomagnetic field.

Furthermore, the effect of the terrestrial magnetic field is felt not only through the anisotropy of the ambient electrons, but also through the induction of an electromotive force by the differential motions of the vehicle. The induced potential gradients between different parts of a spinning rocket or satellite can reach 1 mV cm^{-1} (for example, between a probe placed at the end of a boom and the body of the vehicle), and are modulated by changes in attitude respective to the field lines.

Consequently, the planning of experiments must be able to accommodate quite a large range of values of the reference potential, and its rapid changes within that range. In particular, the choice of potentials applied to probes must allow a considerable margin, of the order of several volts, and the characteristic time for measurement of a parameter must be small compared with that of the most rapid changes to be expected in the potential of the vehicle (in general a fraction of its spin period).

5.2.1.2. *Effects of Measuring Currents*

Another facet of the problem is that the currents collected by probes return to the ionosphere through the surface of the vehicle, so that the impedance of this surface with respect to the plasma is actually an element in the probe circuit. One must therefore ensure that the collected currents do not significantly affect the reference potential. This implies that the ratio between the areas of the vehicle and the probes should be at least an order of magnitude greater than the ratio between the electron and ion saturation currents that can be collected by the vehicle (in general of the order of 10^3 to 10^4; since the surface area of a vehicle is of the order of m^2, that of the probe cannot be more than some cm^2). Precautions of this type are particularly necessary if, as is almost always the case, the vehicle carries several experiments: a large amplitude sweep in the potential of an electrode, made necessary because of uncertainty in the reference potential, can cause, over part of its cycle, large perturbations in the measurements of another probe, by means of a change in the potential of the vehicle.

Another type of interference between experiments can also appear through the biasing of the plasma sheaths surrounding the vehicle and the electrodes. These sheaths can come to encroach upon each other because of changes in the applied or floating potentials. This is one of the reasons why many probes, particularly those intended for measurement of electrons, are kept far from each other, and from the surface of the vehicle, by means of supporting booms. This is also why the collecting surfaces of probes must be surrounded by 'guard' electrodes at the same potential; these can be, for example, a ring around a circular planar probe, or the end of the supporting boom of a spherical probe. Naturally, the area of the guard electrodes must be taken into account in the problem of return current.

5.2.2. WAKE EFFECTS

We have studied the effect of the plasma on the vehicle. We have seen that the latter behaves like a floating Langmuir probe, whose equilibrium potential is modulated by the attitude but stays slightly negative, except possibly in sunlight at very high or very low altitudes when the photoelectric effect dominates or under energetic particle irradiation. The vehicle is surrounded by a sheath, generally depleted in electrons, whose thickness is of the order of the Debye length in the medium. The motion of the vehicle with respect to the plasma, which increases the flux of collected ions by a factor of the order of the ionic Mach number, causes the potential of the vehicle to be generally less negative than it would be in the absence of motion. We can presume that the sheath loses its symmetry; but we must examine in more detail the effect of the moving vehicle on the distribution of the charged particles surrounding it, to see how well they represent the unperturbed ambient plasma.

5.2.2.1. *Ion and Electron Wakes*

As we have indicated, below about 80 km of altitude, since the mean free path in the medium is less than the Debye length and also less than the characteristic size of the instruments, the flow of ions and electrons is strongly coupled to that of neutrals, which in turn is determined, at supersonic velocities, by the presence of a shock wave and all the thermodynamic and hydrodynamic perturbations that accompany it (rise in temperature, boundary layer, stagnation point, etc.).

On the other hand, above some 120 km, and particularly in the case of orbiting satellites, the flows of charged particles and neutral molecules are totally uncoupled, and collisionless models are appropriate. Study of the motion of the neutrals nonetheless gives an idea of the problem: since the velocity of the vehicle is usually well above the thermal velocity of the molecules, these are swept up, and form themselves in front of the vehicle into a zone of compression, and behind it into a partial vacuum that is then progressively filled in because of thermal agitation. The two regions are analyzed differently. While the wake is essentially determined by the cross-sectional area of the vehicle and the temperature of the gas, the population of the front region depends on the reflecting properties of the vehicle's surface (these problems of gas-surface interaction are very poorly understood).

The behavior of the ions is, to first approximation, identical to that of the neutrals, since their thermal velocities are comparable, but account must be taken of the important differences that contribute to simplify things at the bow, but complicate them at the stern of the vehicle:

– while the neutrals are reflected, the ions are neutralized when they reach the surface of the vehicle; therefore there is no zone of ionic compression,

– the cross-sectional area that causes the wake to form is different from the case of the neutrals, since it is that of the sheath at the floating potential of the vehicle, and

– it is not only thermal motion that causes the wake behind the vehicle to be filled in, but also Coulomb effects due to collective interactions between ions and electrons.

Since, in the vehicle's frame of reference, the velocity distribution of the electrons is as a rule isotropic, one could believe that, to a first approximation, there would be no wake effect in the electron gas. This is not true, since the behavior of the electrons will in large measure be determined by the potential distributions. Firstly, the majority of electrons will be reflected from the plasma sheath; this is a consequence, except when the plasma density is so small that the photoelectric effect dominates, of the necessary net equilibrium of charged flux at the surface of the vehicle. Secondly, the flux of photoelectrons coming from the vehicle will tend to increase the electron concentration in its neighborhood over that in the unperturbed medium. Finally, the ion wake will be the site of complicated, and perhaps oscillatory, plasma phenomena (Liu, 1969).

This discussion, of course, applies not only to the body of the rocket or satellite, but also to any object of size comparable to or greater than the Debye length, such as solar panels, and particularly to the probes themselves, if they are at the end of booms. All the currents received by electrodes or sampling grids will thus be modulated by these shadowing effects when their orientation with respect to the velocity vector of the vehicle changes. This is why it is almost always necessary in analyzing the data to know this attitude. Generally, results obtained when the sampling occurs within a wake are eliminated, or, when the attitude of the vehicle can be controlled, the sampling zone is always kept at the bow. Otherwise, except in the case of spherical probes, the 'angle of attack' (the angle between the vector velocity and a direction characteristic of the probe, e.g. the normal to a plane probe or orifice) must have been monitored for a correct interpretation of the measurements. In the case of the ions, these problems, while complicating the analysis, do not cause great error in the results and can even sometimes be put to good use. In the case of the electrons, where the role of these effects is much hazier, it is difficult to be so positive. However, failure to correct for them in the data reduction will always be recognized when the final results contain a modulation correlated with changes in attitude.

5.2.2.2. *Outgassing*

A problem connected with that of wakes is the risk of pollution of the medium, presented by the emission of gases by the vehicle or the instruments. In decreasing order of importance, the sources of pollution are motor exhausts still hot after the end of propulsion, leaks in pressurized reservoirs, and porous materials used as insulators or shock absorbers. The pollution gases do not put charged particles into the medium, but, by augmenting collisions and favoring certain ion-molecule reactions, they can change the nature of the charge-carriers and decrease the electron density near the vehicle. A drastic example of such an occurrence is the 'hole' observed in the ionosphere after the launching of large rockets, like the Saturn Apollo flights. This is one of the reasons why, in planning *in situ* measurements the greatest care must be attached to cleanliness (in the sense of vacuum techniques).

It is not unusual that the vehicle be surrounded by a 'cloud' of water vapor coming from the outgassing payload. In the case of an orbiting satellite, if sufficient precautions have been taken, this outgassing is quickly exhausted and becomes negligible after several days. In the case of sounding rocket, it is not troublesome at low altitudes, where the effect of collisions confines it to the wake, but it must always be taken into consideration at the apex of the trajectory.

In the preceding section we dwelled on several of the many difficulties of studying the ionospheric plasma *in situ*. Despite the difficulties, these methods have become indispensable both because of their intrinsic value, and because of the geographical coverage furnished by satellite-borne instruments:

– for the description of the fine structure of the spatial distribution of ionization and electric fields,

– for the measurement of the energy distribution of the electrons,

– for the identification of the chemical nature of the ions.

We shall now discuss the characteristics of the different techniques used to achieve these objectives.

5.3. Plasma Probes

We thus designate all simple probes based on the Langmuir characteristic. Recall that there are three distinct regions in this characteristic (Figure V.1):

(1) The region of applied potentials that is positive with respect to the plasma, repelling ions and giving a negative, electronic, saturation current;

(2) The region of applied potentials that is slightly negative with respect to the plasma, in which a small change in potential corresponds to a large change in current; and

(3) The region of applied potentials that is strongly negative with respect to the plasma, repelling electrons and giving a positive, ionic, saturation current.

Actually, it is not very practical to use one single probe to measure correctly the complete characteristic, because of the disproportion between the electronic and ionic saturation currents. If one wishes to measure the latter with good precision (region 3) a probe with a large surface area (some tens of cm^2) is needed, but such a probe cannot be positively charged in order to reach region 1 without drawing a large current that considerably perturbs the reference potential of the vehicle. Inversely, with a small probe designed to collect a reasonable electron current, the ion current in region 3 is small with respect to the noise.

From another point of view, it should be remembered that in region 3, phenomena occur that lead to complications in the interpretation of the recorded signal: the flux of incident ions changes with the orientation of the probe to the vector velocity; the photoelectric current emitted by a sunlit probe changes with its orientation to the direction of the Sun; an excess electron current is present if suprathermal electrons, able to jump the applied potential barrier, are present in the medium. This is why the interesting results are all mainly in regions 1 and 2 of the characteristic, i.e. are limited to electron measurements. Even in this case, we have seen that useful information can only be obtained at those times when the probe is not in the wake.

5.3.1. MEASUREMENT OF ELECTRON AND ION CONCENTRATIONS

Let us examine region 1 of the characteristic. We have an electrode, small compared with the reference surface (that of the vehicle), charged positively with respect to it (several volts), and preferably held at the bow. Along the trajectory, it collects a current proportional to the thermal flux of electrons cutting the surface of the sheath surrounding it. It is difficult to get accurate absolute values of the ambient electron concentrations from the probe current, since we have only estimates for:

– the electron temperature,

– the potential of the vehicle (which determines the potential of the probe with respect to the plasma and, consequently, the thickness and area of the sheath),

– the amplitude of the photoelectric effect on the probe, and

– the role of the magnetic field in the collection of electrons by the sheath.

But if large fluctuations are observed in the collected current that are not correlated with attitude changes of the vehicle, we can be sure that they are not due to these various factors, but to fluctuations in the electron concentration of the medium traversed. Figure V.2 shows a succession of vertical profiles obtained this way in the *E* region; the probe in this case was the insulated end point of the nose cone of a small sounding rocket probe. The altitude resolution is limited only by the bandwidth of the telemetry system, due account being taken of the upward motion of the rocket. This allows the thin layers of ionization known as sporadic *E* layers to be clearly put into evidence, along with their downward motion in the course of the night.

In contrast with Figure V.2, we show in Figure V.3 the ion current collected by a rocket-borne planar probe operating in region 3 of the characteristic. We see clearly the preponderant effect for ion collection of the orientation of the probe as the rocket undergoes its motion 'à la Poinsot'. The modulation almost vanishes at the apex, when only the horizontal velocity of the vehicle remains to make the ion velocity distribution anisotropic.

In general then such d.c. probe measurements are not very reliable or accurate, and consequently, many investigators have turned to radio-frequency types of *in-situ* probes. There exists a large variety of these probes including resonance rectification probes, short antenna capacitance probes, short or long antenna impedance probes, plasma resonance probes, upper hybrid resonance probes, and plasma capacitance probes using a pair of spaced grid electrodes.

Most of the antenna type probes involve somewhat complex theory with consequent doubts regarding the accuracy of the calculations of electron density. There have been cases of experiments which showed a very good agreement with other measurements of electron density (such as ionosonde measurements) but other re-

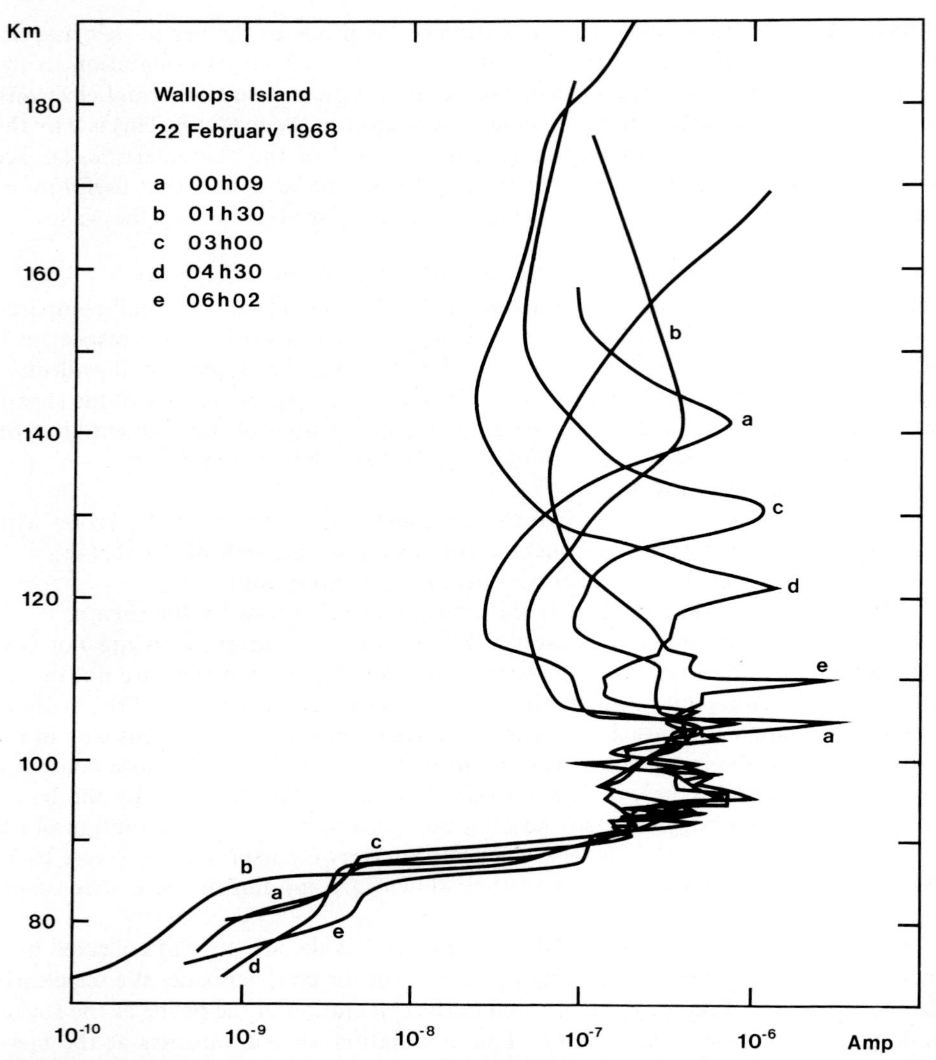

Fig. V.2. Vertical profiles of electron concentration in the *D* and *E* region of the ionosphere, measured with a positively biased Langmuir probe in a succession of nighttime rocket flights (local time of launchings indicated). The fine structure of ionospheric layering is well shown, with a permanent 'ledge' at 85–90 km, and thin 'sporadic *E*' layers moving downward (Smith, 1970).

sults of comparison have been unsatisfactory. The same situation exists in the case of resonance rectification probes.

The upper hybrid resonance probe and the plasma capacitance probe seem to be both in a more satisfactory state and electron density measurements by these probes may be accepted with some confidence. It is probable that the upper hybrid resonance probe gives a measuring accuracy of about $\pm 3\%$ under favorable conditions and this may be better than any other measurements with which it might be compared. Because of the possible proximity of other resonances it does not lend

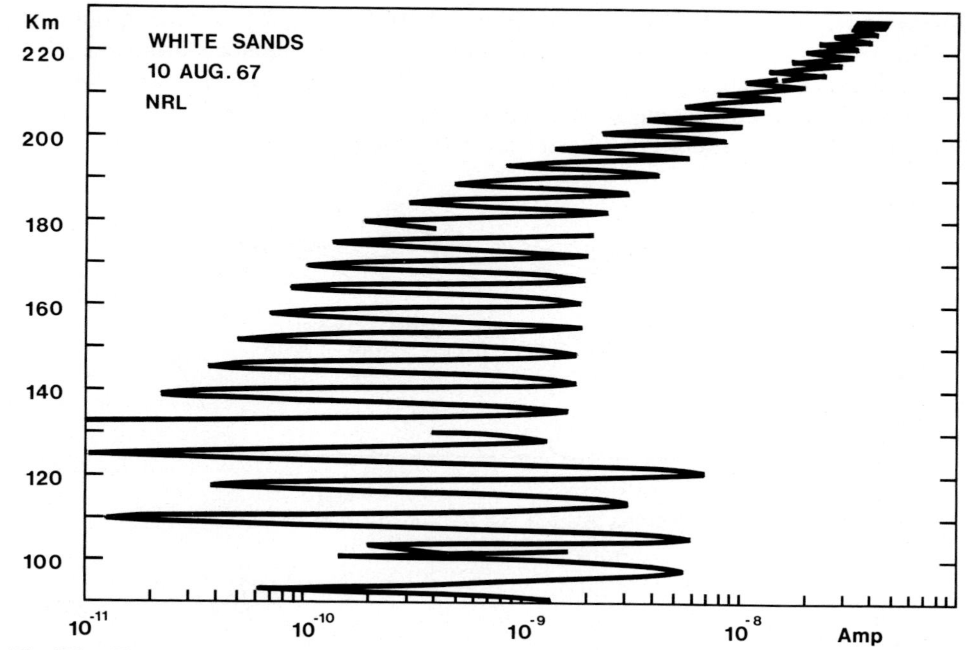

Fig. V.3. The measured ion current on a side looking negatively biased Langmuir probe onboard a spinning rocket is strongly modulated by the varying angle of attack. In principle, this could lead to a measurement of ion temperature (courtesy of C. Y. Johnson).

itself readily to on-board simple data reduction and electron collision damping effects render it inapplicable to the lower regions of the ionosphere.

The plasma capacitance probe, with an accuracy in the region of 5%, is not so precise as the upper hybrid resonance probe but it can be used over a very wide range of electron densities, from 10^2 to 10^7 cm^{-3}, and with a probing frequency of 40 MHz it is applicable in the lower ionosphere. The data reduction can be readily carried out on-board in real time with obvious advantages (Sayers, 1970).

5.3.2. MEASUREMENT OF ELECTRON TEMPERATURE

We shall now see what can be measured in region 2 of the Langmuir characteristic, that is by scanning the potential applied to the probe. We know that the shape of the inclined region of the characteristic is due to the fact that when the potential V of the probe becomes slightly negative with respect to the plasma, the electrons with energy less than eV are repelled and the collected electron current consequently decreases. For a Maxwellian distribution of electron velocities, the collected electron current varies, in the neighborhood of $V = kT_e/e$ as $\exp(eV/kT_e)$.

Even in the presence of superposed ion and photoelectric currents, since these vary only slightly with V, we see that the logarithmic derivative (the slope of the natural logarithm) of the current as a function of potential is simply e/kT_e and, in theory, gives a direct measurement of the electron temperature T_e.

This method has been employed on board American rocket probes and satellites, with probes of cylindrical geometry (wires) by Nagy *et al.* (1963), Spencer *et al.*

(1965) and Findlay and Brace (1969) (Figure V.4). Devices for extracting onboard (before transmission by telemetry) the slope of the characteristic during the scan, with the aid of an analog system, have been used by British and French probes of planar and spherical geometry (Bowen *et al.*, 1964; Wilson and Garside, 1968; Wrenn, 1969; Clark *et al.*, 1972; Berthelier and Godard, 1972).

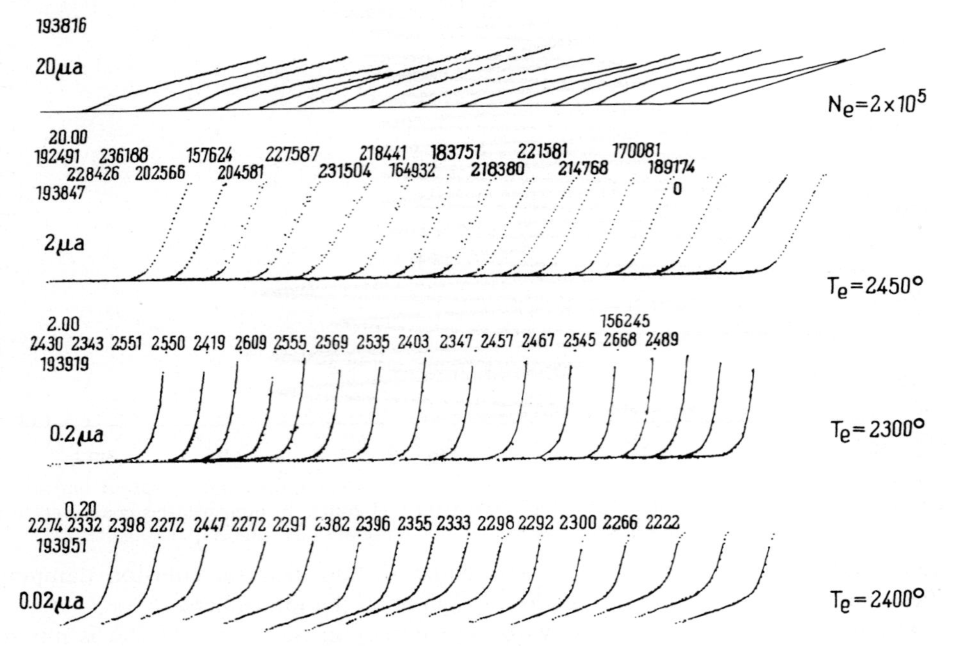

Fig. V.4. Samples of current-voltage characteristics from a satellite borne Langmuir probe, with the inferred electron temperature (Brace *et al.*, 1971).

Unfortunately, this type of measurement, very simple in principle, is handicapped by serious practical inconveniences tied in with hysteresis in the work function of the probe surface. These were revealed from a systematic disagreement with the measurement of the electron temperature obtained by incoherent scatter sounding. Briefly, in the course of the interesting part of the scan, which only extends over some tenths of volts, but during which the collected current changes by more than an order of magnitude, the change in the contact potential of the probe, due to impurities in the crystalline structure of the surface, introduces a distortion in the current-voltage characteristic which always brings about an over-estimate of the electron temperature (Smith, 1972). The solution of this problem lies in the control of the surface state of the probe, through choice of materials and drastic measures of cleanliness (vacuum sealing of a protective bulb opened at the beginning of the measurement), or, alternatively in an increase in the scanning rate to overtake the molecular adjustments at the surface of the probe (Carlson and Sayers, 1970; Hirao and Oyama, 1973). Up to now, these problems have hindered the study of the fine structure of the electron temperature in the lower ionosphere (Oyama, 1976).

5.3.3. MEASUREMENT OF PLASMA POTENTIAL

Note that the determination of the Langmuir characteristic of a probe on board a space vehicle is also a means of determining, with quite good accuracy, the potential of the vehicle with respect to the plasma. It suffices to read off the value of the potential (applied between the probe and the vehicle) that corresponds to the well-marked boundary between regions 1 and 2 of the characteristic, i.e. at the moment when the probe is at the plasma potential (Figure V.1).

With two or more probes spaced a known distance apart, and referred to the same vehicle potential, this is a method for direct measurement of ionospheric electric fields $\vec{\mathbf{E}}$, by measuring the potential difference between points in the plasma through which the spacecraft is moving. Alternatively, the potential difference between floating probes can be directly measured with a voltmeter (Figure V.5).

The electric field derived from such potential difference measurements is $\vec{\mathbf{E}}' = \vec{\mathbf{E}} + \vec{\mathbf{v}}_s \wedge \vec{\mathbf{B}}$ where $\vec{\mathbf{v}}_s$ is the spacecraft velocity. To make this measurement, several space or floating potential probes must be extended on booms from a craft whose aspect behaviour is accurately known. The length of the booms is determined by the precision with which space potential in the plasma can be measured (Cauffman and Gurnett, 1972).

Other methods for direct measurement of ionospheric electric fields, which cannot be attained by the surface charge density method (field mill) because locally the charge is determined by the Langmuir sheath, all involve the measurement of the motion of charged particles, either through the observation of ionizable barium or strontium vapors injected in the medium, or through the analysis of the drift velocity of ions with electrostatic analyzers.

5.4. Electrostatic Analyzers

We thus designate all probes (gridded probes, 'ion traps', etc. . . .) in which the functions of sampling and collecting the charged particles are separated. The first function is in general filled by a grid which extracts from the plasma a current obeying the Langmuir current-voltage characteristic, the second by a plate or 'collector'. Between the two is found a more or less complex analyzer system. The advantage of such systems is to allow the separation of various effects which are mixed in the simple Langmuir probe, and a more detailed access to the energy distribution of the sampled particles. In particular, it becomes possible to analyze the energy distribution of the ions, once the encumbrance of electron and photoelectron currents is removed.

We shall not enter into the details of the many relatively simple systems with plane or spherical geometry that have been used to measure ions and electrons from the satellites SPUTNIK III (Krassovsky, 1960), EXPLORER VIII (Bourdeau, 1961), and ARIEL I (Boyd and Raitt, 1965), or from rocket probes by Sagalyn and Smiddy (1964) and Folkestad (1972) for instance. But we shall show the possibilities of this type of apparatus with the help of only one, developed by Hanson *et al.* (1973) for the OGO VI and ATMOSPHERIC EXPLORER satellites.

The retarding potential analyzer (Figure V.6) is a cylindrical cavity at the satellite potential U_0, with an opening 2 cm in diameter on its forward face. The opening is covered by two grids G_1 and G_2 at the same potential U_0. The sampling therefore

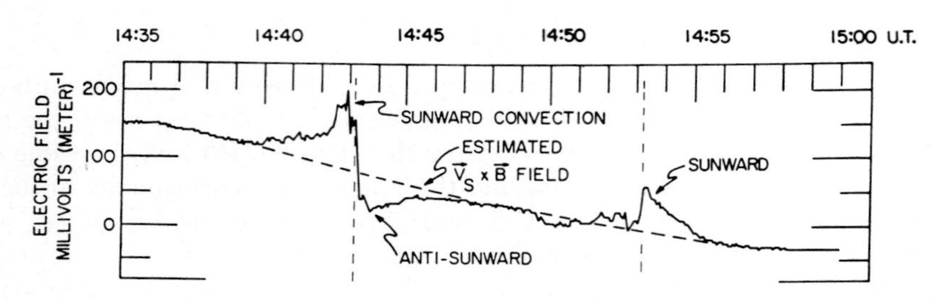

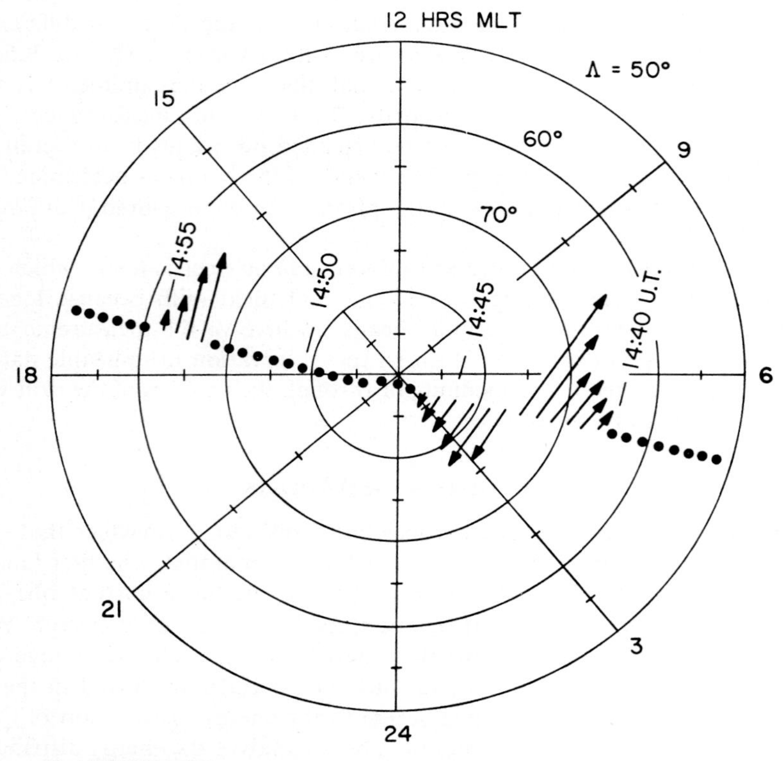

Fig. V.5. Plasma potential measurement across the polar cap on the INJUN 5 satellite. Departure of the observed electric field from the cross magnetic $\vec{\mathbf{V}}_s \times \vec{\mathbf{B}}$ field is interpreted in terms of the convection velocity $\vec{\mathbf{E}} \times \vec{\mathbf{B}}/\mathbf{B}^2$ (Frank and Gurnett, 1971).

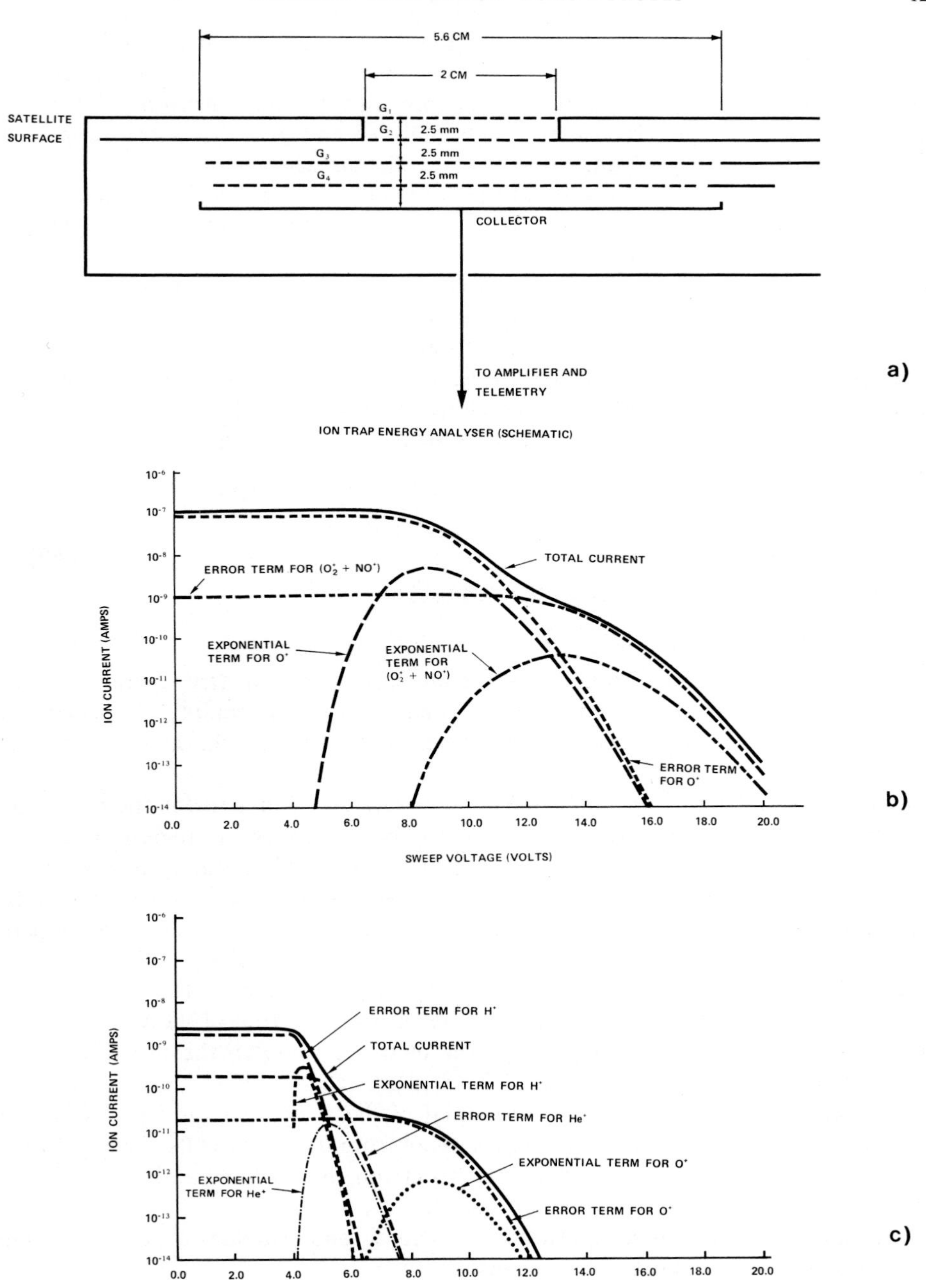

Fig. V.6. Principle of the retarding potential analyzer (cf. text). The middle and lower panels show the interpretation of the variation of collected current as a function of sweet voltage on grid G_3, for lower (O^+, $O_2{}^+$, NO^+ mix) and upper (O^+, H^+, He^+ mix) F regions, respectively.

takes place near the boundary between regions 2 and 3 of the Langmuir characteristic: the flux of sampled particles crossing the sheath of the satellite, then grids G_1 and G_2, is composed of all the ions and those suprathermal electrons with energy greater than eU_0. Grid G_3 is the energy analyzer and grid G_4 is the suppressor, which keeps the electrons from reaching the collector when the ions are being measured, and vice-versa. The collector, at the potential of the vehicle, is connected to the input of an electrometric amplifier that can measure positive and negative currents between 10^{-13} and 10^{-6} A.

For measuring ions, the suppressor is charged to -10 volts (which also has the advantage of repelling towards the collector the secondary electrons that can be emitted through the impact of incident ions and photons), and the potential V of the analyzer is swept periodically from 0 to $+20$ volts in about one second. Under these conditions, the flux of sampled ions reaching the collector is likewise swept and the expression for the collected ion current, as a function of V, is given by:

$$TAev_s \cos\theta \, \Sigma_i \, n_i \, [1/2 + 1/2 \operatorname{erf}(k_i) + (2\pi a_i)^{-1} \exp(-k_i^2)]$$

where $k_i = a_i - b$, $a_i = v_s \cos\theta/\alpha_i$, $\alpha_i = (2kT_i/m_i)^{1/2}$, $b = [e(V + U_0)/kT_i]^{1/2}$

(the function $\operatorname{erf}(x) = \int_0^x e^{-\xi^2} \, d\xi$).

A is the area of the entry grid, T the transmission coefficient (transparency) of the grid system, $v_s \cos\theta$ the component of the satellite velocity normal to the orifice. The index i covers the different species of ion present in the plasma, characterized by their mass m_i.

Figure V.6 shows the current-voltage characteristic calculated for the conditions encountered by a satellite at the top and the bottom of the F region. The terms corresponding to $\operatorname{erf}(k_i)/2$ and $\exp(-k_i^2)$ are drawn for each species of ion. Numerical techniques of 'least squares' type, adjusting these calculations to the measured signals, allow the determination of n_i and T_i, i.e. the density and temperature of each of the principal groups of ions (Hanson *et al.*, 1973).

An ambient electric field of 10 mV m^{-1} gives drift velocities of the particles comparable to their thermal velocities, and the resulting distortion of the velocity distribution is well within the resolution power of this technique, the use of a planar geometry enabling a vector analysis to be made.

For measuring the electrons, the potential of the suppressor grid G_4 is kept at $+20$ volts, and that of the analyzers G_3 is swept from 0 to -20 volts. The apparatus then has the capability of furnishing the spectrum in direction and energy of the ambient suprathermal (photo-) electrons.

Naturally, as always, data on attitude, velocity $\vec{v}_s$, and potential U_0 of the satellite have to be secured by some other means.

5.5. Mass Spectrometers

Even though energy analysis of the ions ends up discriminating them by mass, it is necessary to carry out a finer mass discrimination:

– in the upper and middle ionosphere, in order to determine the relative abundances within the categories of light (H^+, He^+), medium (N^+, O^+) and heavy (N_2^+, O_2^+, NO^+) masses, which are not further separated by simple electrostatic analyzers, and

– in the lower ionosphere, in order to identify correctly the chemical nature of the peculiar species that dominate there (metallic ions, hydrates, negative ions), and which is not known *a priori*.

5.5.1. PRINCIPLE

Fine mass discrimination between the sampling orifice and the collector can be achieved in different ways, but always consists in scanning a parameter of the system whose value determines in a unique way the e/m ratio of those ions that are allowed to reach the collector. The other ions are made to deviate, and end up neutralized at the walls. The deviation is brought about either by the application of a high frequency electric field in resonance with ion motion, longitudinally between transverse grids, as in *linear analyzers*, or else transversely between longitudinal rods, as in *quadrupole mass filters*, or by the use of a permanent magnet that curves the ion trajectories, as in *magnetic analyzers*. In any case, the scanned parameter is usually the acceleration voltage V applied to the ions after they are extracted from the plasma, which can be as high as a kilovolt, and which causes them to enter the analyzer with a scanned velocity $v_i = (2eV/m_i)^{1/2}$. The output signal after dispersion in the analyzer has the shape of a 'mass spectrum', consisting of peaks at those values of the acceleration voltage V_i that correspond to the e/m_i ratios of the ions present, with a spread of width ΔV_i. The ratios $V_i/\Delta V_i$ characterize the 'resolution' or 'resolving power' of the apparatus. It need not be very high except possibly in lower ionospheric studies where ion masses can exceed 100 amu. The amplitude of the current peaks is proportional to the flux of sampled ions and to the transmission coefficient of the apparatus, which can depend on the mass and has to be calibrated in the laboratory.

Again, the sampled flux, and hence the amplitudes of the peaks, like all the ion currents we have considered, depend not only upon the ambient ion concentration,

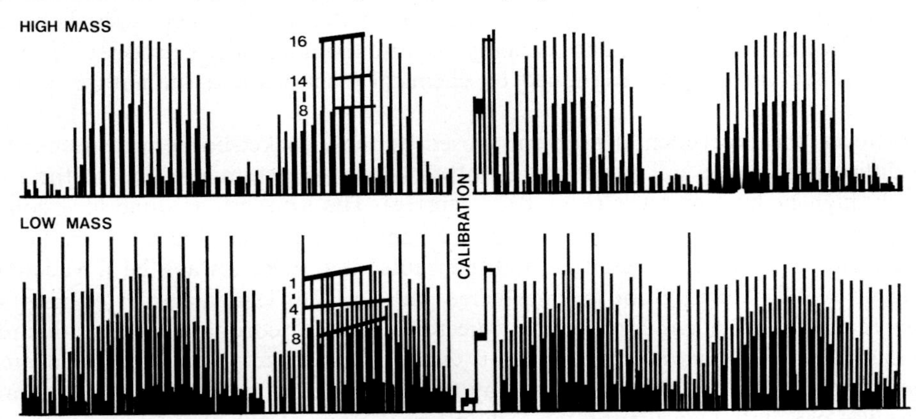

Fig. V.7. Telemetry of the outputs of the magnetic mass spectrometer on board the ISIS 2 satellite, over a succession of about one hundred spectral voltage sweeps, corresponding to a path of 700 km on the orbit. Mass peaks at 16 (O^+), 14 (N^+), 8 (O^{++}), 4 (He^+) and 1 (H^+) are modulated by the varying angle of attack as the satellite spins (Hoffmann *et al.*, 1973).

which is looked for, but also upon the effects of changes in velocity, attitude and potential of the sampling orifice following those of the vehicle.

Figure V.7 shows as a function of time the peak amplitudes in successive mass spectra measured by a mass spectrometer aboard a spinning satellite. As in Figure V.3, the dominant modulation due to the rotating angle of attack is evident, and we again note that the amplitude of this modulation is a function of the mass of the ions, due to the fact that the ratio of the velocity of the satellite to the thermal velocities of the different ions varies as ${m_i}^{1/2}$; this ratio is for example four times greater for O^+ than for H^+.

All these complications in data analysis (Whipple *et al.*, 1974) mean that usually mass spectrometers are used only to furnish the composition of the plasma, i.e. the relative proportions of the different ions, the absolute total density being obtained simultaneously by other methods when possible (probes and/or radiofrequency soundings).

5.5.2. APPLICATIONS

The application of mass spectrometry to the exploration of the ionosphere has proven to be very fruitful (cf. reviews by Istomin and Pokhunkov, 1963; Holmes *et al.*, 1965; Narcisi, 1973). The distribution of the various ions in the upper E and F layers and their variations had been determined aboard rockets and satellites by a number of experimenters, starting as early as 1956–1958, and progress in the mastery of the method has been slow but spectacular.

Figure V.8 shows the latitudinal distribution of ions in the upper ionosphere, measured by satellite. The works of Taylor (1972) and Chappell (1972) have been most important in the study of the phenomena in the plasmasphere and the polar depression of the light ions.

The fundamental contribution of mass spectrometry to the study of the ionosphere has been the measurement by sounding rocket probes of the composition below 120 km, in the lower E and especially the D layer (Narcisi, 1966; Goldberg and Aikin 1971; Krankowsky *et al.*, 1972), firstly because almost all of the other methods have been inadequate for this aim, and secondly because the ionospheric processes in this altitude range are largely determined by chemical reactions and the nature of the ions.

However, the experimental problems posed by these rocket-borne measurements are difficult and have not been fully resolved. They are essentially due to the relatively high neutral pressure (1 to 10^{-5} mmHg). The sampling is done in a zone dominated by collisions and perturbed by the 'ram effect' of the vehicle. Naturally, in the interior of the apparatus, it is indispensable that there always be a vacuum sufficient ($\leqslant 10^{-5}$ mmHg) for the mean free path to be at least an order of magnitude greater than the length of the ion trajectories, and for corona discharges due to the high applied voltages to be avoided *a fortiori*. This makes necessary a vigorous evacuation of the neutral flux entering by the sampling orifice (10^7 to 10^{13} neutral molecules per ion!). Considering the technological limits to the pumping velocity of flyable equipment (cryosorption or titanium sublimation pumps) of about 1000 s^{-1}, this implies as well a very small sampling orifice, of diameter less than a millimeter. The flux of sampled ions consequently becomes very small, and the collector behind

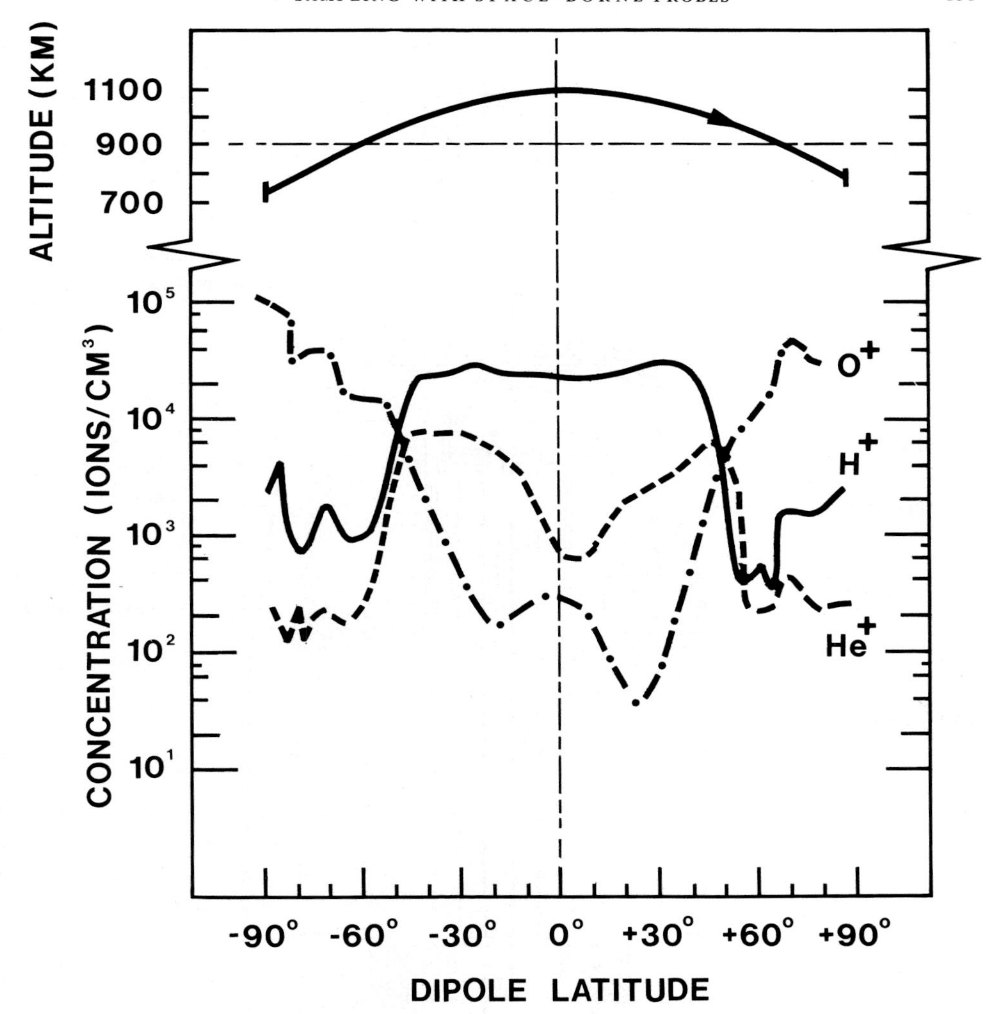

Fig. V.8. The latitudinal structure of the ionic composition measured with a radiofrequency mass spectrometer on board the polar orbiting OGO VI satellite.
Note the altitude variation (upper panel) partly responsible for the observed variation of O^+ concentration, and the marked depletion of light ions in the higher latitudes (Taylor, 1972).

the analyzer must by necessity be an electron multiplier than can furnish, for each incident ion, an avalanche of 10^4 to 10^6 electrons. These avalanches may be integrated by an electrometric amplifier or else counted individually in the form of pulses by a logical circuit, with the advantage of not introducing calibration problems connected with the 'gain' of the multiplier, and of easily permitting the electrical insulation of high voltage collector from the telemetry, which has to remain at the reference potential of the vehicle.

Independent of these technical problems, complicated by miniaturization, which make the mass spectrometers used for these studies both sophisticated and expensive

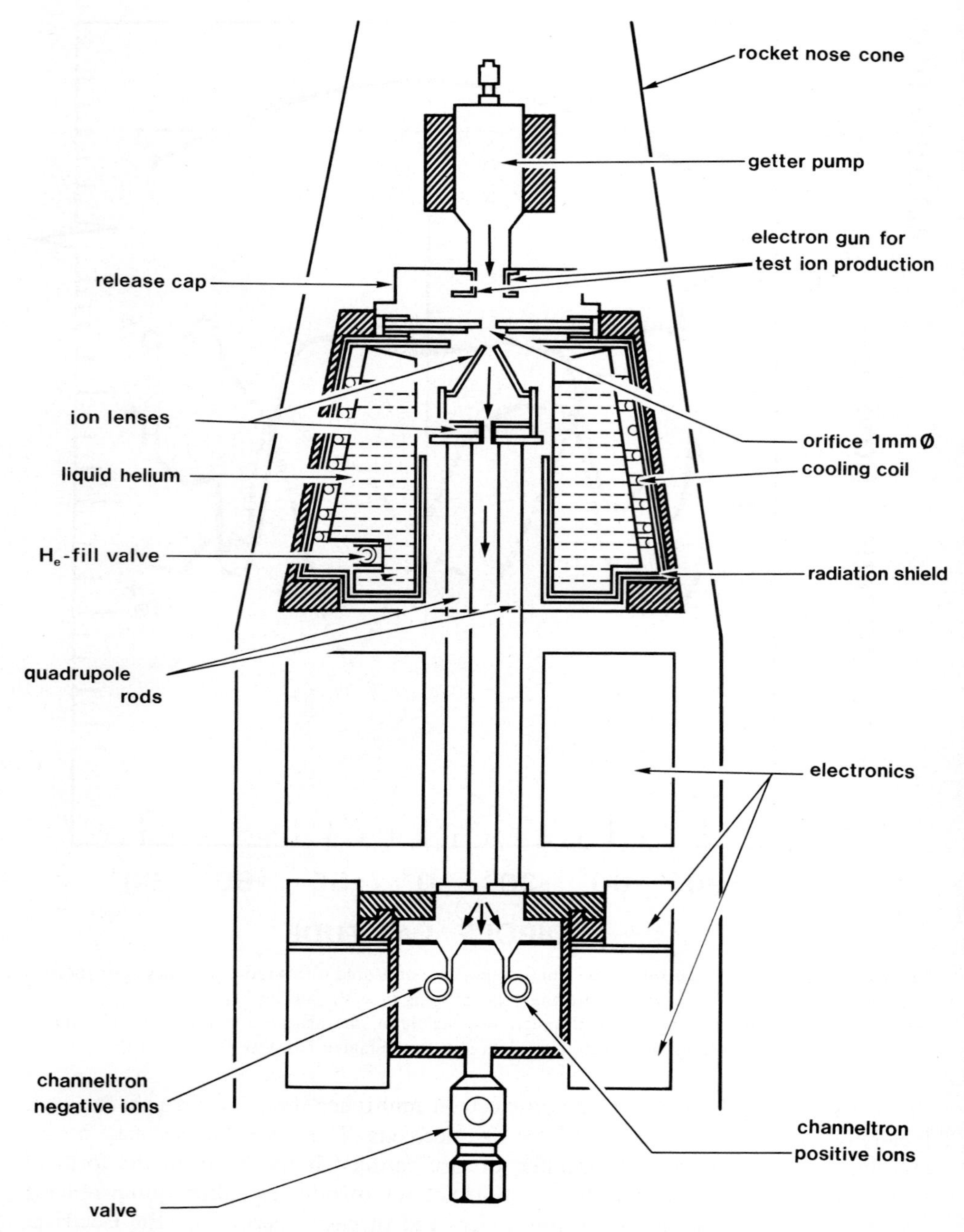

Fig. V.9. Ion mass-spectrometer with cryogenic pump developed at the Max Planck Institut für Kernphysik in Heidelberg for rocket investigations of lower ionospheric composition.
The cap with the test ion source and getter pump is released in altitude for exposure of the orifice. After passage through the quadrupole analyser, positive or negative ions, according to applied voltages, are detected by channeltron multipliers with pulse counting techniques.

instruments (Figure V.9), the difficulties in measurement and analysis come essentially from the effect of collisions suffered by the ions in the immediate vicinity of the sampling orifice. Upstream of the orifice (except possibly near the apex of the trajectory, where the ambient mean free path becomes long enough and the rocket slows down to subsonic vertical velocity), they must pass through a zone of compression of the neutrals, where the temperature strongly increases (shock layer). Downstream, they undergo a quick cooling, owing to the isentropic expansion of the gas entering the evacuated enclosure; at the same time they begin to feel the acceleration field, which increases the energy of the collisions that they can still undergo with the neutrals. Although these processes have little effect on atomic (metallic), or on molecular (NO^+ or O_2^+) ions whose binding energy is sufficiently large, we now know

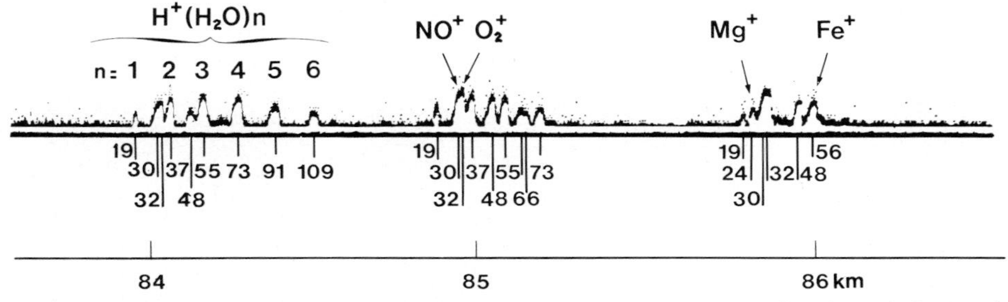

Fig. V.10. Telemetry from the mass spectrometer of Fig. V.9 on a rocket ascending through the *D* region (altitude in km is shown on the lower scale). Four successive positive ion mass spectra are shown, with the mass of each peak indicated, and the main categories of ions identified.

from laboratory work (BURKE, 1972a), that they break up the weakly bound clusters that dominate the ion population in the lower *D* region. For example, Figure V.10 shows the experimental mass spectrum of hydrated protons $H^+ (H_2O)_n$ between 80 and 85 km. It is very probably shifted by fragmentation towards small *n*, in comparison to the real mass distribution, which remains conjectural at the moment.

PART THREE

THE INTERPRETATION OF IONOSPHERIC PHENOMENA

In the laboratory experiment shown here, the conditions of both ion-neutral chemistry and rocket mass-spectrometry in the lower ionosphere are being simulated. A primary plasma is formed in the cavity at the left at a pressure of about 10 mm Hg and reacts with neutrals while cooling to mesopause temperatures as it expands out through a supersonic nozzle into a 0.3 mm Hg vacuum. Product ions are sampled in the stationary supersonic flow through a 250 μm orifice at the tip of the conical cap of the mass-spectrometer enclosure on the right, evacuated to 10^{-6} mm Hg. The photograph was made with a plasma from an electrical discharge exciting nitrogen to optically radiating states, in order to visualize the flow and the (attached) shock layer. In the normal operating mode, the primary ions are obtained under thermal field-free conditions with a radioactive source. The evolution of plasma composition as a function of reaction time is analyzed by varying the distance between nozzle and sampling port. The vector gas yielding the primary ions is N_2, O_2 or Ar, to which are added ppm controlled reactant gases NO, CO_2, H_2O etc.... This experiment demonstrated that, at low temperatures, ions become quite 'sticky' and grow very large by catching trace polar molecules, but that supersonic sampling with a blunt (shock detaching) probe destroys these large clusters. (Courtesy of R. Burke, Centre de Recherches en Physique de l'Environnement, Orléans)

CHAPTER VI

CHEMISTRY OF CHARGE CONSERVATION

In this chapter, we begin the detailed study of the processes responsible for the structure and variations of the ionosphere by considering the interactions of a chemical character between the components of the upper neutral atmosphere and the incident ionizing agents.

In the first section, the basic charge reactional schemes determining the fate of an ion pair (electron + positive ion) in the atmosphere are reviewed, and in the second section, it is shown how these schemes, together with the stratification of the neutral gas, account for the layered structure of the ionosphere.

The following sections describe the specific time-varying phenomena connected with the different ionizing agents.

6.1. Ion Chemical Processes

The sources and sinks of charged particles in the atmosphere involve the following three basic mechanisms:

– the direct production of ion pairs from neutral species;

– inelastic collisions (reactions) between ions or electrons and neutral molecules, which change the nature of the charge carriers without changing the number of charges;

– inelastic collisions (recombination) between ions and electrons through which the charges are neutralized, the resulting neutral species conserving, in the form of chemical or kinetic energy, a part of the initial ionizing energy.

Note that these processes represent only one particular chain of reactions induced by photonic or corpuscular radiation in the upper atmosphere, that in which free charges appear. We shall not systematically follow the side steps, which concern the fate of by-product neutral species and photons, remembering however that the latter are an important clue to these processes through optical aeronomy.

6.1.1. ION PAIRS FORMATION

Chemi-ionization, which is the essential source of ions in flames, could possibly occur in the chemosphere from photodissociated CH_4 and O_2 (Burke, 1972b):

$$CH + O \rightarrow CHO^+ + e^-$$

a scheme not confirmed even though an ion of mass 29 like CHO^+ has indeed been found in the *D* region (Zbinden *et al.*, 1975).

The primary ionizing agents for the ionosphere are all of external origin, in order of importance:

– UV light and X rays propagating from the Sun in a straight line, and only negligibly scattered by the geocorona;

– Charged corpuscules (electrons, protons, alpha particles, etc., solar or magnetospheric, whose trajectories are controlled by the magnetic environment of the Earth; and

– Meteoric dust particles in orbit about the Sun, swept up by the oribiting Earth.

These three ionizing agents are absorbed by the atmospheric gas in various ranges of altitude, with overall peak production of ionization in the lower thermosphere. They have in common the property that their flux changes in time and is spatially anisotropic. The 'map' of the creation rate of ions in the atmosphere is, to a first approximation, fixed with respect to the direction of the Sun; the 24 h proper rotation and annual revolution of the Earth are therefore felt as diurnal and seasonal variations in the ionization rate above a fixed point with given latitude and longitude.

Superposed upon these variations, we find world-wide variations connected to the varying interplanetary (solar and meteoric) conditions encountered by the Earth, with irregular, 27 days, annual and 11 yrs. components. Note that these variations in ionization rates are due not only to the variability of the ionizing agents, but to the variability of the parent neutral atmosphere as well.

The primary ions produced in the Earth's atmosphere essentially correspond to air composition, with accordingly a proportion varying with altitude of N_2^+, O_2^+, and O^+ dominating the lot, plus other trace species like NO^+, He^+, N^+, as well as metallic species stripped from extraterrestrial dust particles in the meteor trails. All in all, the situation is not very different from ion production in air at ground level when for instance a spark occurs between two electrodes. But just as in a spark, primary ion production has little to do with the actual ion composition in the plasma formed, because of the very fast plasma chemistry that takes place.

6.1.2. ION-MOLECULE REACTIONS

The ions created by photo or impact ionization undergo many binary collisions with the molecules of the neutral atmosphere at the level of and below the maximum of primary production in the thermosphere. The importance of the chemical reactions that result from these collisions is obvious when the composition of the primary ionization products, that is the neutral air composition, is compared with that of the bottomside ionosphere in its quasi-steady state (Figure I.10). Even though for example an ion most often produced is N_2^+, only a small proportion of it remains, while one of the most abundant ions, NO^+, must be produced by secondary reactions, since it is not created directly in significant amounts.

Triple (three-body) collisions can considerably improve the chances of a chemical reaction taking place, the third particle playing the role of a catalyst, removing extra momentum. They have a frequency that increases with the square of the atmospheric density, becoming important below the mesopause. They are involved in the formation of cluster ions in the *D* region. They are also responsible for the appearance of negative ions through the attachment of free electrons onto neutral molecules. This attachment process is the reason why free electrons exist only ephemerally (lightning flashes, flames), throughout the lower mesosphere and below, down to ground level, in spite of the constant production of ion pairs by cosmic rays and radioactivity from the Earth. The weak electrical conductivity of the dense air in these subionospheric lower regions is entirely due to positively and negatively charged clusters, the stable end products of chains of rapid ion-neutral reactions (Chalmers, 1957).

6.1.2.1. *Reaction Coefficients and Their Determination*

The driving force of ion-molecule reactions is the electrical attraction between the ions and the induced or permanent dipole of the neutral molecules. This explains why they are much faster than ordinary neutral-neutral chemical reactions.

The quantitative importance of ion-neutral reactions in a given ionospheric situation depends naturally on:

– the concentration and temperature of the reactants (this is a problem of aeronomy); and

– the probability or *coefficient* of the reactions (a matter for chemists).

Even though historically some of the principal reaction schemes were successfully guessed by theoreticians, the paths and coefficients of ionospheric reactions have later been made into the object of detailed experimental laboratory studies. The study of these ion-molecule reactions, of fundamental importance for the disciplines of radiation and combustion chemistry, has furthermore opened new paths in the theory of elementary chemical reactions. A decisive step in its application to space chemistry (planetary ionospheres, and lately interstellar chemistry) dates from the perfecting of experimental devices that allow ions to interact with neutrals under controlled conditions of near thermal equilibrium ('after-glow'). A buffer gas, He for example, is used as a medium for thermalizing the ions and electrons. For a reaction of the type: $X^{\pm} + R \rightarrow Y^{\pm} + S$, the decrease in the reactant ion $X^{\pm}$ is monitored with the help of a mass spectrometer, and the reaction coefficient K is deduced from the continuity equation that defines it:

$$\partial Y^{\pm}/\partial t = Kn(X^{\pm})n(R).$$

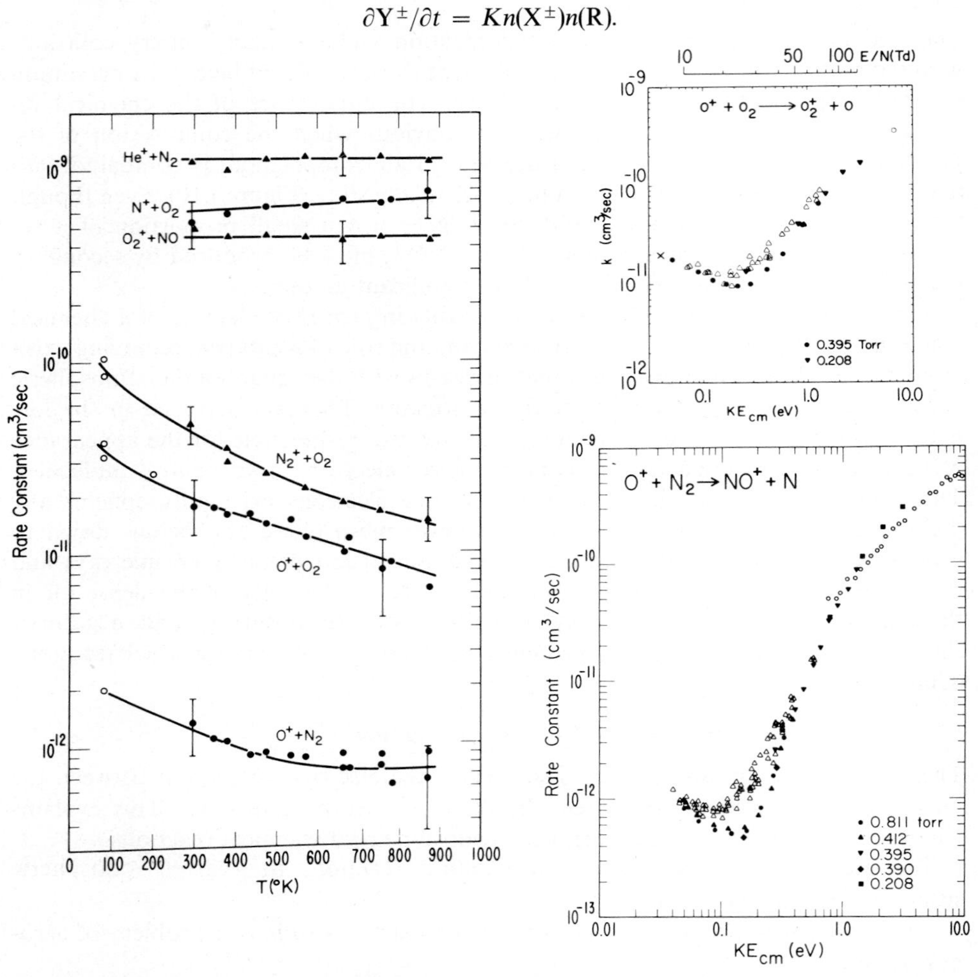

Fig. VI.1. The variation with temperature (left panel) and relative kinetic energy (right panels) of the coefficients for several important ionospheric ion-molecule reactions (Ferguson, 1974). It is seen that the decrease of the charge transfer reaction of O^+ on the dominant atmospheric molecules O_2 and N_2 with increasing thermal energies is followed by a sharp rise at suprathermal energies; this could have a profound effect on the ion composition of the ionosphere when electric fields set ions in motion with respect to neutrals (cf. Figure VI.13).

The most fruitful experimental method, which ressembles the classical techniques of chemical kinetics, consists of injecting the reactant R into the flow of primary $X^{\pm}$ ions, carried along with the buffer gas. Knowledge of the flow velocity and rate of injection of the reactant allows the calculation of K. The change of K with temperature (relative kinetic energy of the reagents) is obtained by heating or cooling the buffer gas or by submitting the ions to a small electric field. This is the 'flowing afterglow' (or 'flow drift') method developed in Boulder by Ferguson and his collaborators (Ferguson *et al.*, 1969), some results of which are to be found in Figure VI.1.

6.1.2.2. *Positive Ion Reactions*

In general, ionized molecules will easily exchange their charge with neutral molecules whose ionization potential is smaller (Table VI.1), i.e. exothermally. For example, this is true of the important reactions:

$He^+ + N_2 \rightarrow N_2^+ + He$

$O^+ + H \leftrightarrows H^+ + O$ (the main source and sink of protons in the upper ionosphere)

(a) $O^+ + O_2 \rightarrow O_2^+ + O$ (main sink for O^+ ions)

(b) $N_2^+ + O_2 \rightarrow O_2^+ + N_2$

(c) $N_2^+ + O \rightarrow O^+ + N_2$.

TABLE VI.1

Electrochemical series for positive ions.
The lower the ionization potential of a molecule, the greater its tendency to lose an electron, and the smaller the possibility of the ion produced to regain an electron (After Denny and Bowhill, 1973).

Ion Produced	Ionization Potential (eV)	Wavelength (Ångstrom)
He^+	24.6	504
N_2^+	15.6	790
H_2^+	15.4	803
N^+	14.5	852
CO^+	14.0	882
CO_2^+	13.8	895
O^+	13.6	910
H^+	13.6	910
OH^+	13.3	930
N_2O^+	12.9	960
O_3^+	12.8	970
H_2O^+	12.6	980
O_2^+	12.0	1030
O_2^+ from $O_2(^1\Delta_g)$	11.1	1118
NO_2^+	9.78	1260
NO^+	9.27	1330
Si^+	8.15	1520
Fe^+	7.83	1595
Mg^+	7.43	1670
Ca^+	6.11	2060
Na^+	5.14	2410

But a reaction can take place through exothermal rearrangement processes similar to ordinary chemical reactions, with the charge ending on a molecule again in general lower down on the scale of ionization potentials. This is the case with other essential ionospheric reactions like:

(d) $O^+ + N_2 \rightarrow NO^+ + N$ (the other main sink of O^+ ions, and a major source of NO^+)

(e) $N_2^+ + O \rightarrow NO^+ + N$ (another major source of NO^+ ions)

$N^+ + O_2 \rightarrow NO^+ + O$

$Mg^+ + O_3 \rightarrow MgO^+ + O_2$ (the main sink for metallic atomic ions).

The more or less previewed set of reactions (a) to (e) was found sufficient to explain the ion composition in the *F* and *E* regions as soon as it was measured; in particular, it accounted for the deficiency in $N_2{}^+$ ions and the dominance of NO^+ ions (Norton *et al.*, 1963).

The discovery of hydrated clusters $H^+(H_2O)_n$ in the *D* region by the team of Narcisi in 1965 was, on the other hand, a complete surprise, until then it had been assumed that mesospheric positive ions were essentially NO^+. For a while, it was debatable as to whether the hydrated clusters were of genuine ionospheric origin. In retrospect, it seems that these ions should have been expected, because they were known to appear in flames and other weakly ionized plasmas in the presence of even minute quantities of water vapor. In fact, it takes very stringent dehydrating techniques for such ions not to appear in air, so that it is not surprising that they dominate the ion composition of the atmosphere up to the mesopause. Basically, this is due to the fact that three body collisions can lead to the formation of 'clustering' bonds with polar molecules, less energetic than the bonds of the constituting molecules, for example:

(f) $O_2^+ + O_2 + M_2 \rightarrow O_4^+ + M_2$

(g) $\left\{ \begin{array}{l} NO^+ + N_2 + M_2 \rightarrow NO^+N_2 + M_2 \\ NO^+ + CO_2 + M_2 \rightarrow NO^+CO_2 + M_2 \\ NO^+ + H_2O + M_2 \rightarrow NO^+H_2O + M_2. \end{array} \right.$

In the above clustering reactions, the 'chaperon' M_2 can be any molecule, and will then be statistically the most abundant neutral molecule, N_2.

Clusters can react with neutrals by the 'cluster switching' mechanism whereby the clustered molecule is replaced by another molecule with larger 'proton affinity', for example:

$$O_4^+ + H_2O \rightarrow O_2^+H_2O + O_2$$
$$NO^+N_2 + CO_2 \rightarrow NO^+CO_2 + N_2.$$

A particularly important case of clustering is 'hydration': that is the tying of the strongly proton affine water molecule, for instance

$$H^+(H_2O)_n + H_2O + M_2 \rightarrow H^+(H_2O)_{n+1} + M_2$$

a reaction which can proceed to very large *n*'s, explaining how ions can in fact become 'condensation nuclei' (Wilson's cloud chamber experiment). Indeed, this may have something to do with the existence of noctilucent clouds in the mesosphere (Witt, 1969).

While an understanding of the cluster formation mechanism now exists, it does not mean that the chain of reactions which lead to the actual equilibrium composition in the *D* region is completely understood. The reason is that whereas a satisfactory scheme exists starting with O_2^+ ions, the precursor ions in the *D* region are believed to be predominantly NO^+ ions. The direct scheme of NO^+ hydration (chain b in Figure VI.2) whose coefficient for the successive steps has been measured in the laboratory at 300°K, is much too slow to result in the observed proportions of the different ions considering the small amount of H_2O in the mesosphere (of the order of 1 ppm). The Boulder group has proposed side stepping chains via the more abundant CO_2 and N_2 reactants, with cluster-switching back to hydration (Figure VI.2, chains c and d), somewhat improving things.

Reactions involving H_2O however have not been studied below 0°C, while mesospheric temperatures can be as low as 120°K. And it is known that three-body reactions have coefficients increasing with diminishing temperature, sometimes appreciably ($T^{-3.2}$ for instance for reaction f; Payzant *et al.*, 1973).

Note at any rate that O 'nips in the bud' the O_2^+ clustering mechanism by breaking the chain at the start with

(h) $$O_4^+ + O \rightarrow O_2^+ + O_3.$$

Figure VI.2 gives the overall schematic of the principal reactions between positive ions and neutral molecules in the atmosphere as we presently understand them. The species are arranged from left to right in order of increasing ionization potentials so that the reactions transfer charge from right to left. Atomic species are on top, molecular species in the middle, and clusters at the bottom. This coincides with the vertical stratification of the dominant final charge-carriers in the ionosphere, as shown in Chapter I and explained in a following section, and the ordinate of the diagram can be thought of in some qualitative way as representing altitude.

Excluded from the above scheme and from the discussion below are doubly charged ions (O^{++} and N^{++} have indeed been observed), and excited state ions, on which little work has been done.

6.1.2.3. *Negative Ion Reactions*

Free electrons, like ions, can react with neutral molecules: when a neutral molecule is electronegative the energy balance is in favor of electron attachment. The importance of triple collisions in allowing such captures comes from the fact that the energy liberated by the attachment of the electron can be carried away by the third 'chaperon' particle, so that the ion can be stabilized.

In reactions involving negative ions the parameter called 'electron affinity' (the energy liberated by electron capture) plays a role comparable to that of the ionization potential in positive ion reactions. This is also the energy that must be given to the negative ion in order to detach the electron.

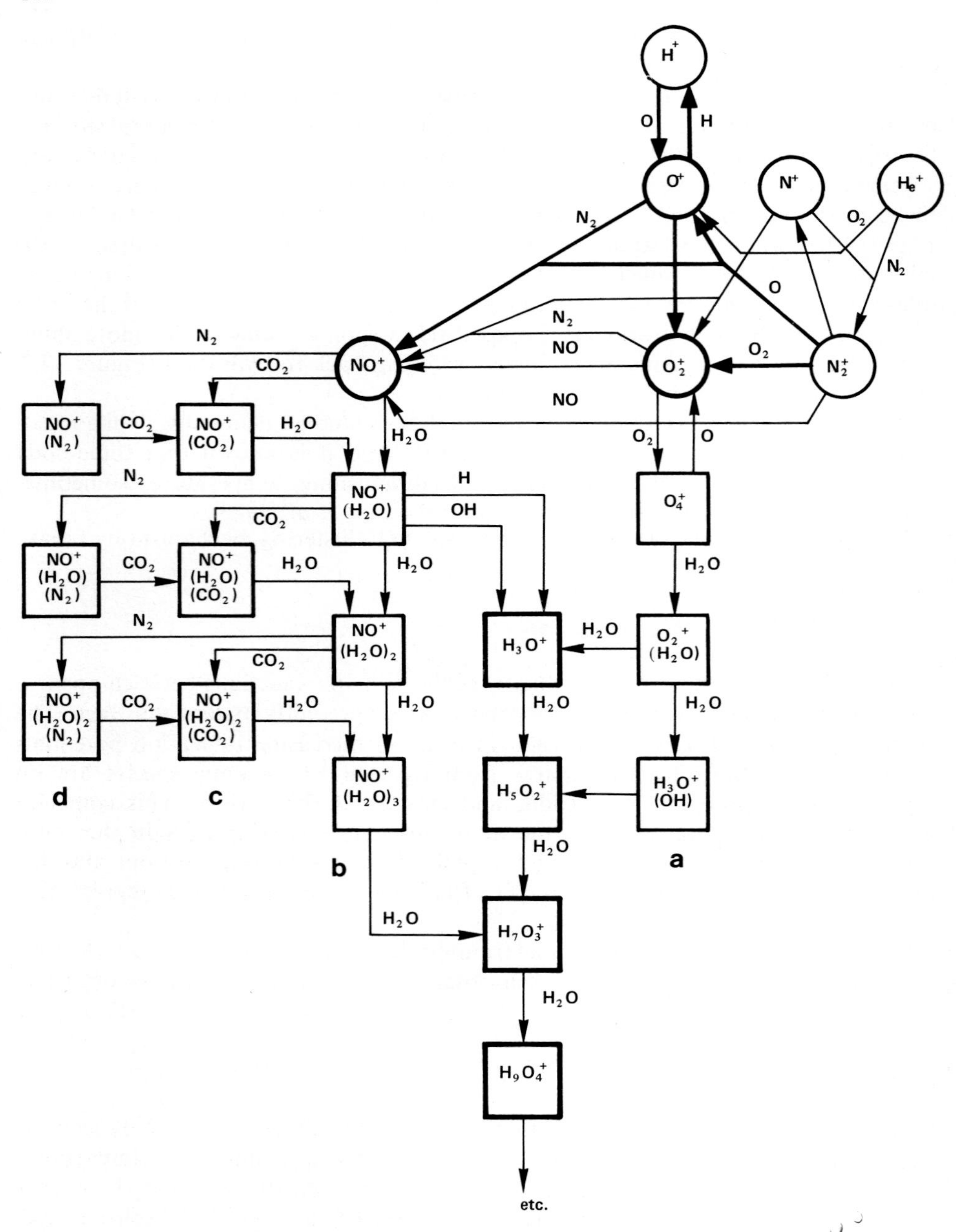

Fig. VI.2 A general scheme of positive ion reactions in the atmosphere. Reaction $A^+ + B \rightarrow C^+ + D$ is noted $A^+ \xrightarrow{B} C^+$. Unclustered species are in circles, clustered species are in squares. The different clustering chains are labelled with lower case letters (cf. text). This scheme governs most ionic processes in air, but the dominant channels and resulting terminal ion species are strongly dependent on altitude, as they vary with concentration ratios, temperature, and density of the neutral constituents.

TABLE VI.2

Electrochemical series for negative ions.
The higher the electron affinity of a molecule, the greater its tendency to gain an electron, and the smaller the possibility of the ion produced to lose the extra electron (After Denny and Bowhill, 1973).

Ion produced	Electron Affinity (eV)	Wavelength (Ångstrom)
NO^-	0.09	140,000
O_2^-	0.48	26,000
CO_4^-	1.22	10,100
O^-	1.47	8,600
OH^-	1.83	6,750
O_3^-	1.96	6,300
CO_3^-	Unknown	
NO_2^-	$1.8 < EA < 3.8$	$<6,900$
NO_3^-	$> EA(NO_2) + 0.9$	$<4,600$

Table VI.2 gives the electron affinities of the principal electronegative constituents in the mesosphere. Except for O_2 they are all trace constituents. Therefore, because of its abundance, O_2 is the substance to which primary attachment preferentially occurs, by the reaction (Chanin *et al.*, 1959)

(j) $e + O_2 + M_2 \rightarrow O_2^- + M_2.$

But just as for positive ions, the primary ions are precursors in a chain of ion-molecule reactions involving:

- charge transfer, for example $O_2^- + O_3 \rightarrow O_3^- + O_2$;
- rearrangement $O_3^- + CO_2 \rightarrow CO_3^- + O_2$;
- clustering $O_2^- + O_2 + M_2 \rightarrow O_4^- + M_2$;

etc. (Figure VI.3).

Indeed it is probable, as indicated by certain recent results (but not shown in Figure VI.3), that the terminal ions are clusters of polar molecules (H_2O, HNO_2, HNO_3) growing through reactions like:

$$NO_3^-(H_2O)_n + H_2O + M_2 \rightarrow NO_3^-(H_2O)_{n+1} + M_2.$$

However supplementary processes enter into the kinetics of negative ions, as there are various ways for them to gain the relatively small energy necessary to detach the extra electron. Although collisions by themselves might not play an important role, they can lead to so-called 'associative detachment' mechanisms whereby the liberation energy is gained from the bond energy of the neutral molecule formed, for instance:

(k) $O_2^- + O \rightarrow O_3 + e^-$

a reaction which again 'nips in the bud' the whole negative ion forming chain. Photodetachment, for example:

$$O_2^- + h\nu \rightarrow O_2 + e^-$$

should be an efficient daytime process as there is a rather large flux of solar photons of wavelength greater than 1200Å, with energy in excess of the electron affinity of attaching species, throughout the lower ionosphere. Finally, and again probably in

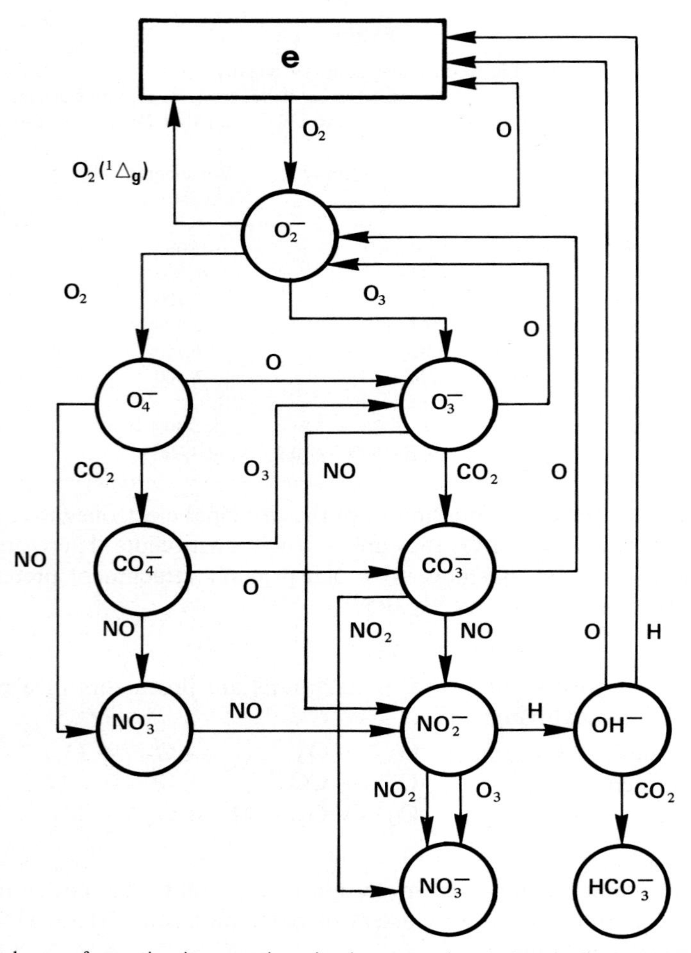

Fig. VI.3. A scheme of negative ion reactions in the atmosphere when hydration is neglected. This scheme starts with the attachment of electrons on the most abundant electron affine molecule, O_2, and involves essentially the cascading of the extra-electron through the electro-chemical series of Table VI.2, against the generally adverse tendency of reactions involving O atoms.

contrast to the case of positive ions, 'photo-dissociation' of negative ions, and of clusters in particular, has an importance which is just beginning to be recognized; for instance $CO_3^- + h\nu \rightarrow CO_2 + O^-$ or $CO_3^- H_2O + h\nu \rightarrow CO_3^- + H_2O$ as it occurs well below the photodetachment energy threshold (Peterson, 1976).

6.1.3. ION-ELECTRON RECOMBINATION

We now come to the second basic ionic mechanism, recombination, in which free charges disappear by the mutual neutralization of a positive and negative charge. Throughout almost all the ionosphere this mechanism operates through the capture of electrons by positive ions. It is only in the lower *D* region (and in the stratosphere and troposphere) that neutralization between positive and negative ions plays the dominant role ('ion-ion' recombination $X^+ + Y^- \rightarrow X + Y$, coefficient α_i about 10^{-7} cm^{-3} s^{-1}, Moseley *et al.*, 1975): because of the much larger concentration of neutrals there, free electrons attach themselves much faster than they can recombine.

6.1.3.1. *Recombinatory Processes*

If the mutual attraction between a positive ion and an electron is to result in recombination, that is in a tied system of lower energy than the dissociated system, the liberated energy must be removed in some way, while the momentum is conserved. In clustering or attachment reactions, which are similar (two particles joining together to form one), we have seen that the participation of a third particle is well-nigh necessary. In the case of recombination, wherein higher energies are liberated, there are other possibilities that allow the reaction to take place during the much more numerous binary collisions.

The transition of an electron from a free state to a bound state by emission of a photon, the inverse of photoionization, which is the only direct recombination mechanism available for atomic ions, is called *radiative recombination.* Although photoemission corresponding to such reactions is indeed observed, the importance of radiative recombination is negligible in the ionospheric charge budget, as its coefficient is only the order of 10^{-12} cm^3 s^{-1}, and throughout the ionosphere the dominant ion-electron reaction is the *dissociative* recombination of molecular ions:

$$XY^+ + e^- \longrightarrow X^* + Y^*$$

whose coefficient α_{XY^+} can exceed 10^{-7} cm^{-3} s^{-1}.

The theory of this process suggests that the recombination coefficient should vary with the energy of the incident electron as $(T_e)^{-1/2}$, but it is known that the coefficient depends as well on the distribution of vibrational states of the ions subject to recombination (Bardsley, 1970).

Note that the atoms produced through dissociative recombination are excited, and fall back into the ground state by emission of a photon. The case of O (1D) atoms coming from the recombination of O_2^+ is particularly important, since it is the source of one of the most intense optical manifestations of the ionosphere, the red line at 6300 Å in the airglow, which is a major means for studying the upper atmosphere (Thuillier and Blamont, 1973).

For the polyatomic clusters, there doubtless exist many ways in which the energy liberated by electron capture can be repartitioned among the degrees of freedom of the clustering bonds to transform the ion-electron system into a stable neutral system. A simple argument due to Reid (1970) approximates an upper limit to the coefficient of recombination, using as a model the simple Coulomb capture of an electron by a 'solid point' such as a large cluster or a small aerosol. For the electrostatic attraction to overcome the thermal energy at a distance r, it is necessary that:

$$e^2/r \geqq kT_e \quad \text{or} \quad r \leqq e^2/kT_e.$$

The capture cross section is πr^2, and if v_e is the (electron) thermal velocity, the recombination coefficient is given by:

$$\alpha \leqq v_e \pi r^2 \qquad (\text{note } \alpha \propto T_e^{-3/2}).$$

Numerically, we obtain $\alpha \leqq 2 \times 10^{-3}$ cm^{-3} s^{-1} at a temperature of 200°K.

This suggests that the higher the order of clustering the easier the recombination, a suggestion that is being confirmed in the laboratory (Leu *et al.*, 1973).

6.1.3.2. *Laboratory Measurements*

Experimental measurement of recombination coefficients cannot be carried out in the same way as in the case of ion-neutral reactions (electrons and ions cannot separately be injected into an enclosure). Therefore, one simply monitors the decay in the electron concentration of a thermal plasma after-glow:

$$\partial n(e)/\partial t = -\alpha_{XY^+} n(e)\, n\,(XY^+).$$

The problem is to insure that the ion population of the plasma is purely XY^+, without parasitic ions with higher recombination coefficients eating up the electrons, and that other processes do not take place. If that is so, $n(e) = n(XY^+)$, and the solution $1/n\,(e) = 1/n_0\,(e) + \alpha_{XY^+}\, t$ gives α_{XY^+} directly as the time derivative of the reciprocal of the electron concentration (Biondi, 1969). One can heat the electrons

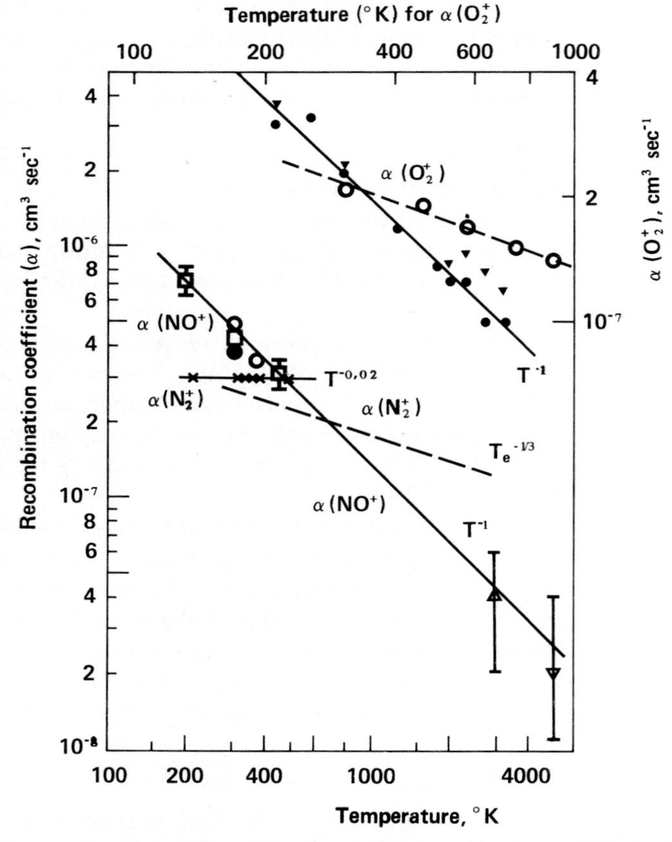

Fig. VI.4. The temperature dependence of the dissociative recombination coefficients of the main ionospheric molecular ions O_2^+, NO^+ and N_2^+. Solid lines are for conditions where $T_e = T_i = T_n = T$, while dashed lines are for conditions where $T_i = T_n = 300$°K, and T_e is varied (Biondi, 1969).

selectively by the application of a high frequency wave, in order to study the change of the recombination coefficient with T_e, or one can heat the buffer gas ($T_i = T_e$) in order to study the effect of the temperature of the ions (Figure VI.4).

In the ionosphere, $T \neq T_i \neq T_e$, and the vibrational temperature need not be in equilibrium with the kinetic temperature. At the moment it does not seem that this question of how recombination coefficients vary with the plasma temperatures has been cleared up. Some doubts therefore subsist about the validity of the application of laboratory results to ionospheric conditions, except perhaps in the *E* region where $T = T_i = T_e \simeq 300$ to 600°K.

6.2. The Chemical Structure of the Ionosphere

The ionosphere exhibits essentially, as briefly described in Chapter I, a vertically stratified structure characterized by abrupt composition changes from layer to layer. Figure V.10 for example illustrates the rapid transition from dominant hydrated clusters to dominant molecular ions near 85 km altitude, while Figure IV.11 shows the transition from dominant molecular ions to dominant O^+ near 150 km.

In this section, we account for this overall permanent vertical chemical structure, tied in with the relatively stably stratified neutral reactants background, while in the two following sections of this chapter, we study the variations in layers density and composition due to the highly fluctuating strength of the ionizing agents.

6.2.1. CHEMICAL EQUILIBRIUM RELATIONS

The quantitative treatment of the budget of charged species involves the coupled equations of continuity of the concentration n_K (cm^{-3}) for each species of ions and of electrons (Equation (I.9)), which we rewrite in the form:

$$\partial n_k(z)/\partial t = \sum P_k(z) - n_k(z) \sum L_k(z) - \vec{\nabla} \cdot \vec{\varphi}_k(z) \qquad \text{(VI.1)}$$

$\sum P_k$ gathers all the sources of production (cm^{-3} s^{-1}), either from direct ionization or from chemical reactions. $\sum L_k$ gathers all the loss frequencies (s^{-1}), either from ion-neutral, ion-ion or ion-electron reactions (these are terms consisting of the product of a reaction coefficient by the concentration of the reactant). $\vec{\varphi}_k$ is the flux due to gravity, pressure gradients, and drifts induced by winds and electric fields.

If production and transport are neglected, the situation is described by the differential equation:

$$\partial n_k/\partial t = - n_k \sum L_k$$

which has the solution

$$n_k = n_{ko} \exp\left(- \sum L_k t\right).$$

Such a situation probably never occurs in the ionosphere, but this expression is of interest because it gives the chemical relaxation time (decay to $1/e$ of initial concentration) for a species *k*, which can be considered very generally as a measure of the average *lifetime* for particles of the species: it is the inverse of the loss frequency. For instance, the lifetime of molecular ions against dissociative recombination on electrons with coefficient α_e of the order of 10^{-7} cm^3 s^{-1}, is $1/\alpha_e n(e)$ which can vary

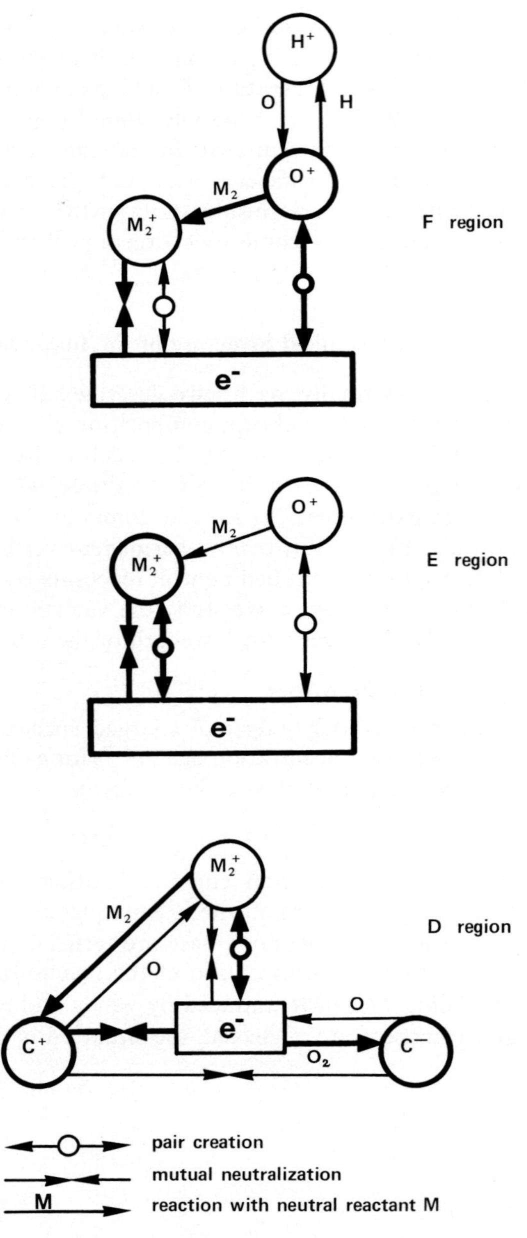

Fig. VI.5. Overall qualitative scheme of ionic equilibria for the main charged species: electrons e^-, positive and negative ion clusters C^+ and C^-, molecular ions M_2^+ (essentially NO^+ and O_2^+), and atomic ions (essentially O^+ and H^+). The dominant channels, indicated with thick arrows, for pair creation, reactions with neutrals, and mutual neutralization, vary with altitude.

from a few tens of seconds when electron concentration is more than 10^5 cm^{-3}, to a few hours when it is below 10^3 cm^{-3}. The lifetime of O^+ against charge transfer to N_2 with coefficient K_d of the order of 10^{-12} cm^3 s^{-1} is $1/(K_d\ n(N_2))$, which increases exponentially with the N_2 scale height, from a value of about one second at 100 km where n (N_2) is about 10^{12} cm^{-3}, to the order of days in the upper ionosphere.

Whenever the lifetime of charged particles is short enough, transport can be neglected, and this is in general the case in the bottom side ionosphere, the 'source' region, where ion-pair production is proceeding at a high rate, as discussed by Walker and McElroy (1966). Except when fast changes in creation rates take place (precipitations; flares or eclipse of the Sun, including rising and setting), the $\partial n_k/\partial t$ term in the continuity Equation (VI.1) also remains negligible. Thus the bottom side ionosphere mostly evolves in a state of quasi-equilibrium, the concentration of each species being determined by the fact that the loss rate, which is proportional to it, balances the production rate. Such conditions are expressed by the relation:

$$n_k(z) = \sum P_k(z)/\sum L_k(z) \tag{VI.2}$$

in words: at chemical equilibrium, concentration is the ratio of the production rate to the loss frequency (or, equivalently, the product of the production rate by the lifetime).

Modeling the concentration of the ionospheric species in chemical equilibrium on the basis of a reaction scheme like Figure VI.2 thus requires computing the height profile of the primary ions creation rates, and the height profile of all the chemical production rates and loss frequencies, with the help of a model atmosphere for the concentration of the neutral ionizable and reactant gases.

Without going into such a detailed procedure, of which examples will be given in Section 3.2 below, for the particular case of photochemical equilibrium in the D and E region, it is possible to account qualitatively for the observed vertical chemical structure of the ionosphere on the basis of the simplified scheme of Figure VI.5, and gross equilibrium relations like Equation (VI.2) for the main categories of charged species, using the data on ion chemical processes which have been reviewed in the preceding section.

6.2.2. CEILING HEIGHTS AND EFFECTIVE RECOMBINATION COEFFICIENTS OF THE MAJOR ION LAYERS

6.2.2.1. *The Ceiling of the Negative Ion Layer*

Negative ions (noted $-$) are formed by three-body attachment of electrons to the most abundant electronegative atmospheric constituent, O_2 (reaction j). Their rate of production is therefore $K_j n(e) n(O_2) n(M_2)$.

They are lost through associative detachment on (reaction k) with a frequency $\bar{K}_k n(O)$, ion-ion recombination on positive ions (noted $+$) with a frequency $\bar{\alpha}_i n(+)$, and photodetachment with a frequency $\bar{\varphi}$ which is constant as a function of altitude because the upper atmosphere is optically thin to the photo-detaching radiation.

(The bar on the coefficients is to indicate that they are a weighted mean of the coefficients appropriate to the various terminal negative ion species.)

According to relation (VI.2), the *ratio*

$$\lambda = n(-)/n(e)$$

of bound to free electrons is then

$$\lambda = K_j n(O_2) n(M_2)/[\bar{K}_k n(O) + \bar{\alpha}_i n(+) + \varphi]. \quad \text{(VI.3)}$$

The numerator is decreasing with altitude like the square of atmospheric density, while the denominator is increasing with altitude. Therefore λ is strongly decreasing with height. The altitude where $\lambda = 1$, that is, above which there are more free than bound electrons, can be computed from Equation (VI.3) and varies like the relative magnitude of the different terms in the denominator.

If values for the primary negative ions O_2^- are appropriate, with $K_j = 1.6 \times 10^{-30}$ cm^6 s^{-1}, $K_k = 3.3 \times 10^{-10}$ cm^3 s^{-1}, $\varphi = 0.3$ s^{-1}, $\lambda = 1$ obtains at the level where

$$n(O_2) n(N_2) \simeq 2 \times 10^{20} n(O).$$

This is about 75 km during nighttime as O recombines in the mesosphere, and about 65 km under the effect of sunlight as n(O) increases due to photodissociation of O_2 and O_3 (Turco and Sechrist, 1970).

The positive ion budget in the region where negative ions are present is

$$n(+) = Q/\,[\bar{\alpha}_e n(e) + \bar{\alpha}_i n(-)]$$

where $Q = Q(e) = Q(+)$ is the rate of creation of ion pairs, and $\bar{\alpha}_e$ the mean dissociative recombination coefficient. Considering that the electrical neutrality of the plasma implies

$$n(+) = n(-) + n(e) = (1 + \lambda) n(e),$$

this budget can be rewritten:

$$Q = \alpha_{\text{eff}} n^2(e) \quad \text{(VI.4)}$$

where $\alpha_{\text{eff}} = (1 + \lambda)(\bar{\alpha}_e + \lambda \bar{\alpha}_i)$.

Below the negative ion ceiling, $\lambda \gg 1$, $\alpha_{\text{eff}} \simeq \lambda^2 \alpha_i$, and no matter how large the ion creation rate Q,$n(e)$ is kept very small by the large value of this effective recombination coefficient.

6.2.2.2. *The Ceiling of the Cluster Ion Layer*

Positive ion clusters (noted C^+) are formed by three-body reactions f and g of the primary molecular ions O_2^+ and NO^+ (noted M_2^+), with O_2, N_2, CO_2, H_2O. The concentration of these neutral reactants is some fraction (mixing ratio) of the molecular nitrogen density, so that the rate of formation of clusters is $\bar{K} n^2(N_2) n(M_2^+)$, with $\bar{K}$ a weighted mean of the reaction coefficients and mixing ratios of the reactants.

They are lost by collisional detachment on O (reaction h) with a frequency $K_h n(O)$, by recombination on electrons with a frequency $\bar{\alpha}_{C^+}\, n(e)$, and by recombination on negative ions with a frequency $\bar{\alpha}_i n(-)$.

Just like the ratio of bound to free electrons, the *ratio of clustered to unclustered positive ions*

$$\mu = n(C^+)/n(M_2^+)$$

given by the expression

$$\bar{K}n^2(N_2)/[K_h n(O) + \bar{\alpha}_{C^+}n(e) + \bar{\alpha}_i n(-)] \qquad \text{(VI.5)}$$

whose numerator is decreasing with altitude and denominator non decreasing, is strongly decreasing with height.

As mentioned above, the problem of the conversion of the primary NO^+ ions into hydrated clusters remains somewhat unclear; but if values for the primary positive ion clusters O_4^+ are appropriate, the ceiling height $\mu = 1$ is reached at the level where

$$n^2(N_2) = 2 \times 10^{17} n(O)$$

that is in the vicinity of 85 km (Cf. Thomas, 1976).

The electron budget above the negative ion ceiling is

$$n(e) = Q/[\alpha_{C^+}n(C^+) + \alpha_{M_2^+}n(M_2^+)]$$

which can be written

$$Q = \alpha_{\text{eff}} n^2(e)$$

$$\text{with } \alpha_{\text{eff}} = [\alpha_{C^+}n(C^+) + \alpha_{M_2^+}n(M_2^+)]/n(e). \qquad \text{(VI.6)}$$

The top of the clustered ion layer is detected as a marked ledge in electron concentration (Figure V.2), a factor 100 jump corresponding to the effective recombination decrease from $\alpha_{\text{eff}} \simeq \alpha_{C^+} \simeq 10^{-5}$ cm^3 s^{-1} where $\mu \gg 1$, to $\alpha_{\text{eff}} \simeq \alpha_{M_2^+} \simeq 10^{-7}$ cm^3 s^{-1} where $\mu \ll 1$ (Reid, 1970). This ledge should properly be considered as the 'floor' of the E region, and is liable to vary, if the above scheme is correct, with the O ledge.

6.2.2.3. *The Ceiling of the Molecular Ion Layer*

In the thermosphere the scale height of O is about twice as large as that of the molecular neutrals O_2 and N_2; from some altitude upwards, there are many more O^+ than M_2^+ primary ions, and the molecular ions are being predominantly produced as a result of the charge transfer reactions a and d of O^+ ions to molecular neutrals, with a frequency $\beta = K_a n(O_2) + K_d n(N_2)$. These secondary molecular ions recombine dissociatively on electrons with a frequency $\alpha_{M_2^+}n(e)$.

Therefore the *ratio of molecular to atomic ions*

$$\nu = n(M_2^+)/n(O^+)$$

is simply the ratio $\beta/\alpha_{M_2^+}\, n(e)$, again decreasing with altitude more rapidly than the square of atmospheric density (that is, O density). With $K_{a,d} \simeq 10^{-12}$ cm^3 s^{-1} and $\alpha_{M_2^+} \simeq 10^{-7}$ cm^3 s^{-1}, the ceiling for molecular ions $\nu = 1$ is accordingly reached where $n(M_2) \simeq 10^5 n(e)$. In the daytime, with $n(e) \simeq 10^5$ cm^{-3}, it lies between 150 and 200 km, and in the night-time, with $n(e) \simeq 10^3$ cm^{-3}, between 200 and 250 km, depending on M_2 scale height, that is on exospheric temperature (cf. Figure I.7). Above this level, the F_1 ledge, the budget of O^+ ions is

$$n(O^+) = Q(O^+)/\beta \quad \text{(VI.7)}$$

and since $n(O^+) \simeq n(e)$, $Q(O^+) \simeq Q(e)$, the budget of electrons is just

$$Q = \beta n(e). \quad \text{(VI.8)}$$

This linear relation linking production rate and electron concentration is the reason why for a long time it was thought that the dominant electron removal process in the F region was 'attachment'.

6.2.2.4. *The Ceiling of the Chemical Equilibrium Domain*

The exponential rise of the concentration of O^+ ions and electrons with altitude (Equations VI.7 and VI.8), proportional to the ratio $n(O)/n(M_2)$, does not continue indefinitely since their lifetime $1/\beta$ also increases exponentially, so that they are eventually redistributed by diffusion.

The exact solution of the continuity Equation (VI.1) with the effect of diffusion (and winds) included will be studied in Chapter VII, but the ceiling of the chemical equilibrium region, or base level for the diffusive upper ionosphere, can be defined in a general way as the level where, during their lifetime, the ions can diffuse a neutral scale height away.

The mean square distance travelled in a time t is $\sqrt{2Dt}$, where D is the diffusion coefficient in O, so that an ion travels a scale height H in a time $H^2/2D$. Since D is proportional to $n(O)$, this characteristic diffusion time increases exponentially with altitude more slowly than the lifetime $1/\beta$ against charge transfer to molecular neutrals, and diffusion sets at the altitude where $1/\beta$ becomes larger, varying between 250 and 350 km depending on exospheric temperature, corresponding in fact to the F_2 peak (Donahue, 1968).

Above the F_2 peak, the O^+ are distributed hydrostatically with their ambipolar scale height $2H$, as discussed in Chapters I and VII. The question of knowing whether the minor F region ions (NO^+, O_2^+, H^+), which all result from charge transfer of O^+, are in chemical or diffusive equilibrium is discussed further in Section 3 below and in Chapter VII.

6.3. Photoionization

In the preceding section, it was demonstrated that the vertical structure of the ionosphere as layered by chemical processes is not too dependent on the mode, and the rate, of primary ions creation.

The general relations governing the ratio of the concentrations for the main ionospheric species (bound to free electrons, clustered ions, molecular to atomic ions) were found to result essentially from the plane stratified structure of the background neutral atmosphere, and the corresponding ceiling heights were established without reference to a particular ionization source.

Similarly, the relations between the ion pairs creation rates Q and the charge concentration n_e in the source region were derived under a set of assumptions valid for any ionizing agent, and were found to depend primarily on the dominant terminal ion species:

$$Q(z) = \begin{cases} \alpha_{\text{eff}}(z)\, n_e^2(z) & \text{for polyatomic ions; and} \\ \beta(z)\, n_e(z) & \text{for } O^+. \end{cases}$$

(When not in steady state, because Q is varying rapidly with time, we can transform those equilibrium relations to the proper continuity equations by adding the time derivative $\partial n_e(z)/\partial t$ on the right hand side.)

Since various ways exist to measure or evaluate $\alpha_{\text{eff}}(z)$ and $\beta(z)$, either from ion composition measurements coupled with laboratory results, or from observations of plasma relaxation times, it follows that the profile $Q(z,t)$ of the total rate of ion-electron pairs creation in the ionosphere and its variations, which are not amenable to direct measurement, can be derived on the basis of available information on the height and time variations of electron concentration. This approach, at least, sets the right order of magnitude for Q; for instance, in the daytime E region with $\alpha_{\text{eff}} \simeq 10^{-7}$ $\text{cm}^3\ \text{s}^{-1}$ and $n_e \simeq 10^5\ \text{cm}^{-3}$, Q is of the order of $10^3\ \text{cm}^{-3}\ \text{s}^{-1}$.

Historically however, the approach has more often been to model $Q(z,t)$ itself, both because of the considerable uncertainty which long obscured the recombination processes and coefficients, and because interest was centered on the tractable solar zenith angle dependence of the bottom side ionosphere.

6.3.1. PHOTOIONIZATION COMPUTATIONS AND MODEL PARAMETERS

In Section 2.2.1. of Chapter I, the basic formulae describing the photolysis of atmospheric gases by solar radiation were recalled. The rate of photoionization of a neutral species K, at a wavelength λ is:

$$Q_K(z,\lambda) = n_K(z)\sigma_K^i(\lambda)\, I_\infty(\lambda) \exp\{-\tau(z,\lambda)\}$$

where σ_k^i is the photoionization cross section, I_∞ the incident photon flux outside the atmosphere, and $\tau(z,\lambda)$ the optical depth of the atmosphere, which is the integral over the ray path of the product of the absorption cross section by the concentration of the absorbers, summed over all absorbing processes.

This expression can be integrated over λ to obtain the rate of creation of any species of primary ions, or summed over K to obtain the contribution of any spectral ray to the total ion-pairs creation rate (Figure VI.6).

The main computational problem to be solved for a complete time dependent treatment is that of the available photon flux at any altitude and wavelength, i.e. of the variations of the optical depth spectrum $\tau(z,\lambda)$ of the overlying atmosphere with zenith angle of the sun, especially near dawn and dusk when the curvature of the Earth has to be taken into account. Chapman (1931) gave the analytical solution for an isothermal atmosphere; Nicolet (1945), Swider (1964), Turco and Sechrist (1970) gave solutions for the more realistic case of an atmosphere with a varying scale height.

Numerical integration can then be performed, but requires a considerable amount of experimental data on the UV solar spectrum, and the neutral atmosphere (temperature, composition, spectra of absorption and photoionization cross sections for each constituent).

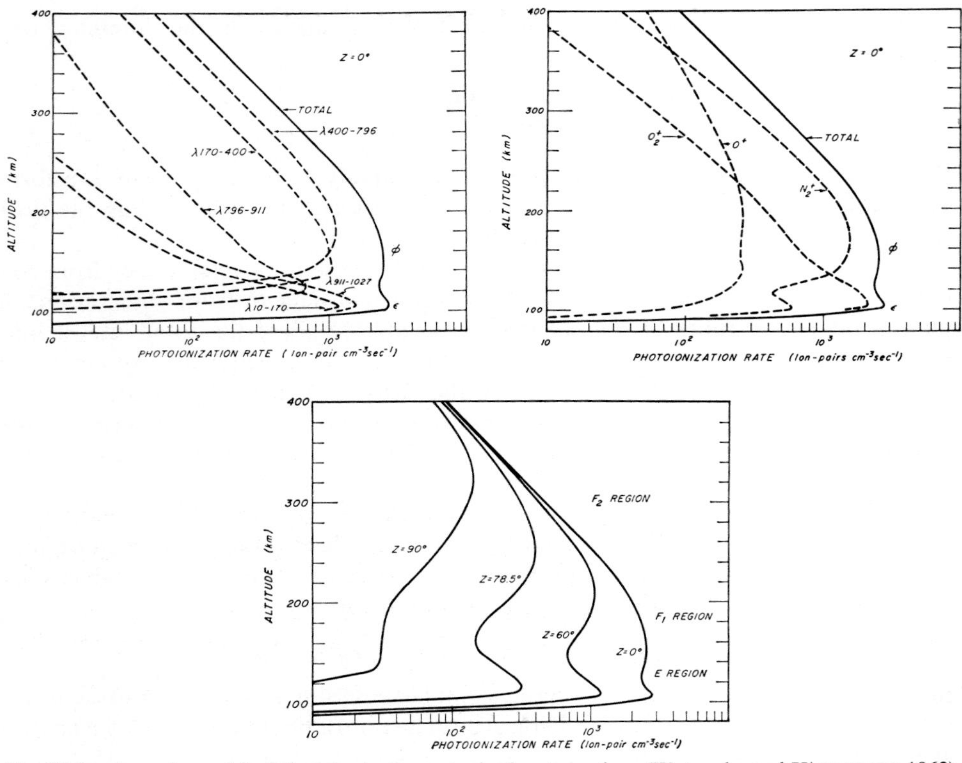

Fig. VI.6. An early model of photoionization rates in the atmosphere (Watanabe and Hinterreger, 1962). The lower panel shows the variation of the total ion-pairs production rate with zenith angle of the Sun. For the Sun at zenith, the upper left panel details the contribution of the different ultraviolet wavelength ranges, and the upper right panel gives the composition of the primary ions produced. In addition to the 10–1027Å range shown here, hard X-rays and the Lyman α-ray at 1215.7Å ionize weakly the atmosphere below the 90 km level.

6.3.1.1. *Daytime*

A major difficulty comes from the fact that the solar photoionizing radiation, as remarked in Chapter I, is distributed in numerous narrow emission lines, with intensity strongly varying over intervals as small as 1Å. Since the cross sections of atmospheric constituents also exhibit a complex structure as a function of wavelength (cf. Dalgarno *et al.*, 1964; Hudson, 1971; Stolarski and Johnson, 1972), it is evident that everything cannot be modeled in detail. Generally the spectrum is divided into rather large bands for which average values of the solar photon fluxes and cross sections data are estimated (cf. for example Oshio *et al.*, 1966).

Another difficulty comes from the fact that the lower wavelength end of the solar spectrum, as measured with space-borne equipment, is not well calibrated. The absolute magnitude of the XUV flux reported by solar workers for medium solar activity has been revised several times (Hinterreger, 1965, 1970). Furthermore the monitoring of its variations with the help of ground observable indices is not to be relied upon faithfully (Manson, 1977; Timothy, 1977; Hinterreger, 1976).

Their complicated and often erratic diurnal and latitudinal behavior is due to the combined effects of photochemistry, diffusion, winds and electric fields on the formation of the F_2 peak, and will be discussed in Chapter VII. The observed long term variations of the mid-latitude F layer on the other hand, although well correlated with solar elevation and activity, seem anomalous:

– Over most of the globe (with the exception of the South Atlantic and South Pacific regions) the noontime peak electron concentration is greater in winter than in summer, in spite of the smaller elevation of the Sun above the horizon. This effect is modulated by the 11 year solar cycle and nearly disappears at the minimum of solar activity (cf. Figure III.3);

– There is also a world-wide semi-annual variation (maximum near the equinoxes) which is masked by the seasonal variation in the Northern hemisphere, but is very evident in the Southern hemisphere; and

– Finally, over the whole globe, the electron concentration at the F_2 peak is 20% greater in December than in June, while the solar flux change due to the eccentricity of the Earth's orbit has an amplitude of only about 5% with a maximum in January.

Actually, these variations fit well with photochemical theory. We have seen that the photochemical equilibrium concentration of O^+ ions has an exponential rise with altitude proportional to the ratio of the concentration of O (from which they are photo created) to that of the molecular neutrals M_2 (to which they are lost with frequency β), up to the level where diffusion sets in, as discussed in Section 2.2 above.

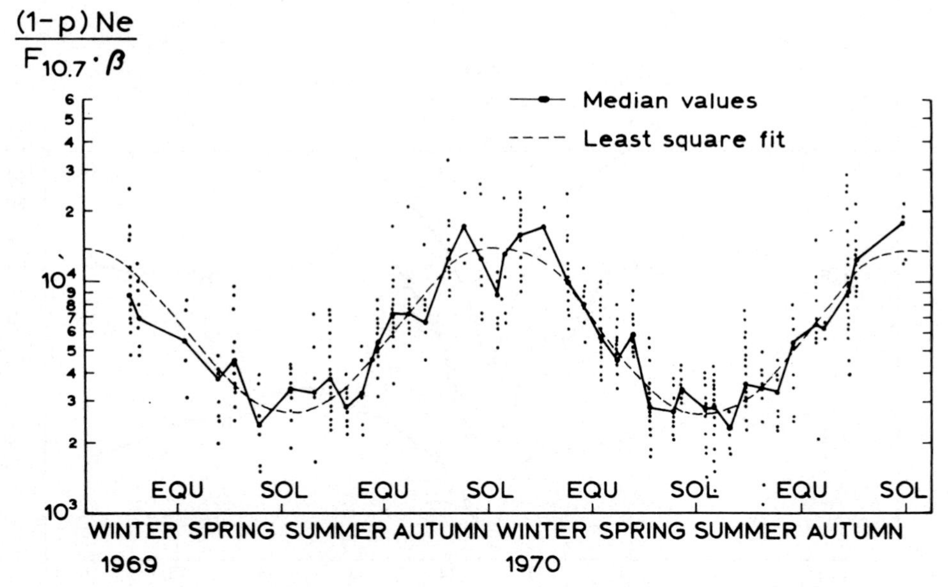

Fig. VI.9. The long term variation of the proportion of atomic to molecular ions at 200 km altitude, as observed with the Saint Santin incoherent scatter sounder, corrected for the effect of variations in the solar photoionizing flux. Since it is proportional to the concentration ratio of O and M_2 (molecular neutrals), this demonstrates that the F region 'winter anomaly', that is to say the fact that the noon peak electron density in the ionosphere is larger the lower the Sun is on the horizon, is the result of a seasonal variation of composition in the background neutral atmosphere (Alcayde *et al.*, 1974).

But the $n(O)/n(M_2)$ ratio in the thermosphere undergoes large changes (cf. Figure I.7) and these overrule the solar variations of creation rates, as Rishbeth and Setty were the first to suggest in 1961.

When for instance the concentrations of N_2 and O_2 decrease with respect to that of O (winter O bulge), the F_2 region O^+ and electron concentration increase, just as happens, as a function of altitude, below the F_2 peak. Such variations have indeed been associated with changes in the proportion of molecular and atomic ions in the F_1 region by incoherent scatter observations (Waldteufel, 1970; Evans and Cox,

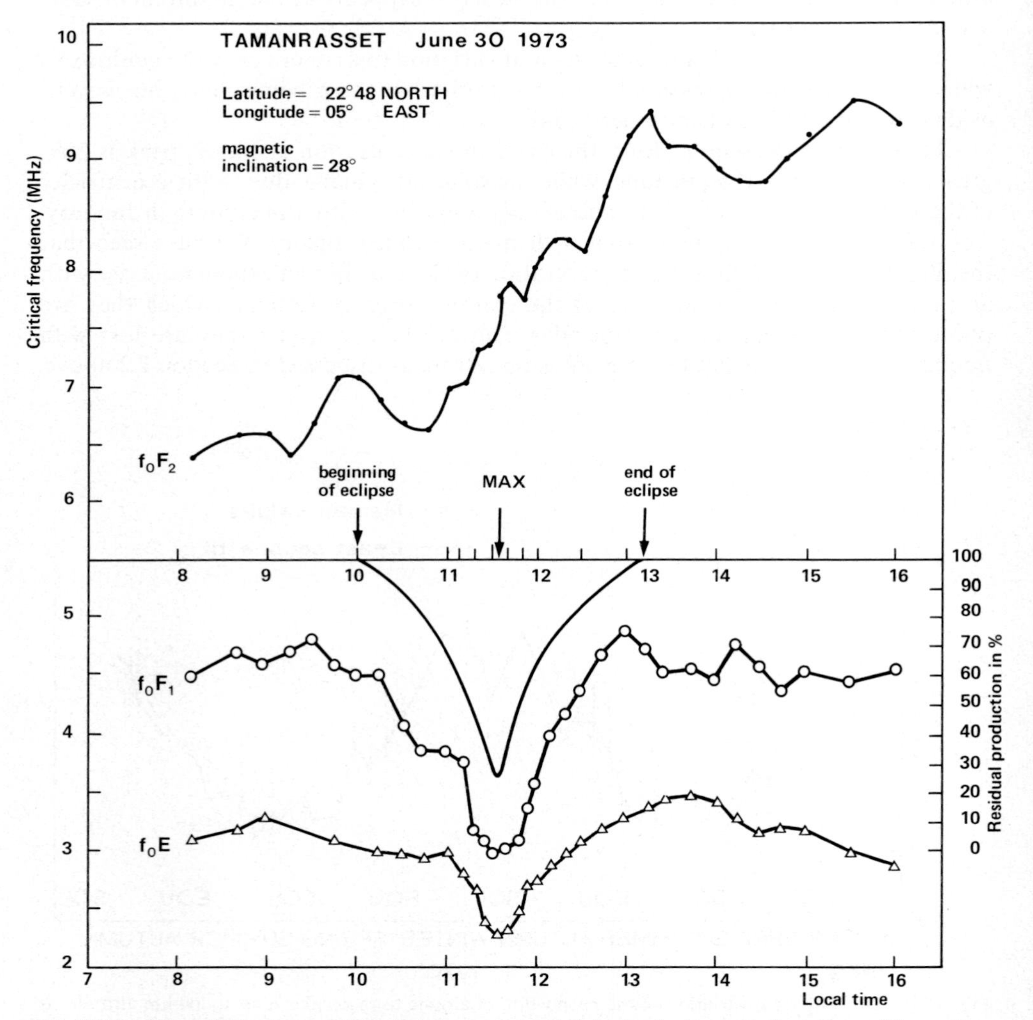

Fig. VI.10. These (hitherto unpublished) data were recorded during the longest solar eclipse of the century by physics students of the University of Constantine (Algeria) during a summer school at Tamanrasset, in the Sahara, in June 1973, using an old French ionsonde on site that had not been in operation for years; a rare example of 'amateur' ionospheric observations. Note that the data were obtained near the tropic, solstice, and local noon, that is, while the Sun was very close to zenith.

1970) and *in situ* ionic mass spectrometric measurements (Figure VI.9). These results have helped to elucidate the changes in composition of the mid-latitude neutral thermosphere, under the influence of changes in conditions at the diffusosphere base level, changes in exospheric temperature, and large scale circulation, described in Chapter I (Alcayde *et al.*, 1974).

6.3.3. PERTURBATIONS OF THE PHOTOCHEMICAL IONOSPHERE

Besides the precipitations and transport phenomena to be discussed below, the regular behavior of the photochemical ionospheric layers can be disturbed either by drastic modifications in the incident ionizing flux (solar eclipses, solar flares) or by strong disruptions in the background atmosphere (thermospheric storms, stratospheric warmings). Special cases, which will not be discussed here, are man made perturbations (nuclear explosions and injection of pollutants like large rocket exhausts). Very special cases, not yet observed, on which it would be interesting to speculate, are the occurrence of novae or supernovae in our Galaxy, liable to irradiate the atmosphere with a far more intense XUV flux than the Sun.

6.3.3.1. *Disruptions in the Solar Radiation*

The effects of *solar eclipses* on the ionosphere, which lend themselves to a straightforward interpretation in the light of recent models, have provided in the past decisive clues for the development of our present understanding of the photochemical ionosphere.

Figure VI.10 exemplifies the direct dependence of the rapidly recombining E and F_1 layers on solar radiation, first demonstrated by Appleton and Chapman (1935). The long lived F_2 peak electrons and ions on the contrary are seen to be unaffected by the photoionization cutoff, although their temperature and equilibrium are perturbed by side effects of the passage of the Moon's shadow through the thermosphere, which is felt as a supersonic cold wave (Evans, 1965; Chimonas and Hines, 1970).

Rocket measurements through the D region during an eclipse have indicated that the cluster ions ceiling is raised by a few km from full Sun to totality at the expense of a large decrease in NO^+ concentration (Narcisi *et al.*, 1972). This observation provided the first direct evidence for the conversion process from NO^+ to water-cluster ions, whose exact scheme is, as we mentioned, still not definitively established.

Solar flares are rather frequent violent eruptions which can raise the local temperature of active regions on the Sun by a factor 2 to 10 with a time scale of a few minutes to several hours. They are visible in Hα light (6563Å) as a brightening of the chromosphere; the coronal X-ray spectrum from the flaring region is quite enhanced, sometimes by several orders of magnitude, as satellite observations have shown (Culhane *et al.*, 1964).

The entire dayside ionosphere is affected, at the level where the atmosphere reaches optical depth unity for the enhanced X-ray spectrum, that is the D and lower E regions (Figure VI.11).

The corresponding enhancement in electron concentration is described under the name of 'sudden ionospheric disturbance' (S.I.D.), and is observed under various forms with most ionospheric techniques (cf. Garriott *et al.*, 1967). The onset of severe

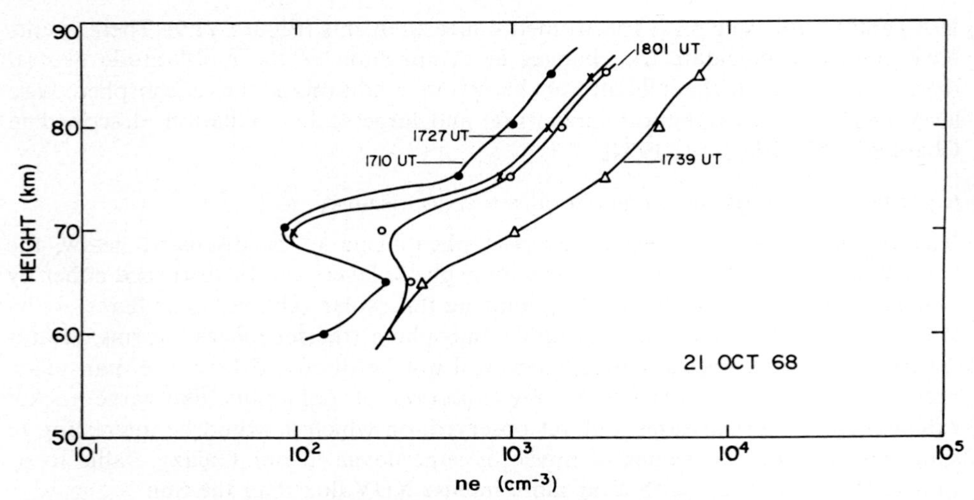

Fig. VI.11. The evolution of the profile of electron concentration in the *D* region during a solar flare, as observed with a wave interaction experiment of the Pennsylvania State University (Rowe, 1970).

D region absorption of H.F. radio communication links is so abrupt that the tendency of operators is to incriminate an equipment failure.

Rowe (1970) has attempted to model the chemistry of the flare disturbed *D* region, but the lack of ion composition measurements, and the uncertainties about the actual NO content, do not allow a precise assessment of the extent to which the quiet regime is qualitatively perturbed by the large enhancement of X-ray induced primary ionization (N_2^+, O_2^+) with respect to the normal Lyman α (NO^+) source.

6.3.3.2. *Disruptions in the Background Atmosphere*

While eclipses and flares represent clear-cut effects of well identified changes in photoionizing rates, the day to day variability of ionospheric photochemical layers is often much larger than that of solar optical indices, and has been interpreted in terms of shifts in the photochemical equilibria brought about by large scale disruptions in the temperature and composition of the neutral upper atmosphere.

Ionospheric absorption of radio waves, that is to say *D* region electron content, is relatively constant during daylight in the summer months, but exhibits large variability in winter time (Denny and Bowhill, 1973). This 'winter anomaly' bears some correlation with the variations of atmospheric temperature in the stratosphere. Large and sudden increases of the temperature at the 30 km level, lasting for periods of the order of a week, called *stratospheric warmings*, are often accompanied by concurrent enhancements in *D* region electron concentration (Bossolasco and Elena, 1963). Attempts have been made to interpret this effect in terms of modifications in the NO content at mesospheric heights (Rowe *et al.*, 1969; Strobel, 1974), but it could result more simply from the sensitivity of the water clustering reactions to temperature, leading to a reduction of the effective recombination coefficient with increased mesopause temperature.

The ionosphere on the other hand responds in a complex way to geomagnetic disturbances, and while many features of the higher latitude storm time ionosphere are due to precipitations and electrodynamical effects, the laggy response of the mid-latitude *F* layer to geomagnetic storms (Figure VI.12) has been demonstrated to follow quite faithfully the $n(O)/n(N_2)$ ratio as disturbed by the dissipation of storm energy (Von Zahn, 1975; Raitt *et al.*, 1975; Prölss *et al.*, 1975).

In addition to the effects of changes in the neutral temperature and composition, electron concentration and ion composition in the photo-chemical ionosphere could

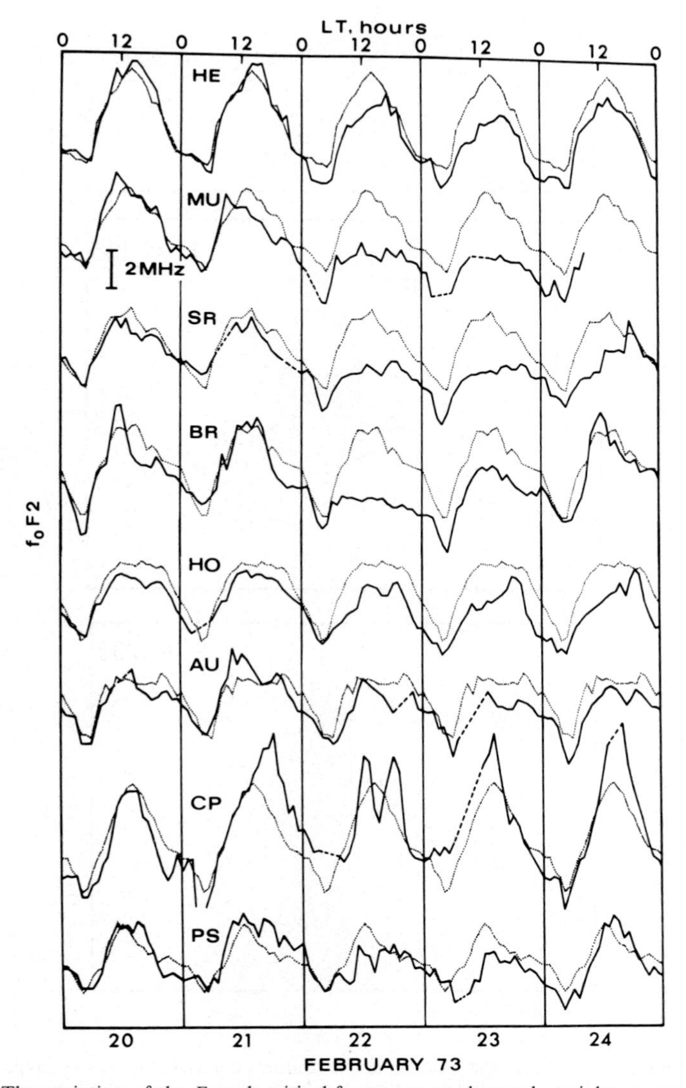

Fig. VI.12. The variation of the *F* peak critical frequency as observed at eight representative ionsonde stations during a series of moderate to moderately strong magnetic storms (solid lines). Also shown are the respective mean values of the peak critical frequency for the whole month. The ordinate scale is linear, with the bar indicating the magnitude of a 2 MHz change in critical frequency (Von Zahn, 1975).

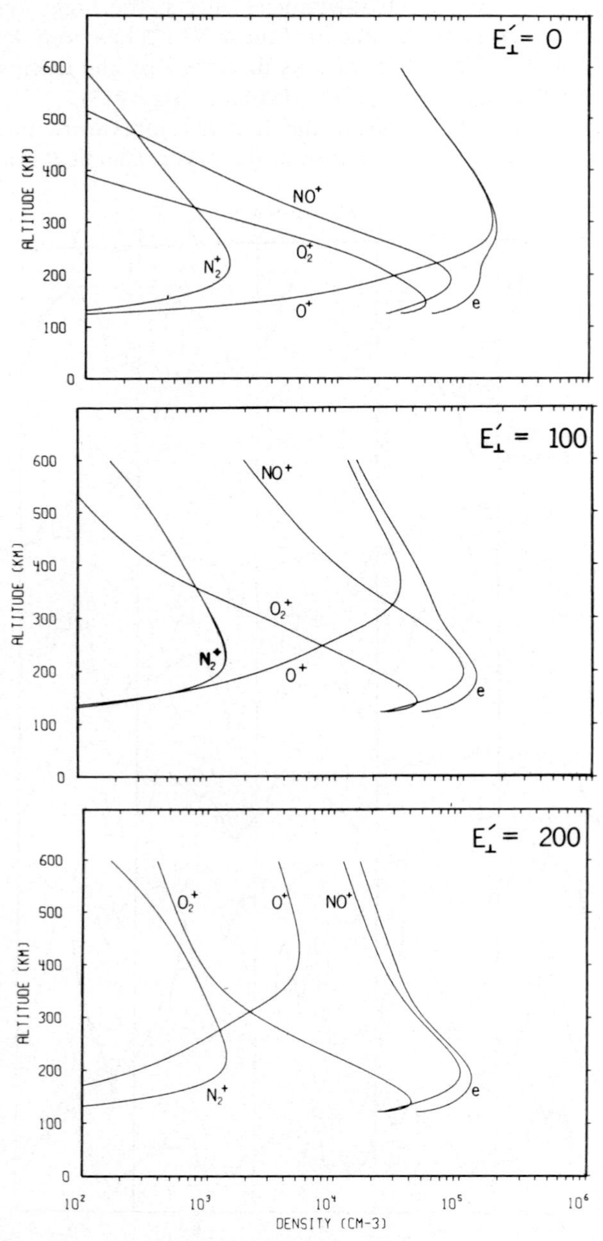

Fig. VI.13. Altitude profiles of ion and electron concentrations computed from the coupled continuity, momentum and energy equations for O^+, O_2^+ and NO^+. As the electric field intensity increases, owing to the rapid enhancement of the coefficient for the charge transfer from O^+ to N_2 with ion energy (cf. Figure VI.1), $\vec{E}_\perp \times \vec{B}$ drifts deplete O^+ in favor of NO^+. For large electric fields ($E_\perp > 200$ mV m^{-1}) NO^+ completely dominates at high altitudes, with a diffusive equilibrium scale height (Schunk *et al.*, 1975).

be drastically affected by changes in the magnitude of the frequency β of charge transfer from O^+ ions as a consequence of the rapid increase of the coefficients of the reactions $O^+ + O_2$ and $O^+ + N_2$ with ion energy. As shown by Schunk *et al.* (1975), ion drifts induced by electric fields act to enhance the reaction rates both through increase in ion temperature as a result of Joule heating (cf. Chapter VIII) and through the relative motion between the ion and neutral gas (cf. Chapter VII).

The rate of the reaction $O_2^+ + N_2 \rightarrow 2NO^+$ (Figure VI.1), in particular, varies as E^4, where E is the electric field intensity, while it varies only linearly with $n(N_2)$, so that doubling the electric field from 50 to 100 mV m^{-1} is equivalent to a factor 16 increase in N_2 density (Figure VI.13). The depletion of O^+ ions and the corresponding enhancement of NO^+ ions may even cause the latter to be the dominant F region ions in the higher latitude ionosphere which at times experiences strong electric fields. These predictions however remain to be borne out by experimental evidence.

6.4. Corpuscular Ionization

High energy charged particles (electrons, protons, alpha particles), from the galaxy, the Sun, and the magnetosphere, are constantly being precipitated in the atmosphere. Their energy is dissipated in numerous collisions with neutral molecules, which are the source of a host of ionization, dissociation and excitation processes through cascading secondary particles and photons emissions. These interactions occur the deeper into the atmosphere the higher the energy of the precipitating particles, and

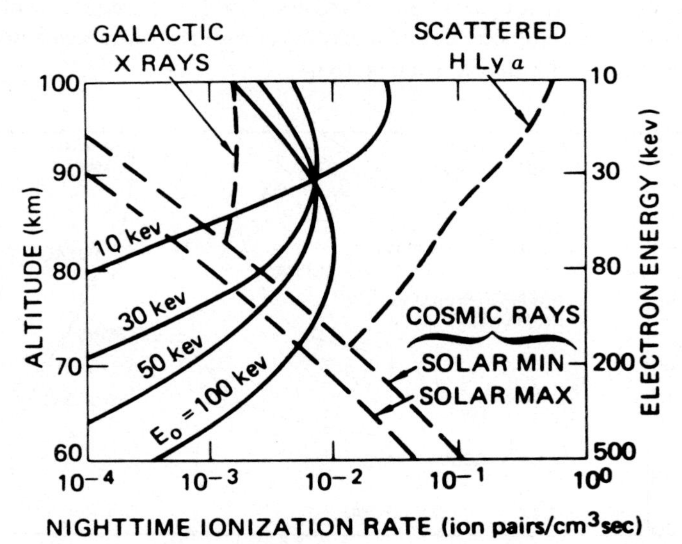

Fig. VI.14. The various ionization sources in the lower latitude night time F region. During daytime, the main production is that of NO^+ by solar Lyman α, with that of N_2^+ and O_2^+ by solar X-rays generally minor. During night time, NO^+ production by solar Lyman α scattered by the geocorona probably dominates, but the fluctuating contribution of precipitating magnetospheric electrons to N_2^+ and O_2^+ production can compete with it. Thick lines indicate the ion pair production rate due to a unit flux cm^{-2} s^{-1} of electrons with energy greater than 40 keV and integral spectral distribution of the form exp $(-E/E_0)$. Fluxes of a few tens of electrons cm^{-2} s^{-1} may be rather common (Paulikas, 1975).

are essentially observed through their consequences: the absorption of radio waves by electrons produced in the collisional region, and the optical radiation of metastable species above the region where they are deactivated by collisions.

The computation of the production rates is difficult, in particular because they require modelling of the incident particles energy and pitch angle spectra (cf. Webber, 1972; Jones and Rees, 1973; Berger *et al.*, 1974; Francis and Bradbury, 1975).

Cosmic rays are permanently, with only a small solar activity dependence (Figure VI.14), the main ionization source below the ionosphere. But within the ionospheric layers, corpuscular ionization only rivals with photoionization sporadically and locally, in the polar caps and in the auroral zones; it may play a marginal role in the lower latitudes at night.

6.4.1. POLAR CAPS EFFECT OF SOLAR COSMIC RAYS

High energy solar events, like severe chromospheric flares, are followed after a delay of 15 min to a few hours by the arrival on the Earth of large fluxes of protons (α particles and electrons to a lesser extent) with energy reaching up to hundreds of MeV. The access of these solar cosmic rays to the atmosphere, where they dissipate their energy at mesospheric levels, is funneled by the geomagnetic field to the higher latitudes. The incident flux is rather constant over the polar caps, with a nearly isotropic pitch angle distribution. The latitudinal cut-off is lower, the higher the energy and the level of magnetic activity, and is characterized by a flux drop of several decades over only a few degrees of latitude; it exhibits in addition a strong local time dependence (McDiamid and Burrows, 1969). The frequency of such events, which can result in a complete 'polar blackout' of radio communications, is typically a few times a year for moderate solar activity.

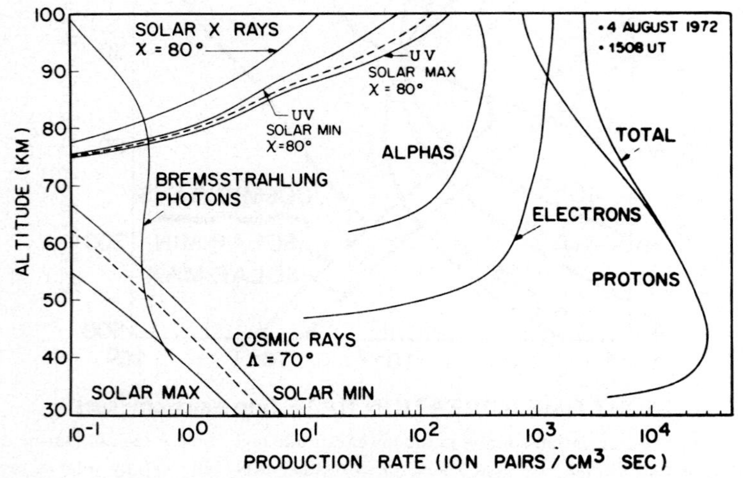

Fig. VI.15. The contribution of solar particles (electrons, protons and alpha particles), and associated bremsstrahlung photons, to the ion pair production rate in the *D* region during a P.C.A. event, as compared to the effect of normal solar XUV radiations and galactic cosmic rays. These production rates were computed from satellite data on the energy and angular spectra of precipitating particles. (Reagan and Watt, 1976.)

Mass-spectrometric measurements during such a polar cap absorption event (P.C.A.) (Narcisi, 1973), have shown that the cluster ions ceiling is lowered by as much as 10 km, as enhanced dissociative recombination favors the earlier members of the ion-molecule reactions scheme starting with the primary impact produced N_2^+ and O_2^+ ions. The large enhancement of mesospheric electron concentration is thus largely due to a reduction in the effective recombination coefficient, as if *E* region chemistry encroached upon the *D* layer. A drastic difference in negative ion composition as compared to quiet conditions is also observed, but remains largely unexplained.

A thorough investigation of the effects of the intense, long lasting, solar proton event of August 1972 has been conducted by Reagan and Watt (1976) using the Chatanika incoherent scatter radar electron concentration measurements, and satellite data on the ionizing solar corpuscular radiation (Figure VI.15); they derived and analysed the variations of the effective recombination coefficient as a function of altitude, solar zenith angle, and ion pair production rate. This work confirmed the role of the sunrise increase and the sunset decay of atomic oxygen, due to the interaction of solar UV and visible radiation with the top of the O_3 layer, on the *D* region ionic chemistry, as suggested by Adams and Megill (1967) and predicted by Turco and Sechrist (1972).

6.4.2. MAGNETOSPHERIC PRECIPITATIONS

6.4.2.1. *Auroral Effects*

In contrast to the large scale and relatively stable *D* region polar cap disturbances due to the direct entry of high energy solar particles, the precipitations in the Earth's atmosphere of magnetospheric ring current and plasma sheet particles during geomagnetic storms (cf. Chapter II) exhibit an intricate spatial and temporal structure, as well as a wide range of energies (from a few hundreds eV to several keV) and pitch angles (from monodirectional to nearly isotropic) (cf. Vallance-Jones, 1977). Their effects are observed with high resolution ionospheric techniques as a highly fluctuating condition of the *E* and upper *D* auroral regions (Figure VI.16).

Ion composition measurements with rocket borne mass spectrometers have been obtained in a number of auroral events (Donahue *et al.*, 1970; Swider and Narcisi, 1970; Krankowsky *et al.*, 1972; Narcisi *et al.*, 1974; Narcisi and Swider, 1976). Despite the expected large variability, there does not seem to be any significant qualitative difference in the ionic chemistry between the auroral and the lower latitude photochemical molecular ions layer. The observed ion composition ratios which are in particular characterized by a somewhat enlarged proportion of NO^+ and O^+ ions have been interpreted in terms of enhancements in production due to impact dissociation and excitation of N_2 (cf. Hyman *et al.*, 1976). It must be appreciated however that the experimental monitoring and the theoretical modelling of auroral conditions is an extremely difficult problem because of the overlapping effects of many dynamical processes with a large range of time constants, and of the effects on the chemical rate coefficients of the excess kinetic and internal energy of the reacting species (cf. Rees, 1975).

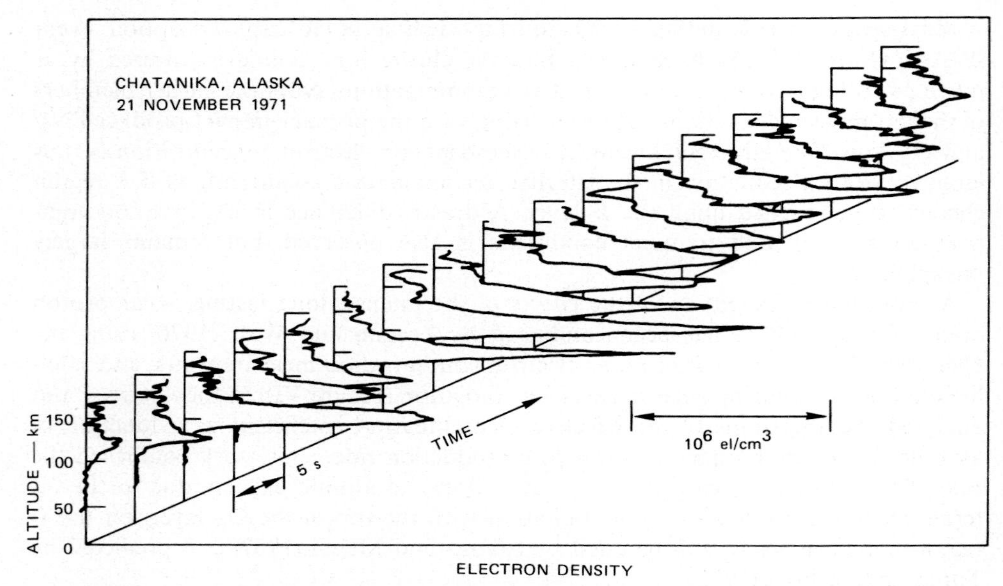

Fig. VI.16. A sequence of electron concentration height profiles in an aurora, taken at 5 s intervals with a 5 s integration time, using the Chatanika incoherent scatter sounder. The dynamic, rapidly changing, auroral *E* region ionization exhibited in this display is not an everynight occurrence at Chatanika, but is observed frequently in the midnight sector. These *E* region enhancements generally occur coincident with the appearance of magnetic activity on the College magnetometer (Leadabrand *et al.*, 1972).

6.4.2.2. *Middle and Low Latitude Effects*

Since magnetospheric energetic particles are injected from higher to lower *L* shells, and since precipitations empty the loss cone (cf. Chapter II), only a slow leak to the atmosphere can be expected from the reservoir of stably trapped protons and electrons in the inner radiation belts.

Paulikas (1975) has reviewed the phenomenology and aeronomy of magnetospheric precipitations at middle and low latitudes. Three types of processes can be distinguished:

(a) – Regions of low magnetic field strength, and in particular the South Atlantic 'anomaly' resulting from the offset and tilt of the geomagnetic dipole with respect the Earth's axis, depress the mirror point altitude of trapped particles, so that many do not survive a complete drift revolution of their bounce line because of increased interaction with the atmosphere. While optical effects are apparently never generated as a result of precipitations in the South Atlantic anomaly, the flux of precipitating electrons may be sufficiently strong to induce localized enhancements of ionization in the *D* and lower *E* regions of the ionosphere after the injection of a fresh population of trapped particles during a storm;

(b) – Wave-particle interactions in the plasmasphere cause trapped electrons' pitch angles to diffuse into the loss cone, and can provide a significant and sometimes dominant source of ionization in the night-time lower ionosphere at middle latitudes (Figure VI.14). No significant optical emission is excited by these precipitations; and

(c) – Precipitations of ring current protons proceed through charge exchange with exospheric H, converting them to energetic neutrals, and allowing them to transfer to low altitudes where they can be again ionized and temporarily trapped or precipitated on low L field lines. This precipitation gives rise to a weak optical emission at or near the geomagnetic equator.

Paulikas concludes that there is no persuasive evidence that energy fluxes approaching anywhere near the level of 1 erg cm^{-2} s^{-1} are deposited into the middle and low latitude atmosphere by such effects.

6.5. Meteor Ionization

The presence of metallic species in the upper atmosphere was first detected by Bernard (cf. Bricard and Kastler, 1944), who identified the Na line in the twilight glow spectrum. Later, lines of K and of ionized Ca were also identified and extensively studied from the ground. The abundance of these vapours was found to exhibit marked geographical and temporal variations (Hunten, 1967).

Early rocket mass-spectrometric measurements confirmed the presence in the lower thermosphere of neutral and ionized metallic atoms and oxides (Istomin and Pokhunkov, 1963). Since then, as many as thirty elements with their different isotopes, oxides and hydrates have been detected in the D and lower E regions by rocket borne ion mass spectrometers (Goldberg and Aikin, 1971; Zbinden *et al.*, 1975).

There was a slow realization in the beginning that these species were of extraterrestrial origin; later, in the 1960's, that this transport was somehow connected with the so called sporadic E layers (Cuchet, 1965); and finally that their chemistry was an interesting aeronomical problem of its own (Ferguson and Fehsenfeld, 1968).

6.5.1. PRODUCTION BY INTERPLANETARY DUST

Vaporization of solid cosmic particles impinging on the atmosphere is a permanent process over the whole globe, only much more intense on the 'front' (sunrise) face of the Earth facing the 'apex' towards which it moves through space with its orbital speed of about 30 km s^{-1} in the solar-stellar frame of reference. The meteoric particles involved in this process are not so much the rare stones predominantly of asteroidal origin that form visible trails (shooting stars), and occasionally reach the ground, as the cloud of interplanetary dust particles of cometary origin with mass less than 10^{-2} g (1 mm in radius) also responsible for the zodiacal light and gegenschein. The larger dust particles, called 'meteors', colliding upon the dense layers of the atmosphere, form kilometers long trails, comparable to that of shooting stars, but much more tenuous and observable only with a telescope or a radar. The trail results from intense heating and vaporization of the surface of the particles, which dissipate their kinetic energy the higher up the faster the particle and the larger the zenith angle of its path. The smallest particles (mass $< 10^{-7}$ g, radius $< 10\mu$m), called micrometeorites, are slowed down without being overly heated and vaporized. In both cases, there first occurs a process in which atoms are torn out by impact with the air molecules ('sputtering'). At the outset, there has been deposited in the decelerating slab of the atmosphere a diffuse gas of sputtered or vaporized atoms whose composition resembles that of the particles (O, Fe, Ni, Si, Ca, Mg, Na, etc.) along with the solid fragments that have survived the thermal shock. The total accretion has been estimated to be 10^2 to 10^4 t day^{-1}.

The average energy input is much less than 1 erg cm^{-2} s^{-1}, negligible in the daytime compared to solar XUV radiation (Lebedinets *et al.*, 1973), but possibly important in the nighttime lower thermosphere.

From the ionospheric point of view, the essential fact is not so much the direct production of ionization by impact, which is less than 0.1 ion cm^{-3} s^{-1} on the average at is maximum near 100 km of altitude (Figure VI.17), as the injection over all the surface of the Earth, between the altitudes of 80 and 120 km, of a vapor of low ionization potential atoms. Only during rare, particularly intense meteor showers, in general associated with the Earth's passing through the debris of a comet, has it been established that the electron production rate increases directly because of the arrival of meteors.

6.5.2. METEOR ION CHEMISTRY

Although sputtering and vaporization of cosmic particles can produce metallic ions directly, the processes responsible for the existence of permanent layers of ionized metallic atoms in the upper atmosphere are photoionization and charge exchange.

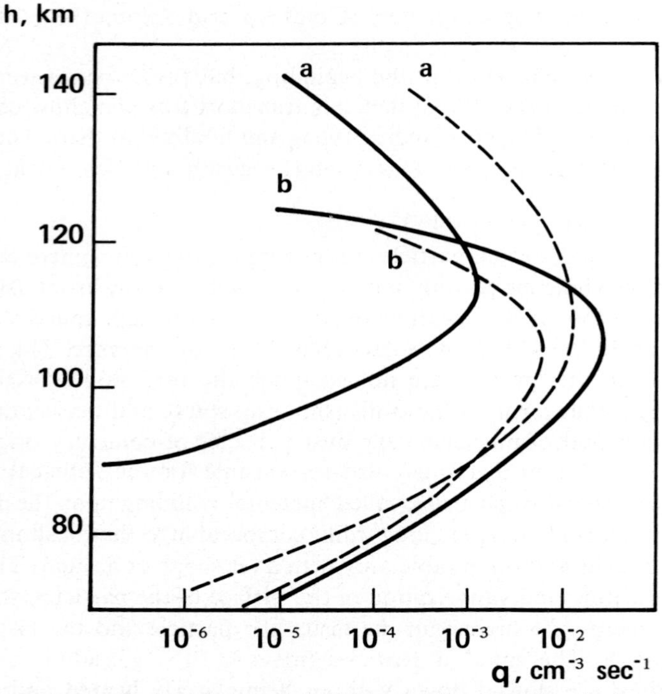

Fig. VI.17. The total contribution of meteoric ionization to the atmosphere is due to: – the smaller particles, micrometeorites, which essentially do not vaporize (a); – the larger particles, which vaporize (b). The ions produced are both: – metallic ions Fe^+, Si^+, Ca^+, Mg^+, etc. (solid lines); – and atmospheric ions O^+, O_2^+, N_2^+, etc. (dotted lines) (from Lebedinets *et al.*, 1973).

Because the metals (noted M below) have lower ionization potentials than atmospheric molecules, these processes are very efficient. For instance reactions like

$$NO^+ + M \rightarrow M^+ + NO$$

can occur in essentially all collisions of metallic atoms with *E* region ions (Rutherford *et al.*, 1971). Conversely, there is no efficient loss process for the metallic ions M^+, down to the level where they can be oxidized by reactions like

$$M^+ + O_3 \rightarrow MO^+ + O_2.$$

Note however that the ratio $n(O)/n(O_3)$ has to be low enough for such reactions to lead to significant amounts of oxides, because the dissociating reaction with O is exothermic:

$$MO^+ + O \rightarrow M^+ + O_2.$$

Three body reactions can lead to higher order oxides, and to hydrates by cluster switching:

$$M^+ + O_2 + N_2 \rightarrow MO_2^+ + N_2$$
$$MO_2^+ + H_2O \rightarrow M^+H_2O + O_2.$$

The compound ions can recombine dissociatively with electrons, regenerating the neutral metal or oxide.

The general scheme is thus a sort of merry-go-round between the different states of the metallic elements, with downward vertical transport of the ions by 'dynamo' drifting and upward diffusion of the neutrals playing a dominant role (Gadsden, 1970). The metals and metal compounds are always found layered, with what seems to be permanent strata below 100 km and patchy sporadic strata above, except at the magnetic equator (cf. next chapter).

Ion composition measurements have shown that for elements with similar chemical behavior, the abundance ratios agree with meteoritic (solar system) abundances. On the other hand the relative abundance of the various chemical families of meteor ions reflects the role of the chemical processes going on: alkali ions (Na^+, K^+) are overabundant, a probable effect of their particularly low ionization potentials; the refractory elements (Al^+, Ca^+, Ti^+, Mg^+) are underabundant; the non-metals S^+ and Si^+ are absent in the lower levels, a probable consequence of their larger O affinity (Zbinden *et al.*, 1975).

An even more dramatic clue to the chemical kinetics in the meteor ion layers is apparent in Figure VI.18, where it can be seen that there is a pronounced rarefaction of the normal ionospheric ions NO^+ and O_2^+ in the main meteor ion stratum. If it is assumed that their production rate is unchanged, a decrease in molecular ion concentration by such a large factor (say 10) is due to an enhancement by the same factor of their loss frequency by some charge exchange process. Their undisturbed loss frequency is $\alpha^+ n(e) = 10^{-3}\ s^{-1}$, so that the enhanced loss frequency due to charge exchange reactions of type:

$$\begin{matrix} NO^+ \\ O_2^+ \end{matrix} + X \rightarrow X^+ + \begin{matrix} NO \\ O_2 \end{matrix} \quad (\text{rate } x)$$

is $\Sigma x n(X) \simeq 10^{-2}\ s^{-1}$.

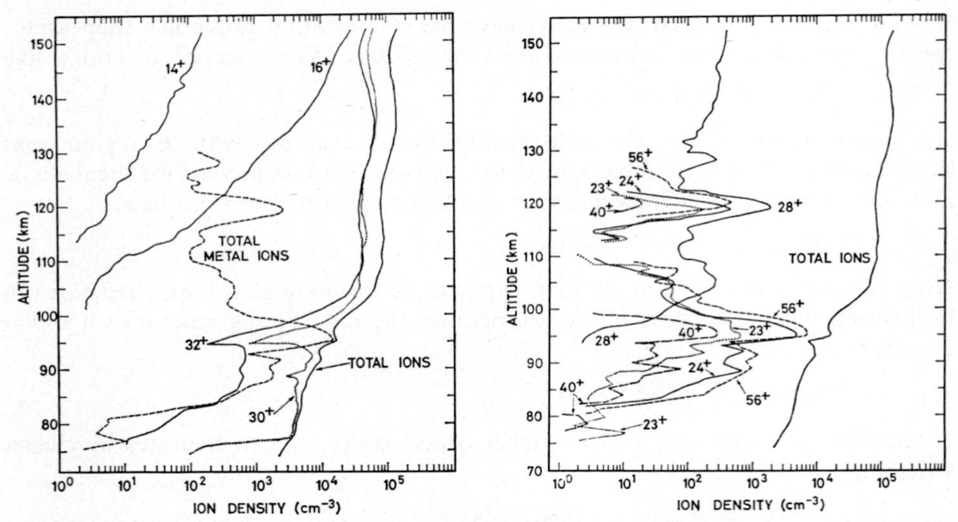

Fig. VI.18. The concentration profile with height of the major *E* region ions (left panel), and of the major metallic ions (right panel), as measured with a cryopumped mass-spectrometer in Sardinia on December 14, 1971, at 13 h local time, during the Geminid meteor shower (Zbinden *et al.*, 1975). Ions are identified by their atomic mass: NO^+ (30); O_2^+ (32); Na^+ (23); Mg^+ (24); Si^+ (28); Ca^+ (40); Fe^+ (56). Note that the main metallic ion layer, between 90 and 100 km, does not show up in the total ion profile, because of the corresponding valley in molecular ion concentration.

The obvious candidates for the reactants X are the (unseen) neutral metals and their oxides M, MO, MO_2 (reaction for instance). The rates of loss of the product ions and the charge exchange coefficients x have to be quite large if the concentration of these neutral species does not exceed the ion concentration, as suggested by optical observations.

Note that if the above scheme is correct, the neutral meteor species catalyze the photoproduction of NO in the lower thermosphere, as it is a by-product of the ion molecule scheme starting with photoionization of N_2:

$$N_2^+ + O_2 \rightarrow NO^+, + MO_x \rightarrow NO$$

with a yield of some 10 molecules cm^{-3} s^{-1} at the peak.

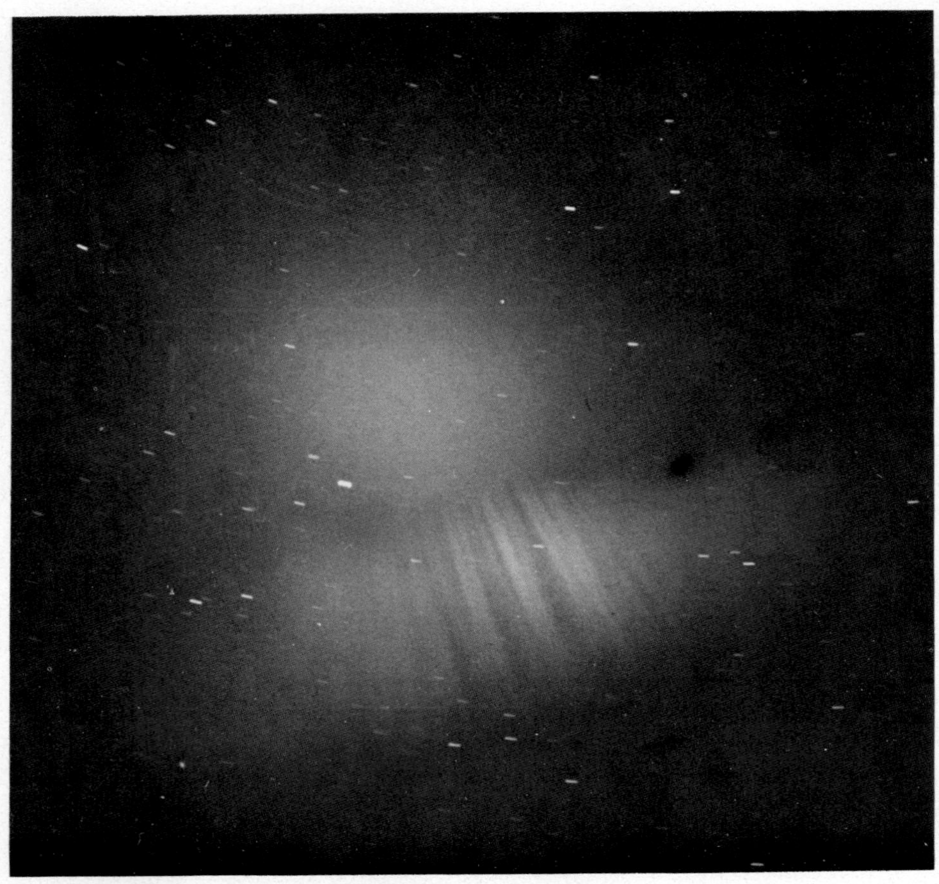

This photograph, taken on August 6, 1967, at Fort Churchill, Canada, permits us to visualize the decoupling of neutral and ionized gas motions in the upper atmosphere. Metallic vapors were released into sunlight from a rocket at a height of 260 km. They were then filmed from three sites on the ground against the starfield in the dark dusk background sky. Neutral Sr atoms glow blue and drift slowly with the neutral wind while diffusing smoothly into a quasi-spherical cloud. Ba atoms are photoionized by solar UV light with a time scale of a few tens of seconds and glow purple. This ion cloud drifts rapidly (southwestward), indicating the existence of a (northwestward) electric field. It is subject to substantial shear, and formation of magnetic field aligned irregularities, due to $\vec{\mathbf{E}} \times \vec{\mathbf{B}}$ plasma instability (cf. Francis and Perkins, 1975). (Courtesy of G. Haerendel, Max Planck Institut für Physik und Astrophysik)

CHAPTER VII

ELECTRODYNAMICS OF MOMENTUM TRANSFER

In the preceding chapter, we discussed photoionization, impact ionization and chemical processes which control production and recombination of the various ion species. The present chapter is devoted to the processes which control the motions of ions and electrons in the ionosphere.

The first section deals with the general rules governing the bulk motion of ions and electrons. In the second section, ionospheric electric fields and currents are discussed in relation to neutral winds and magnetospheric convection, and their effect on the triggering of ionospheric instabilities is briefly presented. The last three sections deal with the influence of transport processes on the distribution of long lived ionospheric particles, first in the diffusive upper ionosphere which is totally structured by such processes, then in the F layer where both chemistry and transport play a role, and lastly in the sporadic E layers, where metallic ions of meteoric origin are stratified by neutral wind shears within the photochemical lower ionosphere.

7.1. The Bulk Motions of Ions and Electrons

The geomagnetic field, through the Lorentz-Laplace force, compels charged particles, in the absence of other forces, to describe helices around its field lines. But we have noted before that in the weakly ionized and dense ionospheric plasma this cyclotron motion is counteracted by collisions with neutrals and Coulomb interactions between charges, so that the trajectories of individual particles are intractable, and a hydrodynamical approach must be used to describe their statistical collective behavior.

7.1.1. COLLISION FREQUENCIES

Let us consider the average transfer of momentum by collisions between the particles of a gas 1 (mass m_1) and those of a gas 2 (mass m_2). It can be shown that, provided the relative velocity V between the two gases is small compared with their thermal velocities, the mean momentum lost by a particle of gas 1 has the form $m_1 \nu_{12} V$. The quantity ν_{12}, which depends on the law of interaction between particles 1 and particles 2, has the dimensions of a frequency and is often called the collision frequency. More exactly, it is an effective collision frequency, valid for momentum transfer, and cannot be identified with the average number of collisions per second without using very restrictive and unrealistic hypotheses about the nature of the collisions.

The momentum lost by gas 1 is gained by gas 2, so that if n_1 and n_2 are the respective numerical densities, we have in general:

$$n_1 m_1 \nu_{12} = n_2 m_2 \nu_{21}.$$

We saw in Chapter I that the use of very simple models for the collisions, such as that of billiards balls, allow us easily to obtain approximate values for the partial collision frequencies of the different types of interaction. More rigorously, for two gases, 1 and 2, having Maxwellian velocity distributions at temperatures T_1 and T_2, integration of the momentum exchange over the collisional impact parameter (the impact parameter is the shortest distance of approach between two particles in the absence of interaction), and over the velocities of the two types of particles, leads to the formula:

$$\nu_{12} = \frac{m_2}{m_1 + m_2}\frac{4}{3}n_2\left(\frac{8k}{\pi}\right)^{1/2}\left(\frac{T_1}{m_1} + \frac{T_2}{m_2}\right)^{1/2} K^3 \int q(g) g^5 e^{-Kg^2} \mathrm{d}g \qquad \text{(VII.1)}$$

where $K = (2k(T_1/m_1 + T_2/m_2))^{-1}$,

and $q(g)$ is the cross section for momentum transfer, which depends on the relative speed g between the two interacting particles. This cross section is obtained from the differential scattering cross section $\sigma(g, \chi)$, where χ is the angle of deflection of the relative velocity, by the formula

$$q(g) = \int_0^\pi 2\pi\sigma(g,\chi)(1 - \cos\chi) \sin\chi \, d\chi. \tag{VII.2}$$

The above value of ν_{12} is similar to that obtained by Banks (1966a), aside from the factor $m_2/(m_1 + m_2)$, which is practically equal to 1 if gas 1 is composed of electrons and gas 2 of heavier neutrals and ions. The disparity comes from a slightly different definition of the collision frequency used by Banks and other authors, who write the motion term in the equation of transfer in the form $m_1 m_2/(m_1 + m_2)\nu_{12} V$, replacing m_1 by the reduced mass $m_1 m_2/(m_1 + m_2)$.

Under the crude approximation of collisions between rigid spheres, Equation (VII.1) gives an effective frequency ν_{12} which is $\frac{4}{3}\frac{m_2}{m_1 + m_2}$ times the number of collisions per unit time. (Using the same definition as Banks for ν_{12}, leads therefore to a factor 4/3.) This correcting factor shall be kept in mind when comparing crude models of collision used in Chapter I with more elaborate ones.

First of all, let us examine the case of collisions between electrons or ions on the one hand, and the various species of neutral particles on the other. Taking care not to forget the factor $m_2/(m_1 + m_2)$, which is different from 1 in the ion-neutral case, we can find in Banks (1966,a,b) expressions for the partial collision frequencies that are the fruit of a meticulous labor of synthesis between theoretical calculations and laboratory measurements of the quantities related to momentum transfer, such as plasma conductivity at very high frequencies, or electron drift velocity under the influence of a steady electric field.

Taken together with a model of the neutral atmosphere, these expressions can be used to determine the overall collision frequencies, ν_{en} (electron-neutral) and ν_{in} (ion-neutral). Progress in the physics of collisions or in the knowledge of the neutral atmosphere could lead to the revision of these values, but it seems improbable that they are wrong by an order of magnitude.

The frequency ν_{ei} of electron-ion collisions can be calculated in a very similar way: $\sigma(g,\chi)$ is then the Rutherford cross section for Coulomb collisions, deduced from the elementary laws of mechanics. The only difficulty comes from the incorrectness of the description of Coulomb interactions as a series of binary encounters. We know that for distances short compared to the Debye length λ_D the particles really can be treated individually, but that it is quite a different matter at distances large compared with λ_D, where collective behavior dominates. A development of this argument leads to the assumption that there are no binary interactions at distance greater than λ_D, so that the integration over the impact parameter should extend from 0 to λ_D rather than from 0 to infinity (the integration over χ in Equation (VII.2) is equivalent to an integration over impact parameter, since χ depends on the impact parameter alone when the relative velocity is fixed). Furthermore, the necessity to take account of the collective nature of the long-range interaction is made evident when we notice that integration up to infinity would lead to an infinite value of the cross-section, which cannot be physically correct. The cut-off of the collision integrals at λ_D causes the appearance in all the results of a factor ln Λ, where, aside from a factor 2π, Λ is simply the number of particles in a sphere of radius λ_D. In the ionosphere, ln Λ does not vary much, remaining between 13 and 15. The fact that Λ is large is a justification of the rough way in which the screening effect due to collective behavior has been introduced, in the form of a sharp cutoff: changing the cutoff distance by a factor of two would change Λ by a factor of 2, hence ln Λ by 0.7 which is only 5% of its value.

The result of the calculation described above is:

$$\nu_{ei} = 3.8 \times 10^{-6} \ln \Lambda n_e T_e^{-3/2}. \quad \text{(VII.3)}$$

7.1.2. ANALYSIS OF THE MOMENTUM TRANSFER EQUATION

The hydrodynamical equation of momentum transfer for particles of species k can be written, when viscosity is neglected, under the form:

$$n_k m_k \partial \vec{\mathbf{V}}_k/\partial t + \vec{\nabla}(n_k k T_k) = n_k m_k \vec{\mathbf{g}} + n_k \varepsilon e(\vec{\mathbf{E}} + \vec{\mathbf{V}}_k \times \vec{\mathbf{B}}) - n_k m_k \sum_{l \neq k} \nu_{kl}(\vec{\mathbf{V}}_k - \vec{\mathbf{V}}_l). \quad \text{(VII.4)}$$

The left hand side includes the inertial term and the effect of diffusion due to a gradient in the partial pressure, and the right hand side describes the effect of the force fields, and of collisional coupling with other species l present.

In the first two terms on the right hand side we made explicit the role of gravity and of the electromagnetic field (ε is 0 for neutrals, +1 for singly charged positive ions, and −1 for electrons). To good approximation, $\vec{\mathbf{B}}$ is the terrestrial magnetic field, the modifications of the field due to ionospheric currents, even if measurable, always being negligible from the point of view of the Lorentz-Laplace force $\vec{\mathbf{V}} \times \vec{\mathbf{B}}$, which, it should be noted, plays the role of a 'brake' on the average motion of the particles, except in the direction of the lines of force. $\vec{\mathbf{E}}$, on the other hand, is the total 'self-consistent' field in whatever reference frame we have defined the velocities, and must include the polarization fields corresponding to space charges resulting from the particles motion; therefore in contrast to $\vec{\mathbf{B}}$ and $\vec{\mathbf{g}}$, it is not in any way a given 'applied' field.

The last term, which includes the 'drag' between the various particles species, has here been expressed in terms of the difference between the species bulk velocities. Note that this friction term can be decomposed into two terms,

$$+ n_k m_k \sum_{l \neq k} \nu_{kl} V_l \text{ and } - n_k m_k \nu_k V_k, \text{ where } \nu_k = \sum_{l \neq k} \nu_{kl};$$

these terms are respectively a driving term and a braking term. This decomposition might seem arbitrary, since only the difference $\vec{\mathbf{V}}_k - \vec{\mathbf{V}}_l$ has any importance in collision phenomena. Actually, the reference frame is arbitrary, but as soon as one has been chosen we have the right to talk of a 'driving' term and a 'braking' term, just as we talk of a 'driving' electric field. It must be remembered however, that these terms, like the velocity and electric field, depend on the chosen frame. We only have to consider relatively steady motions and can take $\partial \vec{\mathbf{V}}/\partial t = 0$ in the transport equation. We can obtain a quantitative criterion for the validity of this simplification by defining a 'characteristic period' τ of the motion. The order of magnitude of the first term $nm\, \partial V/\partial t$ is then $nmV\tau^{-1}$. The order of magnitude of the collisional braking term is at least $nmVv$, and that of the Lorentz braking term is at least $nmV\,\omega_B$ (using the definition of the gyrofrequency $\omega_B = eB/m$). Thus the motion can no longer be considered as steady if τ^{-1} is not small compared with v or ω, i.e. if we are dealing with a time scale well under a second. We therefore exclude discussion of this type of motion, which in general is caused by the excitation of waves (radio frequency propagation or plasma instabilities), and shall consider either quasi-stationary motions or slow oscillations, where the acceleration is, at all times, exactly and instantaneously balanced by the combined braking due to collisional and cyclotronic effects.

A final form of the equation of transport, in which all the terms that do not depend on the velocity $\vec{V}_k$ are on the left, while those that do depend on it are on the right, can thus be written

$$\frac{1}{n_k}\vec{\nabla}(n_k kT_k) - m_k\vec{g} - \varepsilon e\vec{E} - m_k \sum_{l \neq k} v_{kl}\vec{V}_l = \varepsilon e(\vec{V}_k \times \vec{B}) - m_k v_k \vec{V}_k. \quad \text{(VII.5)}$$

The former terms are the *causes* of the motion in the frame considered (that is in the frame where $\vec{E}$ and $\vec{V}$ are defined): pressure gradients, gravity, electric fields, driving by friction. We shall in the next section identify them generically by a driving 'force' $\vec{F}$ in order to study the general character of the motion.

7.1.3. THE MOBILITY AND CONDUCTIVITY TENSORS

Thus, for the moment, we do not concern ourselves with the causes of the motion, which we consider to be 'self-consistent' and given, and represented by the single quantity $\vec{F}$. The general equation of transport (VII.5) is then:

$$\vec{F} = \varepsilon e\vec{V} \times \vec{B} - mv\vec{V} \quad \text{(VII.6)}$$

or

$$\vec{V} = \frac{1}{mv}\vec{F} + \frac{\varepsilon e}{mv}\vec{V} \times \vec{B}.$$

7.1.3.1. *Mobilities*

We project the vectors along axes parallel and perpendicular to $\vec{B}$, and perpendicular to $\vec{B}$ and $\vec{F}$:

$$\vec{V} = \{V_{//}, V_{\perp}, V_{\Lambda}\};\ \vec{F} = \{F_{//}, F_{\perp}, 0\};\ \vec{B} = \{B, 0, 0\};\ \vec{V} \times \vec{B} = \{0, V_{\Lambda}B, -V_{\perp}B\}.$$

The decomposition of Equation (VII.6) along these axes (with $eB/m = \omega_B$) gives:

$$\vec{V} = \left\{\frac{1}{mv}F_{//}, \frac{v}{m(v^2 + \omega_B^2)}F_{\perp}, \frac{-\varepsilon\omega_B}{m(v^2 + \omega_B^2)}F_{\perp}\right\} \quad \text{(VII.7)}$$

From this expression we can immediately draw the following conclusions (rediscovering the mean motion of a single particle, which is indeed described by an equation identical to Equation (VII.6)):

– If $\vec{F}$ is parallel to $\vec{B}$ ($F_{\perp} = 0$), the bulk motion is parallel to the force, and the velocity is limited only by collisions: $\vec{V}_{e,i} = \vec{F}/m_{e,i}v_{e,i}$.

– If $\vec{F}$ is perpendicular to $\vec{B}$ ($F_{//} = 0$), the bulk motion has two components one parallel to $\vec{F}$, the other perpendicular to both $\vec{F}$ and $\vec{B}$, the ratio between the two velocity components being $-\varepsilon v/\omega_B$. This means that $\vec{V}$ is parallel to $\vec{F}$ if $v \gg \omega_B$, and perpendicular to $\vec{F}$ if $v \ll \omega_B$. The motion of the ions and electrons takes place in the plane perpendicular to $\vec{\mathbf{B}}$, and it is easy to compute that their velocities $\vec{\mathbf{V}}_i$ and $\vec{\mathbf{V}}_e$ are inclined with respect to $\vec{\mathbf{F}}$ by the respective angles α_i and α_e, where $\alpha_{e,i} = \tan^{-1}(\omega_{Be,i}/\nu_{e,i})$ (Figure VII.1). Note that Equation (VII.6) implies that the tip of the perpendicular velocity vector for electrons and ions as a function of altitude rotates as a chord on a circle, with diameter F/eB.

More generally (any $\vec{\mathbf{F}}$), the solution to Equation (VII.6) can be written in the

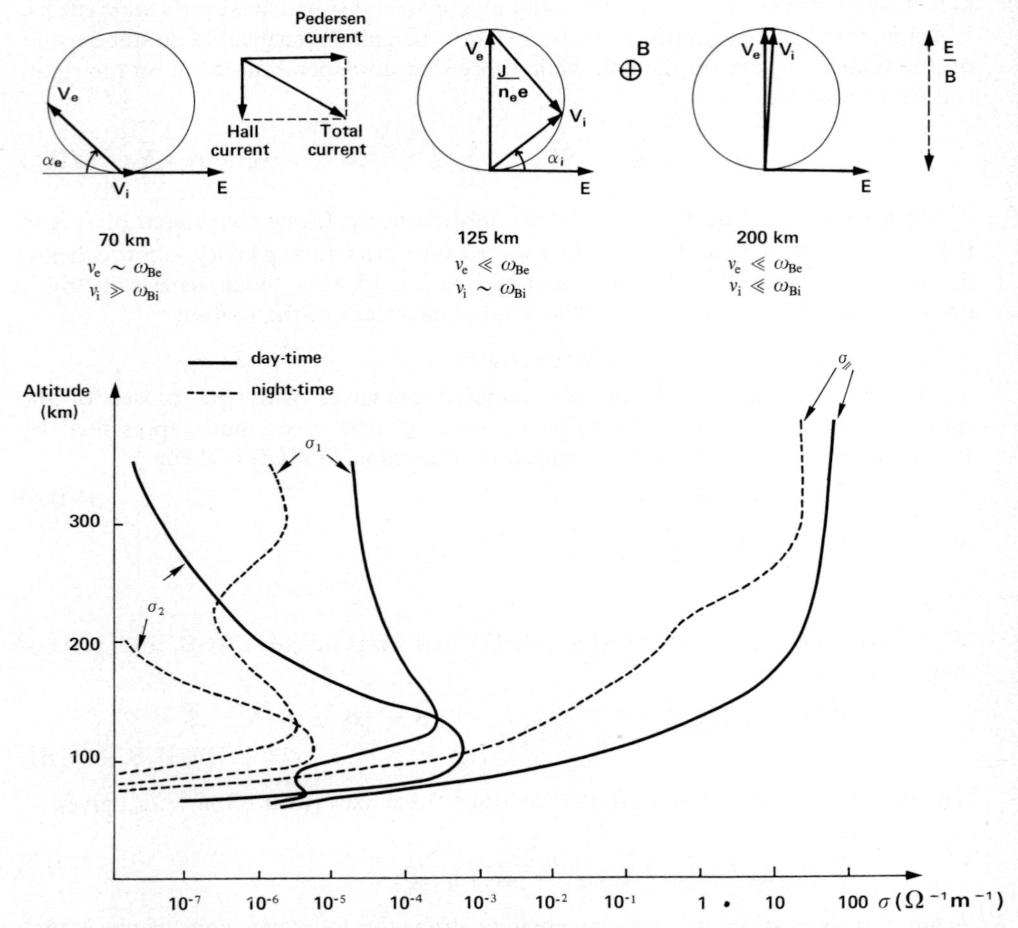

Fig. VII.1. Because of collisions with neutral molecules (frequency ν), and of cyclotron rotation (frequency ω_B), a force (here an electric field $\vec{\mathbf{E}}$) perpendicular to the magnetic field $\vec{\mathbf{B}}$ (here into the plane of the figure) causes the response bulk velocities of ions and electrons in the ionosphere to be strongly height dependent. The vectors $\vec{\mathbf{V}}_e$ and $\vec{\mathbf{V}}_i$ remain chords on a circle of diameter E/B; but the angles α given by $\tan^{-1}(\nu/\omega_B)$ correspondingly vary from 0 in the neutrosphere ($\nu \gg \omega_B$) to $\pi/2$ in the magnetosphere ($\nu \ll \omega_B$). The corresponding current $\mathbf{J} = n_e e(\vec{\mathbf{V}}_i - \vec{\mathbf{V}}_e)$ is strongest in the intermediate dynamo region. With a model for the electron concentration n_e, the vertical profiles of the components of the conductivity tensor relating $\vec{\mathbf{E}}$ to $\vec{\mathbf{J}}$ can be computed. While the longitudinal conductivity $\sigma_{//}$ in the direction parallel to the magnetic field lines keeps increasing with height, the Pedersen (σ_1) and Hall (σ_2) conductivities perpendicular to the magnetic field maximize in the E layer.

tensorial form $\vec{V} = \bar{\bar{\mu}}\vec{F}$, where $\bar{\bar{\mu}}$ is the 'mobility tensor'. In a Cartesian frame independent of $\vec{F}$, but whose x axis is parallel to $\vec{B}$, this tensor is:

$$\bar{\bar{\mu}} = \begin{matrix} 1/m\nu & 0 & 0 \\ 0 & \nu/m(\nu^2 + \omega_B^2) & \varepsilon\omega_B^2/eB(\nu^2 + \omega_B^2) \\ 0 & \varepsilon\omega_B^2/eB(\nu^2 + \omega_B^2) & \nu/m(\nu^2 + \omega_B^2). \end{matrix} \quad \text{(VII.8)}$$

Its components transform by rotation into those associated with any system of axes, independent of $\vec{B}$, that might turn out to be convenient in a given situation, e.g. vertical-horizontal, North-South and East-West axes. The relation $\vec{V} = \bar{\bar{\mu}}\,\vec{F}$ will still hold, provided $\vec{F}$ is transformed in the corresponding way. In any case, if $\omega_B \ll \nu$, that is, in the neutrosphere, the tensor reduces to the scalar mobility $\mu = 1/m\nu$.

We know that the friction force due to a neutral atmosphere having a bulk velocity $\vec{U}$ is $\vec{F} = m\nu\vec{U}$. From Equation (VII.8), we deduce the velocity $V_{//}$, projection of $\vec{V}$ in the direction of the magnetic field:

$$V_{//} = (m\nu/m\nu)U_{//} = U_{//}.$$

Therefore to a neutral wind of velocity $\vec{U}$ always corresponds a motion of the ionized species parallel to the magnetic field, with a velocity equal to the projection along $\vec{B}$ of $\vec{U}$. The components of $\vec{V}$ perpendicular to $\vec{B}$ are much smaller in the F region where $\nu \ll \omega_B$.

7.1.3.2. *Conductivities*

Electric fields appear both as the causes and the consequences of the motion. But there is nothing to stop us from examining formally the relationship between an existing field $\vec{E}$ and the motion of ionospheric electrons and ions, that is, between electric field and current.

In general, if the electrons and ions are subject to a force $\vec{F} = \varepsilon e\vec{E}$, then, within the set of approximations we have been using, the electron velocity will be:

$$\vec{V}_e = -e\bar{\bar{\mu}}_e\vec{E}$$

and that of the (singly charged) ions will be:

$$\vec{V}_i = e\bar{\bar{\mu}}_i\vec{E}$$

where $\bar{\bar{\mu}}_e(m = m_e, \nu = \nu_{en}, \varepsilon = -1)$ and $\bar{\bar{\mu}}_i\ (m = m_i, \nu = \nu_{in}, \varepsilon = 1)$ are respectively the electron and ion mobility tensors, $\bar{\bar{\mu}}$ being defined in the tensorial expression above. The resulting current is:

$$\vec{J} = n_e e(\vec{V}_i - \vec{V}_e)$$

n_e being the electron or ion density (the medium is globally neutral). Then

$$\vec{J} = \bar{\bar{\sigma}}\vec{E}$$

where the 'conductivity tensor' $\bar{\bar{\sigma}}$ is given by:

$$\bar{\bar{\sigma}} = n_e e^2(\bar{\bar{\mu}}_i + \bar{\bar{\mu}}_e). \quad \text{(VII.9)}$$

One can distinguish three types of conductivities. The longitudinal conductivity $\sigma_{//}$ relates the current along the magnetic field to the parallel component of the electric field. For the component of $\vec{E}$ perpendicular to $\vec{B}$, the current component parallel to $\vec{E}$, and the corresponding conductivity are called Pedersen current and Pedersen conductivity (σ_1), while those in the direction perpendicular to both **E** and **B** are called Hall current and Hall conductivity (σ_2). Using the coordinate system in which Equation (VII.8) has been written, $\sigma_{//}$ is the first diagonal term of the conductivity, σ_1 is the common value of the two diagonal terms, σ_2 is the absolute value of the off-diagonal terms.

The behavior of electrons and ions at typical altitudes is plotted in the upper panel of Figure VII.1 and the resulting typical conductivities are given in the lower panel.

For a typical altitude of 70 km, the velocity of ions is very small; the current is mainly carried by electrons and the Hall current dominates over the Pedersen current. In the dynamo region, where ν_i and ω_{Bi} are comparable, the Hall current is mainly carried by electrons; the contribution of ions is of opposite sign and reduces the total Hall current; the Pedersen current carried by ions is maximum and equals the Hall current ($\sigma_1 = \sigma_2$) when $\nu_i = \omega_{Bi}$ (for a fixed value of electron concentration). Above 200 km, electrons and ions essentially move together with the velocity $\vec{V} = \vec{E} \times \vec{B}/B^2$: the corresponding current is almost zero. The small difference between their velocity is almost parallel to $\vec{E}$ and the Pedersen current dominates over the Hall current ($\sigma_1 \gg \sigma_2 \approx 0$). The actual values of Pedersen and Hall conductivities depend of course upon the electron concentration. The values plotted on Figure VII.1 have been derived by assuming typical day and night-time electron concentration profiles.

The longitudinal conductivity (parallel to the magnetic field) is equal to $n_e e^2/m_e \nu_e$, since $m_i \nu_i \gg m_e \nu_e$. As the collision frequency between electrons and neutral particles approaches zero when altitude increases, the longitudinal conductivity tends to increase indefinitely with altitude. However, it is never really infinite, because of electron-ion Coulomb collisions. We can take these into account by introducing into Equation (VII.6) for the electrons a term $-m\nu_{ei}(\vec{V}_e - \vec{V}_i)$, where ν_{ei} is the collision frequency given by Equation (VII.3). The projection parallel to the magnetic field of the equation thus modified is written:

$$0 = -e\vec{E} - m\nu_{en}\vec{V}_e - m\nu_{ei}(\vec{V}_e - \vec{V}_i).$$

When ν_{en} approaches zero, there results:

$$\vec{V}_e - \vec{V}_i = -e\vec{E}/m\nu_{ei} \qquad \text{(VII.10)}$$

from which the current is

$$\vec{J} = (n_e e^2/m\nu_{ei})\vec{E}. \qquad \text{(VII.11)}$$

From Equation (VII.3), ν_{ei} is proportional to n_e; consequently, in this case, the longitudinal conductivity no longer has any dependence on the electron concentration n_e.

In all cases, magnetic lines of force can be considered as good electrical conductors. As a result (Hines, 1963), the two ends of the same line of force (these are said to be magnetically conjugate points) must be practically at the same potential, within, say, 0.1 V, which is negligible compared with the difference in potential between two non-conjugate points which, as we shall see, can reach several thousand volts.

Of course, these conclusions are false when, as sometimes happens, phenomena of instability or turbulence induce a decrease in the parallel conductivity in certain regions of the magnetosphere. The study of these phenomena remains an open subject.

7.1.4. SELF-CONSISTENT MOTION

We have examined so far the relationship between the bulk motion of electrons and ions and the 'driving force'. The driving force however may differ from the applied force trough feedback phenomena.

When an electric field is imposed on the ionosphere, electric charges start moving and there is no reason for the corresponding current to be divergence free. Polarization charges therefore accumulate, causing an additional electric field; equilibrium is reached when the total field (applied field plus polarization field) creates a divergence free current. This phenomenon is illustrated by the 'Cowling conductivity'.

A similar phenomenon arises in friction with neutral particles: the velocity of the neutral particles is not an 'applied' velocity, and depends on the reaction of the ionized species ('ion drag'). Here again, feedback loops occur, and the corresponding solutions to the equations of motion must be 'self-consistent'.

7.1.4.1. *The Cowling Conductivity*

Let us assume that an electric field $\vec{\mathbf{E}}$ in the East-West direction is applied to the equatorial E region. Since the magnetic field is horizontal there, the result of this is a vertical Hall current whose value is

$$J_z = \sigma_2 E$$

σ_2 being the Hall conductivity (off-diagonal term in the conductivity tensor). But at lower altitudes the vertical conductivity is zero because of the low electron concentration and because of collisions, and at high altitudes it is zero because the magnetic field is horizontal. Thus the vertical current causes an accumulation of charges which in turn creates a vertical electric field, and equilibrium will be reached when this field cancels the current (Figure VII.2). Let E_p be the vertical polarization field and σ_1 the Pedersen conductivity (one of the two equal diagonal terms in the conductivity tensor). At equilibrium we must have

$$J_z = 0 = \sigma_2 E + \sigma_1 E_p, \quad \text{or} \quad E_p = -\sigma_2 E/\sigma_1.$$

But on the other hand the field E_p causes a Hall current of size $-\sigma_2 E_p$ in the East-West direction, so that the total East-West current at equilibrium is

$$J = \sigma_1 E - \sigma_2 E_p, \quad \text{or} \quad J = (\sigma_1 + \sigma_2^2/\sigma_1)E.$$

The factor $\sigma_1 + \sigma_2^2/\sigma_1$, called the Cowling conductivity, is thus the true ratio of the current to the applied field, and its value is much greater than σ_1 at an altitude of about 110 km.

This increase in conductivity basically explains the existence of the 'equatorial electrojet', a current that flows during the day in the E region, from East to West, along the magnetic equator, producing the large variations in the magnetic field observed at stations with small magnetic latitude. A similar phenomenon develops in the

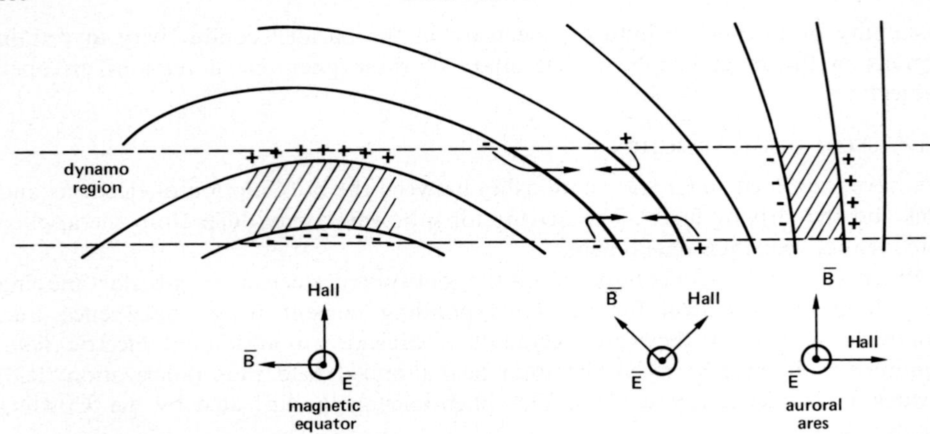

Fig. VII.2. When Hall currents due to an applied $\vec{E}$ field (out of the plane of the figure) cannot flow because a conductive region is bounded in the $\vec{E} \times \vec{B}$ direction by strong negative conductivity gradients, polarization charges appear, and *their* Hall effect raises the effective Pedersen current (parallel to the applied field). This happens permanently at the magnetic equator because of the field geometry, and sporadically in auroral arcs because of a local increase in conductivity by precipitations, explaining the existence of the equatorial and auroral 'electrojets'.

auroral belts as described in Section 7.2.2.2 below. Besides its intrinsic interest, this phenomenon clearly illustrates the role of polarization electric fields induced by space charges, which is basic to an understanding of the ionospheric electrical system.

7.1.4.2. *Couplings between Plasma and Neutral Motion*

The action of neutral winds on the plasma will be discussed throughout the rest of this chapter. However, it must be kept in mind that the thermospheric neutral wind system itself is dependent on the velocity and concentration of the ions through the 'ion drag' phenomenon. When the velocity of the plasma $\vec{V}$ is different from the velocity $\vec{U}$ of the neutral particles, the friction force exerted by the plasma on the neutrals is equal to $(m_i\nu_i + m_e\nu_e)n_e(\vec{V} - \vec{U})$ per unit volume, unit time. Since $m_i\nu_i$ is much greater than $m_e\nu_e$, only the force exerted by the ions is important here. The vertical component of this friction force is small compared with the gravitational forces on the atmosphere. It is balanced by a slight change in the vertical gradient in density and hence in the vertical gradient in pressure. On the other hand, the horizontal component of the friction force acts freely upon the neutral atmosphere, until it moves with the same horizontal velocity as the ions.

7.1.4.2.1. *Ion-drag*. The friction force between neutral particles and ions plays a fundamental role in thermospheric dynamics. In the absence of any frictional force, the neutral wind would result from the equilibrium between the pressure gradient and the Coriolis force $2\vec{\Omega}\Lambda\vec{U}$, where $\vec{\Omega}$ is the rotation vector of the Earth, and $\vec{U}$ the velocity of the neutral wind. In this case, $\vec{U}$ would be perpendicular to the pressure gradient and the wind system would be geostrophic (parallel to the isobars) as happens at meteorological altitudes. If on the contrary, the Coriolis force is negligible compared to the frictional force (this is the case in the daytime F region), the situation is completely different.

Let us assume first that ions are motionless; the frictional force is $-m_i \nu_i \vec{\mathbf{U}}$; the velocity $\vec{\mathbf{U}}$ is therefore parallel to the pressure gradient and the wind blows from the hottest point (in the daytime hemisphere) to the coolest one (in the nighttime hemisphere). Actually the mean velocity of the ions $\vec{\mathbf{V}}_i$ is not zero. As we shall see, their velocity perpendicular to the magnetic field depends upon the existing electric fields, while their parallel velocity depends upon the ambipolar diffusion velocity (Section 3.1 below) and on the parallel component of the neutral particles velocity $\vec{\mathbf{U}}$. The actual frictional force $-m_i \nu_i (\vec{\mathbf{U}} - \vec{\mathbf{V}}_i)$ depends upon $\vec{\mathbf{V}}_i$ and thermospheric winds cannot be computed from the pressure gradients, without a good knowledge of ion velocity and concentration (Rishbeth, 1972).

During daytime, when the frictional term dominates over the Coriolis force, the vertical diffusion velocity is negligible but typical electric fields induce a change of 30% in the computed neutral velocity. During nighttime, thanks to the reduced electron concentration, ion drag is no longer the dominant effect and becomes comparable to the Coriolis force. However, the large variations of the diffusion velocity may play a certain role. In any case, the ion drag term is proportional to the electron concentration n_e and any neutral wind computation is therefore strongly dependent on the geographical and vertical plasma distribution.

The frictional force between ions and neutrals may also influence the behavior of the ionosphere, through a more sophisticated self-consistent process: motions of charged particles drive motions of the neutral species which in turn affect the bulk velocity of electrons and ions. Since neutral particles are more numerous than charged particles, their acceleration will always be small and in order for equilibrium to be reached (equality of the horizontal components of $\vec{\mathbf{U}}$ and $\vec{\mathbf{V}}_i$), a long time is required, of the order of an hour in the daytime F region, several hours in the nocturnal F region. (At lower altitudes the time constants are still greater, and this process is practically negligible.)

A feedback mechanism is then produced, because, according to the results of Section 1.3, the neutral wind induces in ions a velocity parallel to $\vec{\mathbf{B}}$ of magnitude $U \cos I$ which is added to the initial velocity of the ions, and in the final equilibrium state it is the horizontal projection of the resulting velocity V_i that must be equal to U. Under certain conditions, the neutral atmosphere influences the velocity of ions through this mechanism. We shall give some examples.

7.1.4.2.2. *Effect of an Applied Electric Field in the East-West Direction.* Assume an initial situation where $\vec{\mathbf{U}} = 0$. The combined $\vec{\mathbf{E}}$ and $\vec{\mathbf{B}}$ fields cause, with a time constant of the order of a second, a motion of ions:

$$\vec{\mathbf{V}}_{i0} = \vec{\mathbf{E}} \times \vec{\mathbf{B}}/B^2$$

with an horizontal component situated in the meridian plane. This component slowly accelerates the neutral atmosphere, which in turn reacts upon the ions, as we have mentioned. When the horizontal speed of the neutral atmosphere is U_1, the speed of the ions is $\vec{\mathbf{V}}_{i1} = \vec{\mathbf{V}}_{i0} + |U_1| \cos I \dfrac{\vec{\mathbf{B}}}{|B|}$; their vertical velocity is decreased and their

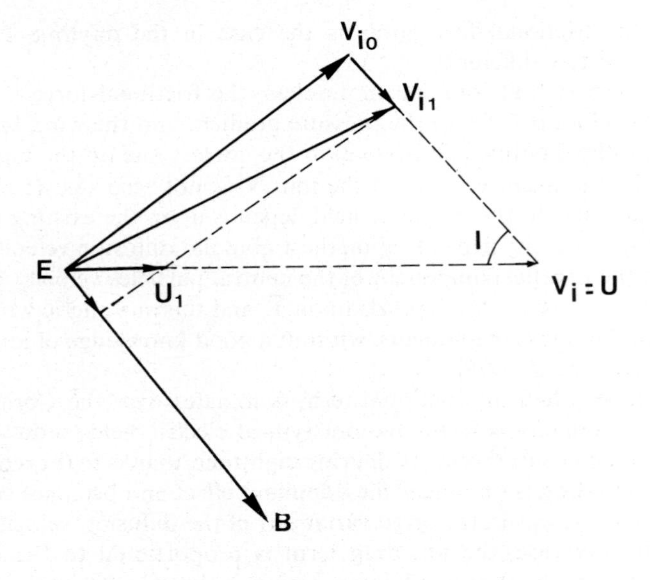

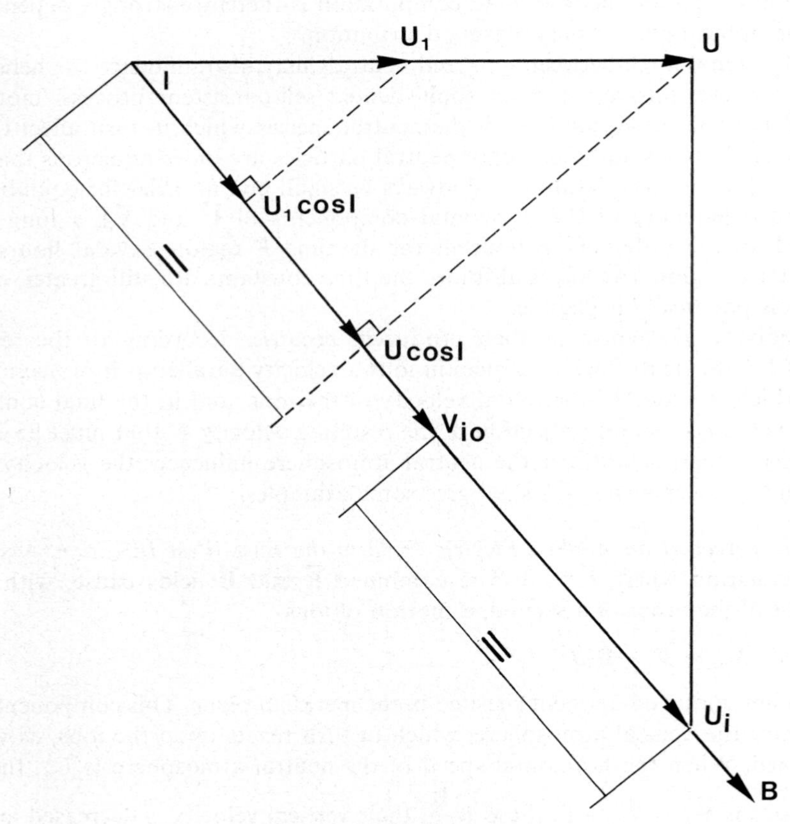

Fig. VII.3. See text.

horizontal velocity is increased. Equilibrium is reached (Figure VII.3a) when ions and neutral particles have the same horizontal speed U which is given by the equation:

i.e. $$V_{i0} \sin I + U \cos^2 I = U$$
$$U = V_{i0}/\sin I.$$

The vertical component of V_i is then zero.

We see that after a time long enough for equilibrium to be attained, a horizontal electric field in the East-West direction causes a horizontal motion of the ions, not a motion perpendicular to the magnetic field. Even in times that are not long enough for this equilibrium to be established, the induced motion of the neutral atmosphere can appreciably change the response of the ions to such a field.

7.1.4.2.3. *Diffusion of Ions.* Let us now assume that $\vec{\mathbf{U}}$ is again initially zero and that the ions diffuse along the magnetic field with initial velocity $\vec{\mathbf{V}}_{i0}$. As before, the neutrals are set into motion and induce in the ions a supplementary parallel velocity $U \cos I$ (Figure VII.3b). The equilibrium state corresponds to the equality of the horizontal component of the ions velocity and the velocity of the neutral atmosphere

$$(V_{i0} + U \cos I) \cos I = U$$

hence $$U = V_{i0} \cos I/\sin^2 I$$

and $$\vec{\mathbf{V}}_i = \vec{\mathbf{V}}_{i0}/\sin^2 I.$$

The reaction of the neutral atmosphere has the effect of increasing the velocity of diffusion, the more so when I is smaller. This phenomenon might therefore be very important in the equatorial regions. However, here again we encounter the problem of the time necessary to reach equilibrium, and Kohl (1965) was able to show that this time varies as $1/\sin^2 I$, so that in the equatorial regions this equilibrium state is in fact never attained.

7.2. Ionospheric Electric Fields and Currents

As shown by Figure VII.1 the Pedersen and Hall conductivities exhibit a strong maximum in the daytime E region, the 'dynamo region', while the parallel conductivity is large at every altitude above 100 km (with the exception already mentioned of possible magnetospheric anomalous conductivity). The dynamo region is therefore the only place of the ionosphere where currents can flow perpendicular to the equipotential magnetic field lines. It is thus part of any current loop in the Earth's environment. The driving forces of the currents are both the motion of neutral particles in the dynamo region itself which is the dominant term at middle and low latitudes, and the motion of solar particles which is felt mainly in the polar ionospheric regions, connected to the outer magnetosphere.

Ground based magnetic field observations, which reveal the existence of ionospheric currents, had been interpreted in terms of equivalent current sheets, assumed to be flowing horizontally in the E region. Magnetometers onboard rockets and satellites have investigated whether the currents are actually flowing horizontally or whether field aligned currents (so-called Birkeland currents) are also flowing, which is the case in the high latitude regions as could be expected since the driving force there is not in the ionosphere.

These current systems are linked to the electric fields distribution through the conductivity tensor. But Lorentz invariance gives a zero order approximation to the fields: Lorentz invariance requires the electric field to vary, in the presence of a magnetic field, by addition of a $\vec{\mathbf{V}} \times \vec{\mathbf{B}}$ term when passing from one frame of reference to another one, moving with a velocity $\vec{\mathbf{V}}$ relative to the first one. A natural frame of reference is the one rotating with the Earth, in which the neutral atmosphere is motionless in first approximation. If the electric field is zero in this frame of reference, the value of the corresponding 'corotational' field is of the order of 15 mVm^{-1} in the magnetospheric frame of reference fixed relative to the Sun, in the equatorial region. Near the poles, this field obviously vanishes. Actually, tidal motions induce a departure from perfect corotation of the neutral atmosphere. Assuming that the electric field is zero in the frame of reference locally attached to the neutral particles, implies in the corotating frame of reference a field of 1.5 mVm^{-1} for typical wind velocities of 50 ms^{-1}. In the auroral and polar zones, on the other hand, the convection electric field originating in the flow of the solar wind on the magnetopause has an order of magnitude of 50 mVm^{-1}, and dominates over other fields in both frames of reference.

Direct electric field measurements are not very sensitive (Chapter V) and results have been obtained in the high latitude regions only, where the fields are stronger. Other measurements of electric field are based on the observation of the motion of ions perpendicular to the magnetic field, either ionospheric ions observed by incoherent scattering (Chapter IV), or artificially seeded photoionized barium clouds observed by optical methods, or ducts of whistler propagation tracked by radiogoniometry.

7.2.1. THERMOSPHERIC WINDS AND THE DYNAMO SYSTEM

As mentioned in Chapter I the neutral atmosphere, at *E* region altitude, is subject to tidal motions, excited mainly by thermal effects from the Sun. The neutral gas couples to the ionized species through collisions to set them in motion, causing currents and polarization electric fields in a way somewhat analogous to what happens in a dynamo machine, hence the old name 'dynamo system'.

Like the diurnal motions of the neutral atmosphere, the resulting currents are organized in a more or less permanent pattern fixed with respect to the direction of the Sun (Figure VII.4). A given point on the Earth's surface rotates underneath this pattern of currents, and this is felt as the *Sq* and *L* diurnal variations in the terrestrial magnetic field at this point (upon which are superimposed the effect of other currents).

7.2.1.1. *Dynamo Theory of the Diurnal Magnetic Field Variation*

A neutral wind of velocity $\vec{\mathbf{U}}$ acts as a driving force $m\nu\vec{\mathbf{U}}$ as described in Section VII.1. The resulting drifts are different for ions and electrons in the *E* region. The current thus produced need not satisfy the equation div $\vec{\mathbf{J}} = 0$. At any point where div $\vec{\mathbf{J}} \neq 0$, electric charges accumulate just like we have described when discussing

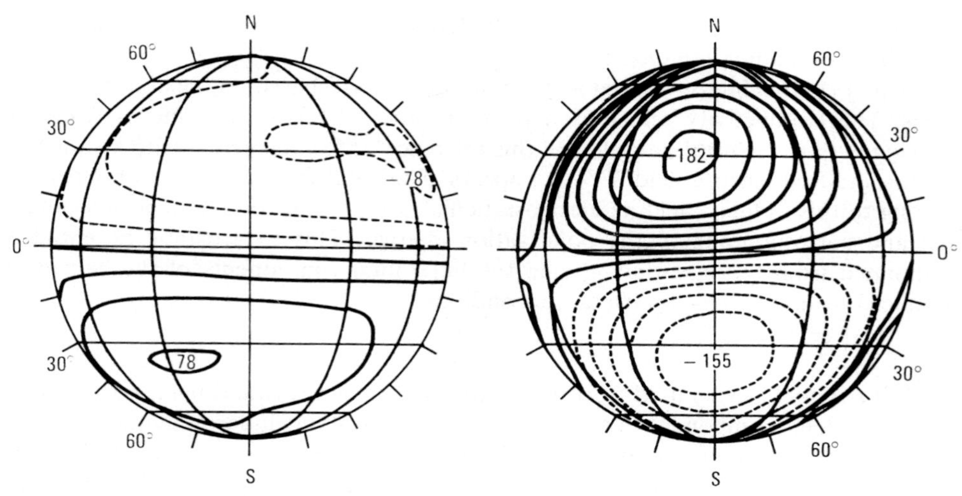

Fig. VII.4. The average current sheet flowing in the dynamo region required to account for the solar quiet diurnal variation of the Earth's magnetic field measured by the network of ground magnetometers at the equinox during the International Geophysical Year (Matsushita, 1965). Dayside (right) and nightside (left) of the dynamo region are represented in geographic latitude coordinates. The figures give the total currents in kA around each vortex. It must be imagined that the Earth rotates under this current pattern, whose magnetic field then reproduces for any site the observed diurnal oscillation of the magnetograms.

Cowling conductivity. A space charge polarization field is set up, and continually adjusts so that the current flow is non divergent.

In the first analyses of the problems, the wind $\vec{\mathbf{U}}$ was assumed to be independent of height throughout the conducting region, and the equation for the height integrated current was written in terms of height integrated conductivities (Baker and Martyn, 1953). However, there are strong indications that the actual wind velocity varies sufficiently with height for such a method to fail accounting for the tidal dynamo.

Dynamo calculations must take into account another constraint due to the large conductivity along field lines through the magnetosphere; the electric potential must be virtually equal at all points on a given field line above the dynamo region. *E* region electrostatic fields which result from polarization effect are thus mapped into the *F* region, where they act on the plasma, causing it to drift with the velocity $\vec{\mathbf{E}} \times \vec{\mathbf{B}}/B^2$. Moreover, any potential difference which might arise between magnetically conjugate points at the top of the dynamo layer tends to be removed by a flow of current along the field line linking the points; this however by no means implies that the current flow between conjugate hemispheres is negligible. The usual assumption that dynamo currents are horizontal must be based on different grounds. The vertical dimension of the dynamo region is small compared with its horizontal extent. Horizontal scale lengths being typically 100 times greater than vertical ones, the nondivergence of the current density implies that the vertical current density must be of the order 1/100 of the horizontal current density. Although this statement does not imply that the total vertical current is negligible compared to the horizontal one, this approximation is usually made in order to simplify the calculations. It is valid, if we assume that the Earth's magnetic field is a dipole with its axis antiparallel to the rotation axis and that the wind and conductivity fields are symmetric with respect to the equator (equinox conditions).

Tarpley (1970a) has developed a computational model based on the above assumptions. The conductivity tensor is derived (Chapter 7.1.3) in a frame of reference moving with the neutral particles. In this reference frame, the driving force vanishes, but an induced electric field $\vec{\mathbf{V}} \times \vec{\mathbf{B}}$ appears. The total electric field is obtained by adding to this induced field an electrostatic field, written as the gradient of a potential and corresponding to the polarization charges. This electrostatic potential is eliminated from the equations giving the horizontal components of the current $\vec{\mathbf{J}}$ (deduced from $\vec{\mathbf{J}} = \bar{\mu}\vec{\mathbf{E}}$), by using the condition

$$\vec{\nabla} \cdot \vec{\mathbf{J}} = 0.$$

Such a model can be used to compute the electric currents and fields corresponding to various components of the neutral winds system: lunar gravitational tide (Tarpley, 1970a), solar thermal tides (Tarpley, 1970b; Richmond *et al.*, 1970). The neutral winds used in such calculations are theoretical models adjusted to the available observations from meteor radars or incoherent scatter radars.

The conclusion of such studies is that the observed neutral winds may indeed account through dynamo theory for the quiet time behavior of ionospheric fields and currents as interpreted from ground based magnetic field observations:

– the *Sq* current system which represents the solar daily component of these variations is caused mainly by the diurnal tidal mode (1, −2) driven primarily by direct solar heating in the thermosphere;

– the *L* current system, having an amplitude smaller than *Sq* and a period equal to half an lunar day (24.8 h), as found in ocean tides, is caused by the gravitational lunar tide.

While the semi-diurnal tide has only a minor action on the currents as seen on ground level magnetograms, both the diurnal (1, −2) mode and the semi diurnal (2, 4) mode must be taken into account simultaneously to obtain a (limited) agreement between results from dynamo computations and the observed electric fields in the ionosphere (Figure VII.5).

7.2.1.2. *The Equatorial Electrojet*

Near the magnetic equator, the terrestrial magnetic field is horizontal in the dynamo region, and this special geometry allows the accumulation of space charges which causes the Cowling conductivity as described in Section 1.4.1.

The equatorial electrojet, detected as a strong latitudinal anomaly on magnetograms, is mainly driven by the electrostatic field produced by the charge distribution resulting from the *global* tidal electromotive force. At the magnetic equator the local wind-induced electric field is vertical because both the neutral particles velocity and the magnetic field are horizontal. This field sets up a charge separation leading to a polarization field that tends to cancel the total electric field. Thus if the cancellation were complete, the neutral air motion at the equator itself would make no contribution to the electrojet current.

However, Kato (1973) has shown that this cancellation is not as complete for curved geomagnetic lines, as when straight horizontal lines are considered. As a

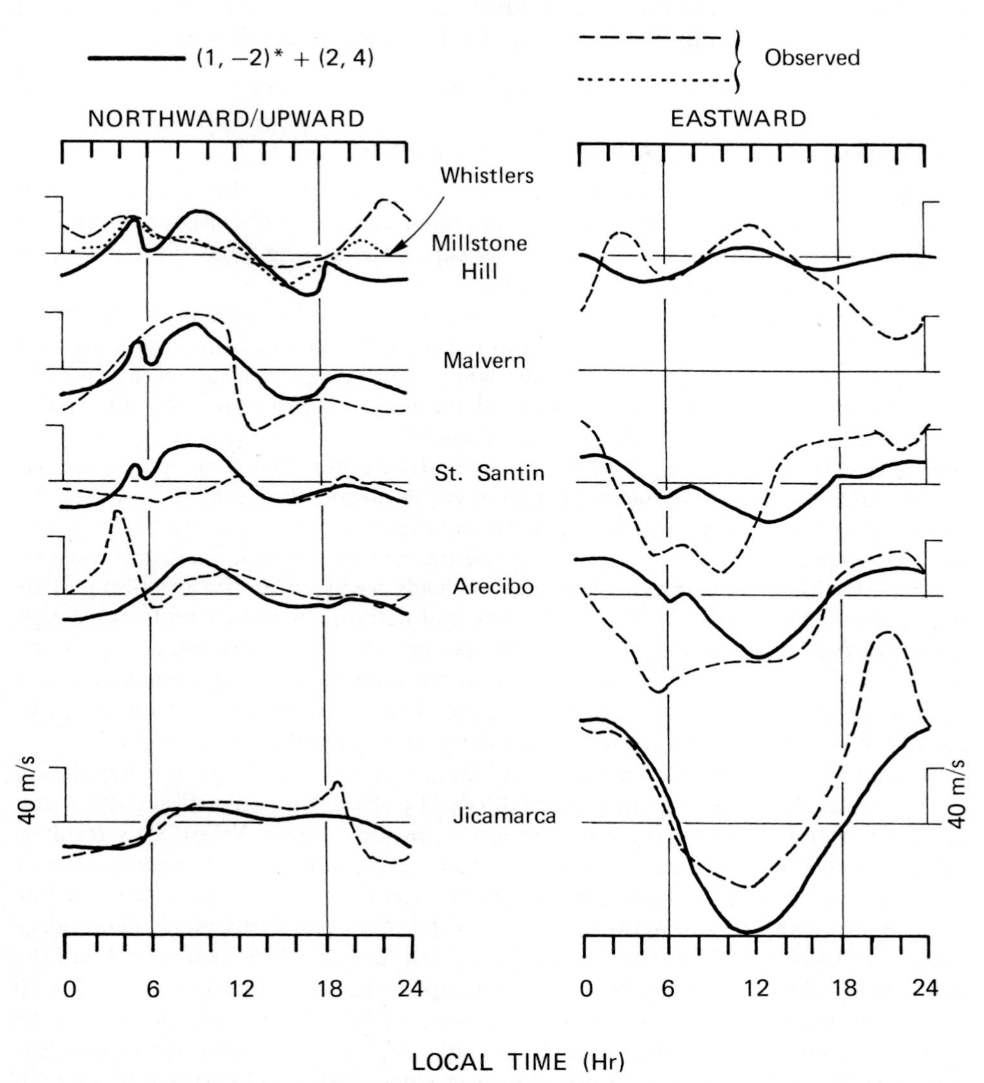

Fig. VII.5. Comparison between modelled (full lines) and observed (interrupted lines) diurnal variation of wind induced E region ion drifts. The observations are from the low and middle latitude incoherent scatter facilities (cf. Table IV.1) and from whistler ducts tracking. The model shown, giving the best fit to the experimental data, includes the diurnal (1, −2), and semi-diurnal (2, 4) modes of the thermal solar tidal wave.

consequence the local neutral winds may directly contribute to driving the equatorial electrojet. Nevertheless, they are less efficient than in mid-latitudes and, as a first approximation, the equatorial electrojet should rather be considered as driven by the global dynamo system in the extra-conducting strip which exists during daytime at the magnetic equator due to Cowling conductivity, as explained in Section 1.4.1.

7.2.2. MAGNETOSPHERIC CONVECTION AND AURORAL ELECTRODYNAMICS

7.2.2.1. *Penetration of the Magnetospheric Convection*

On the magnetopause, as explained in Chapter I, a potential distribution is generated by processes associated with field line merging (less likely, viscous drag). This electric field, directly mapped into the polar cap, is the driving force of the magnetospheric and ionospheric convection system.

The problem is actually self-consistent: an assumed electric field in the magnetopause controls the motion and distribution of protons and electrons which are also subject to gradient and curvature drifts, when their energy is large (cf. Chapter I). The field aligned currents flowing between the magnetosphere and the ionosphere are determined by the average value over each flux tube of the divergence of the perpendicular current. These field aligned currents are closed by perpendicular ohmic currents in the conducting E region of the ionosphere, and the ionospheric conductivity tensor being known there, the ionospheric field is determined. But this field is mapped in the magnetosphere along the conducting field lines and must be consistent with the magnetospheric electric fields assumed at the beginning. This requirement closes the system of equations and determines the currents and fields, provided the value of the electric potential on the polar cap boundary is set. Assuming such a boundary value implies that the solar wind action is equivalent to a voltage generator without internal resistance. This conventional hypothesis looks reasonable within our present poor knowledge of the merging phenomena.

Using such an approach, Vasyliunas (1972) has shown that the proton population of the plasma sheet and the ring current drastically affects the convection pattern and virtually excludes convection from the low L region (Figure VII.6). This result is valid for a steady state situation. For time-varying phenomena the screening effect of the ring current is not efficient and magnetospheric convection penetrates to low latitudes as in the right diagram of Figure VII.6. Numerical time dependent calculations by Jaggi and Wolf (1973) have shown that just after a sudden increase of the convection field, at a time t_0, an electric field appears at low latitudes. In the first 10 min of the relaxation, most of the antisolar electric field that existed initially in the nighttime ionosphere has been eliminated, although little has happened to the dayside electric field in that short time. However, some electric fields still exist even 2h after t_0: the dayside field has dropped by half, whereas the nightside field has dropped an order of magnitude.

Penetration of electric fields to middle latitudes is experimentally observed for periods of the order of a few hours (Evans, 1972, 1973; Testud *et al.*, 1975). The variations of the Earth's magnetic field at the equator under disturbed conditions (DP 2 variations) are likewise interpreted (Nishida, 1968) in terms of penetration of the magnetospheric convection. The observed time constants fit the calculated ones,

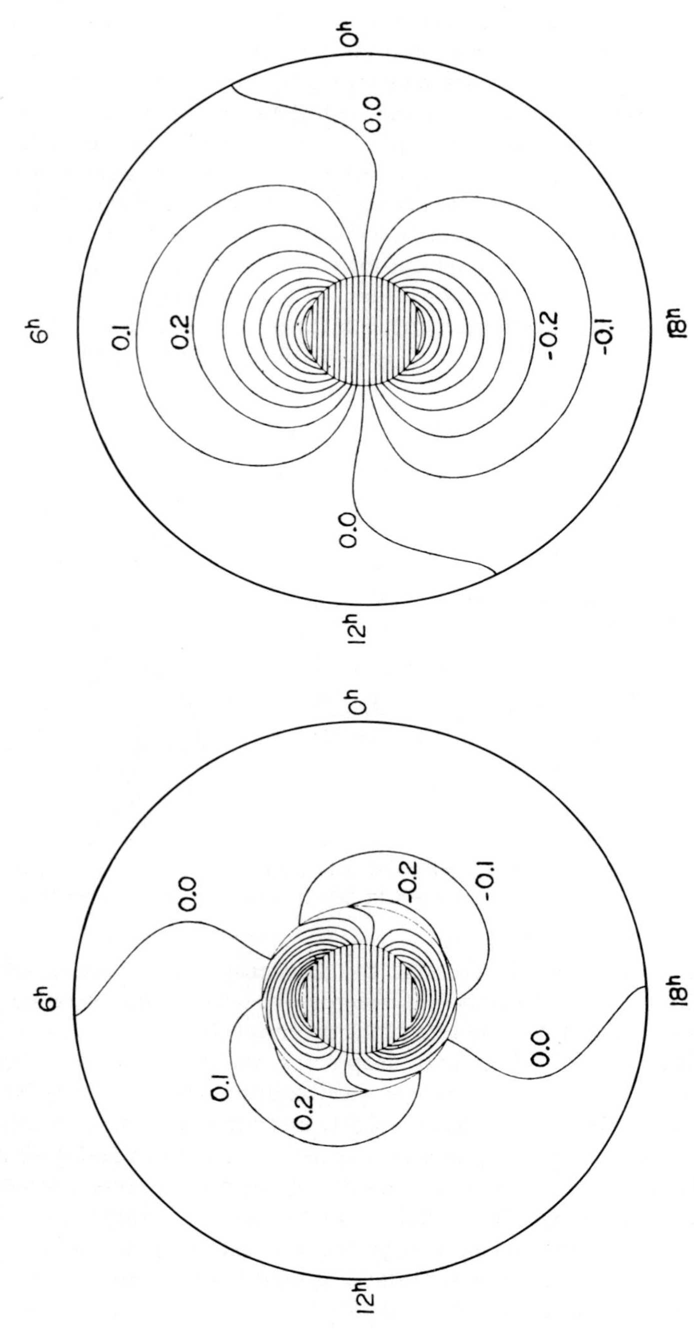

Fig. VII.6. Theoretical hemispherical pattern of magnetospheric convection (electric field equipotential or magnetic field tubes and thermal plasma perpendicular motions) for quiet (left) and disturbed (right) situations, as conditioned, outside the polar cap, by the trapped particles' conductivity (Vasyliunas, 1972).

particularly at night, but quantitative time dependent models of the magnetospheric-ionospheric convection are still beyond us.

The permanent control of the high-latitude ionosphere by magnetospheric convection is, on the other hand, clearly established. Electric field measurements with orbiting polar satellites (cf. Figure V.5) and with the Chatanika radar (Figure IV.9) have established the primary features of plasma convection at low altitude over the auroral zones and polar caps. The most prominent and persistent feature is the occurrence of an abrupt reversal in the convection velocity and electric field at about 70° to 80° magnetic latitude: the convection is anti-solar over the polar cap, while it is directed toward the Sun in the adjacent lower latitude 'auroral belt'.

Gurnett and Frank (1973) have shown that the electric field reversal corresponds to the boundary between open magnetic field lines which connect into the solar wind, on the poleward side of the reversal, and the closed field lines, which connect to

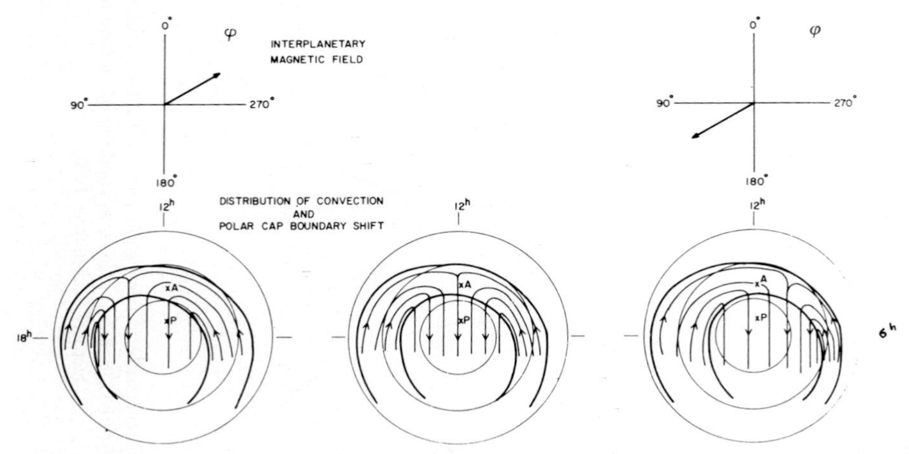

Fig. VII.7. Polar cap electrical equipotentials and convection patterns, as modulated by the variations of the azimuth φ of the interplanetary magnetic field (Heppner, 1972) (cf. Figure II.2).

the opposite hemisphere without crossing the magnetopause, on the equatorward side of the reversal. The changes in potential across the evening and morning auroral belts are frequently unequal, but the sum of the potential changes for the two auroral belts corresponds closely to the potential drop across the polar cap which may vary from 20 kV to over 100 kV. The anti-sunward convection over the polar region presents a large variability, ranging from a nearly uniform electric field in some cases to very asymmetrical profiles, with large differences between dawn and dusk in other cases. Heppner (1972) has reported a correlation between these convection patterns (Figure VII.7) and the longitudinal angle φ of the interplanetary magnetic field. Maximum convection in the northern high latitudes on the morning side corresponds to a maximum convection in the southern high latitudes on the evening side and a value φ in the sector between 90° and 180° (φ is 0 in the Sun direction). Such a correlation implies that the transfer of momentum from the solar wind is most effective where the interplanetary field is parallel to the field direction in the outer most regions of the magnetosphere, and least effective where the two fields are antiparallel.

This two-cell convection pattern which can be inferred from the ionospheric measurements is consistent with the generation of magnetospheric electric fields through magnetic merging as described in Chapter II. Major departures from this gross average behavior include multiple reversals and irregular fluctuations, particularly in the local evening, possibly corresponding to a turbulent convection pattern.

7.2.2.2. *Auroral Electrojets and Polar Currents*

In the past, it had been found convenient to describe observed polar magnetic disturbances by an equivalent current system assumed to be located on a spherical shell (the ionosphere) and which would give rise to the polar perturbations (Figure VII.8). Such a current system is simply a means of expressing geomagnetic disturbances and should not be confused with the actual current system. The equivalent system was first analyzed by using a Fourier transform of the D component of the field along circles of constant dipole latitude. This analysis, which produces the so-called SD

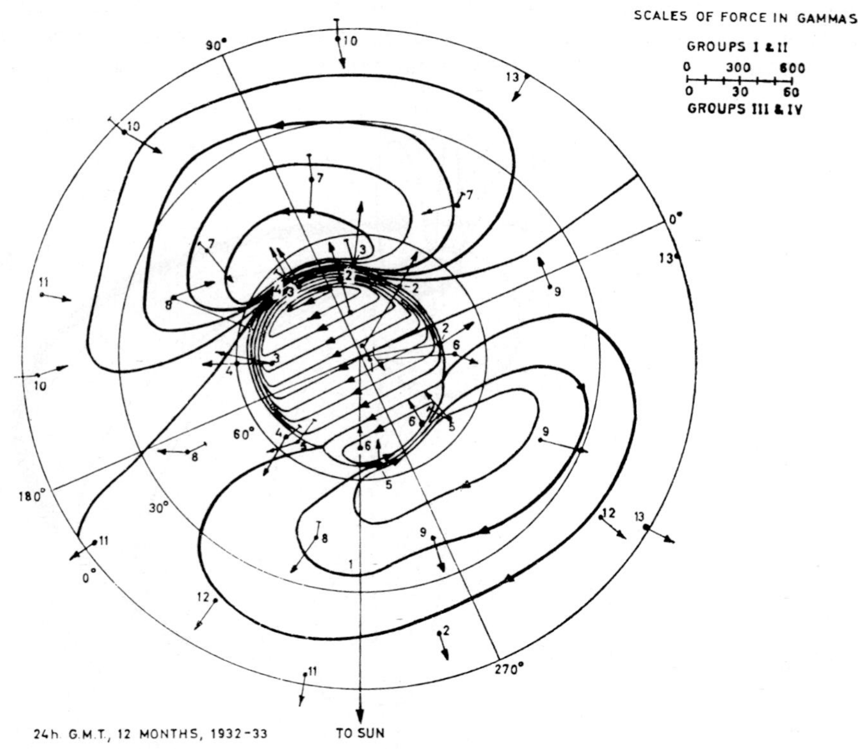

Fig. VII.8. The 'equivalent' dynamo region current sheet over the polar cap required to account for an average magnetic disturbance of the Earth's magnetic field in the high latitudes (Silsbee and Vestine, 1942). Contrary to the equivalent current sheet giving the *Sq* variation (Figure VII.4) this representation is not physically significant, because the actual current system is in fact three dimensional.

current system, implies that this current system is fixed with respect to the Sun, and the Earth rotates underneath. The SD current intensity is supposed to remain constant for 24 h. An auroral zone station observes the daily variation because it moves under different parts of the SD current system.

Our present view is that a current system which has some resemblance with the SD system appears intermittently with a life-time of 1 to 3 h and the Earth rotates under such a situation. This shows the impulsive and intermittent nature of the polar magnetic substorm. The main contributor to this intermittent current system is the auroral electrojet which is observed to flow along the auroral oval westward in the nightside sector. Weaker eastward electrojets are also observed.

It gradually became clear that these phenomena are associated with the triggering and decay of a three-dimensional current system involving currents flowing horizontally in the conducting auroral E region, the so called 'Birkeland currents' flowing along the magnetic lines of force between the ionosphere and the outer magnetosphere, and the 'ring current' of trapped protons near $L = 6$. Chapman had already noticed in 1952 a correlation between the intensity of the local auroral electrojets and that of the ring currents felt as a world wide magnetic disturbance. Magnetic observations made by polar orbiting satellites and by geostationary satellites have shown conclusively that electric currents do flow in and out from the auroral ovals.

A possible model of the three-dimensional current system involved has been presented by Fejer (1963), and by Akasofu and Meng (1969). It is assumed that the

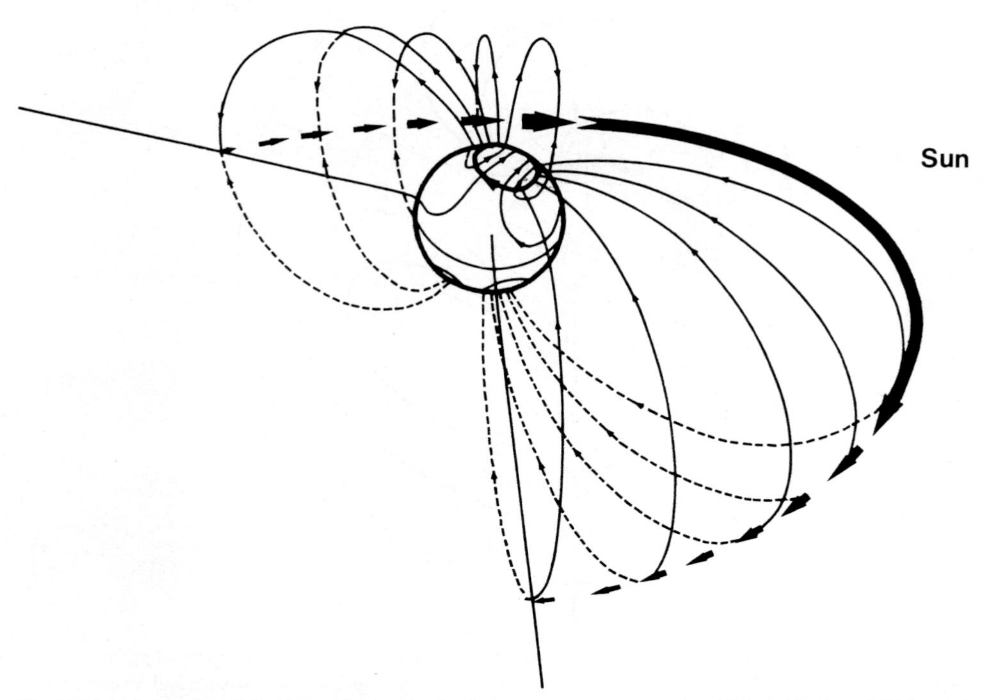

Fig. VII.9. A scheme of the three-dimensional substorm current system in which field aligned Birkeland currents connect the auroral electrojets to the asymmetric ring current (Akasofu and Meng, 1969).

asymmetric growth of the magnetospheric ring current resulting from the injection of hot plasma is an immediate cause of the substorm current. The current flows into the auroral ionosphere from the morning sector of the ring current, and flows out to the ring current in the evening sector, after rounding the auroral oval (Figure VII.9).

Such models are over simplified. The exact relationship between polar ionospheric currents and the magnetospheric currents is far from being clearly understood. The magnetosphere acts on the ionosphere by injecting particles and by setting boundary values for the fields mapped out along magnetic field lines. Particle precipitations increase ionospheric conductivity, and also cause enhancements of the neutral temperature, inducing changes in the thermospheric wind system which in turn may generate electric fields and currents through the dynamo effect. Conversely, ionospheric fields and currents may generate a strong feed-back on magnetospheric convection (Coroniti and Kennel, 1973). A detailed discussion of these transient processes is outside the scope of the present monograph.

The particle precipitations which cause the field aligned currents and the visual auroral displays also produce an increased ionization in the auroral dynamo region and therefore an increased conductivity. This extra-conductive strip is elongated along the auroral ovals. Something akin to the Cowling conductivity phenomenon develops: large currents cannot flow horizontally to the lesser conductive neighboring regions, and polarization charges accumulate both poleward and equatorward of the auroral ovals causing the phenomenon described in Section 1.4.1. Notice that the auroral polarization charges could flow along the magnetic field lines, and that their effective accumulation requires either an anomalous longitudinal resistivity, or a symmetrical situation in the northern and southern hemispheres connected by closed field lines. The importance of Cowling conductivity is thus a common feature of the auroral electrojets and of the equatorial electrojet, even though the accumulation of charges is due to completely different causes.

7.2.3. ELECTRICALLY DRIVEN IONOSPHERIC INSTABILITIES

We have seen that the presence of electron concentration irregularities has been detected by radio experiments (spread F on ionograms, radio-star and satellite beacon scintillations, coherent radar echoes 10^7 to 10^8 times more powerful than the incoherent scatter returns which correspond to thermal fluctuations) as well as by *in situ* measurements. They are often correlated with the existence of large currents (auroral and equatorial electrojets) and of large electron density gradients.

7.2.3.1. *Plasma Instabilities*

Plasma are subject to 'instabilities', phenomena in which irregularities of electron concentration grow exponentially up to very large values for which limiting nonlinear phenomena appear. A literature survey on plasma instabilities (Auer *et al.*, 1970) lists no less than 145 instabilities! The terminology is highly confusing and any classification leads to a certain degree of overlapping. However, the number of mechanisms is very large and we shall discuss only those which have been claimed to play an important role in ionospheric irregularities formation.

A plasma consisting of two interpenetrating streams of charged particles is unstable, if the relative mean velocity of the two streams is large enough. As in a traveling wave tube, particles travelling at a velocity close to the wave velocity interact with the wave; if more particles are decelerated by the wave than are accelerated, the wave gains energy and grows. This is the two stream instability. Farley (1963) has shown that for ionospheric conditions in the presence of the Earth's magnetic field and of collisions with neutral particles, waves propagating in directions nearly orthogonal to the magnetic field can be excited if the component of electron drift velocity in the direction of propagation is somewhat greater than the ion thermal velocity. This phenomenon may thus happen in the auroral and equatorial electrojets.

The $\vec{\mathbf{E}} \times \vec{\mathbf{B}}$ instability (also called either collisional drift mode or gradient-drift instability or neutral drag mode) was first investigated by Simon (1963). As the ionospheric plasma is only weakly ionized, the ions drift more slowly than the electrons under the action of an electric field $\vec{\mathbf{E}}$, since the ion-neutral collisions are more effective than the electron neutral collisions. Charge separation results which causes an additional electric field $\vec{\mathbf{E}}'$ and consequently an $\vec{\mathbf{E}}' \times \vec{\mathbf{B}}$ drift. Instability occurs when this drift is in the direction of decreasing density and this will happen only if there exists a density gradient, to which the applied electric field has to be parallel: the electron density must increase in the same direction as the electron potential energy increases. Moreover the electric field must exceed a critical value. This instability is likely playing a role both in E and F region irregularities.

The interchange or Rayleigh-Taylor instability corresponds to the interchange of neighboring flux tubes, which decreases or increases energy, and, therefore enables or prevents growth of the perturbations. This instability occurs when a force independent of charge acts on the particles, normal to the magnetic field, such as gravity or curvature of the magnetic field: electrons and ions drift in opposite directions. The resulting charge separation produces an electric field $\vec{\mathbf{E}}$ and this new field $\vec{\mathbf{E}}$ another drift ($\vec{\mathbf{E}} \times \vec{\mathbf{B}}$ drift) which amplifies any perturbations. Such a gravitational (sometimes called 'flute') instability is quite similar to the collisional drift instability. However, the drifts are proportional to electron and ion masses, whereas in the $\vec{\mathbf{E}} \times \vec{\mathbf{B}}$ instability the collisions play an essential role in producing unequal drifts in the presence of an electric field. This instability is important in the equatorial spread F phenomena, but also in the behavior of plasmapause.

The theoretical study of instabilities follows the same lines as the wave study described in Chapter III. The complexity of the calculations is increased by the sophistication of the velocity distribution function which is to be taken into account for a description of the instability. An instability occurs when the waves obey a dispersion relation, which implies that the frequency has an imaginary part: the wave amplitude grows exponentially from the thermal fluctuations to a large observable value. If the growth rate is not large compared to the frequency, it is possible to define a phase velocity and a wave vector $\vec{\mathbf{k}}$ on one hand and a group-velocity on the other hand, as discussed in Chapter III. Among the possible waves, a radio scattering experiment selects (Chapter IV) the one having the $\vec{\mathbf{k}}$ vector satisfying the Bragg condition (hence the wavelength of the observed irregularities depends upon the frequency of the radio-wave). The phase velocity $\vec{\mathbf{v}}_\phi = \omega\vec{\mathbf{k}}/k^2$ associated with the $\vec{\mathbf{k}}$ vector may be completely different from the group velocity which describes the motion of the irregularity, i.e., the way in which the ionospheric region affected by the irregularity varies with time.

7.2.3.2. *Equatorial and Auroral E Region Instabilities*

The equatorial *E* region instabilities have been divided in two classes from the radar data. Recent *in-situ* rocket measurements made in India (Prakash *et al.*, 1971) are consistent with the existence of two classes of irregularities. Class 1 which corresponds to narrow spectra of the backscatter return propagates perpendicular to the magnetic field at the ion acoustic velocity and presents a very flat wave number spectrum; it is attributed to the two stream instability (Farley, 1963) associated with the equatorial electrojet. Class 2 corresponds to broader frequency spectra and the amplitude decays rapidly with decreasing wavelength; it seems to travel at the drift velocity of the electrons and is easily excited for wavelengths longer than a few meters even for an electron drift velocity below the ion-acoustic speed, i.e., below the two stream instability threshold (Farley and Balsley, 1973); it is customarily associated with the gradient-drift instability. The 'gradient-drift' plasma instability, arising in the presence of an ionization gradient perpendicular to the current flow, seems quite capable of generating long wavelengths (tens of meters), but not short wavelengths. However, when the large-scale irregularities become strong, they alter the local parameters within the electrojet sufficiently to fulfill the conditions for instability at small wavelengths (Sudan *et al.*, 1973). Several experimental properties of the irregularities of classes 1 and 2 are still awaiting a satisfactory explanation.

Part of the discrepancy could be due to the over-simplification of the models of instabilities which are used and the final understanding may well require an elaborate computer simulation (Farley, 1974). Farley and Fejer (1975) have thus suggested that the gradient drift term might play a role even in type 1 irregularities, changing the value of the threshold of the two stream instability.

The auroral *E* region irregularities might be due to irregular beams of ionizing precipitated energetic particles. This explanation might be valid on some rare occasions, but almost all the observations are consistent with some sort of plasma instability (Unwin and Baggaley, 1972). The two-stream instability explains (Gadsden, 1967) the behaviour of certain types of radio aurorae, characterized as being diffuse.

Some *in-situ* simultaneous observations of electric fields and plasma density fluctuations are also consistent with the two stream instability (Kelley and Mozer, 1973). Other types of radio aurorae characterized as discrete can be explained by the gradient-drift instability, which predicts, at least quantitatively, their observed behavior (Unwin and Baggaley, 1972). The two classes of phenomena discovered in the equatorial *E* region can also be identified in the auroral *E* region (Tsunoda, 1976) and this is in agreement with the suggestion that the two same instabilities are basically responsible for the observed irregularities. Thus, the use of the word aurora in the context of radio aurora does not imply that this phenomenon is identical to optical aurora, although both phenomena are sometimes observed simultaneously.

7.2.3.3. *Spread F*

The predominant feature (Herman, 1966) in geographic distribution of spread *F* occurrence is the existence of two areas of maxima (Figure VII.10): one at the equatorial

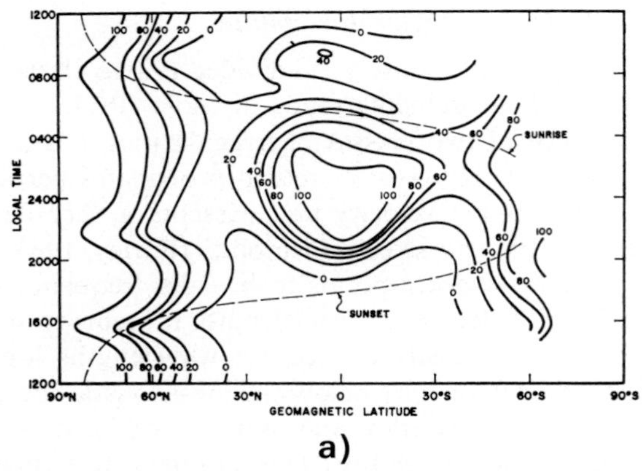

Fig. VII.10 (a). Map of the frequency of occurrence of spread *F* (in %) from the Alouette topside sounder (Calvert and Schmidt, 1964). Bottomside sounding had revealed the nighttime equatorial maximum, but not the permanent polar.

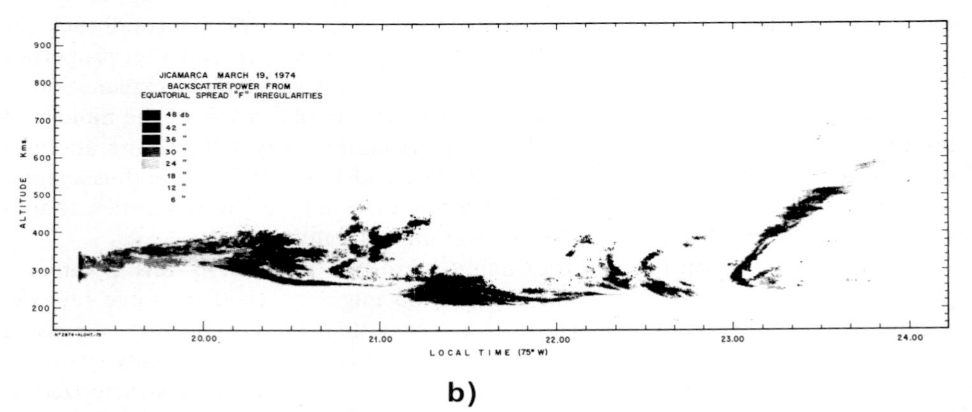

Fig. VII.10 (b). The amplitude of the coherent echoes observed at the Jicamarca incoherent scatter station, plotted in grey scale in an altitude–time diagram shows the evolution of the structure of ionospheric irregularities responsible for equatorial spread *F* (courtesy of R. F. Woodman).

region, but during nighttime only, the other at polar latitudes. Very little spread *F* is observed in the region between 20° and 40° geomagnetic latitude.

The equatorial *F* region irregularities have been extensively observed by using not only ionosondes, but also V.H.F. radars (Figure VII.10), scintillation and *in situ* measurements. Their spatial power spectrum extends from a few meters to several kilometers and varies more or less as k^{-2}, k being the wave number $2\pi/\lambda$. These irregularities are highly elongated along the magnetic field and they are observed only at night and occur over a wide range of altitudes, from 250 km to above 1000 km. They appear only when the equatorial *F* region is high, i.e. the bottom of the layer must be above 300 km. An intriguing association of the irregularities with the presence of Fe^+ ions has been reported by Hanson and Sanatani (1971). It has been suggested (Hanson *et al.*, 1973) that the conductivity structure caused by the long-lived metallic ions leads to large-scale (> 10 km) irregularities and sets up gradients in the electric field and/or plasma pressure that drive other instabilities to produce smaller-scale irregularities.

Farley *et al.* (1970), have concluded that backscatter observations of equatorial spread F are inconsistent with the most popular theories of the origin of the irregularities. A recent theoretical work by Hudson and Kennel (1975) concludes that both the collisional drift mode and the interchange (Rayleigh-Taylor) mode contribute to the electron concentration irregularity spectrum, the Rayleigh-Taylor mode dominating at long perpendicular wavelengths, and the drift mode at short perpendicular wavelengths.

The theoretical interpretation of the equatorial F region irregularities is far from being completed and some of their properties are still waiting for a convincing explanation (Farley, 1974).

At temperature latitudes, spread F is not a frequent phenomenon and occurs mainly during nighttime (Figure VII.10). Near the dip pole spread F occurs almost 100% of the time both day and night during winter. During summer, spread F occurrence frequency remains very high at night but becomes 50 to 60% around noon. Shimazaki (1962) has established that the high-latitude spread F phenomena have a positive correlation with auroral activity, although it appears to exist even during periods of weak auroral activity.

The F region irregularities are also observed by H.F. backscatter (4 to 40 MHZ) (Cornelius, *et al.*, 1975), sometimes simultaneously with satellite scintillation (Möller and Tauriainen, 1975). Using simultaneous particle and plasmapause measurements from a polar orbiting satellite, Pilkington *et al.* (1975), have identified two main regions of irregularities, one coinciding with the position of the average auroral oval, and the other lying close to the plasmapause.

Direct ionization from the precipitated electrons in the high-latitude regions has been quoted as a possible cause of the observed irregularities. Plasma instabilities are likely playing a major role on most occasions, although the discussion on their exact nature is still going on. Perkins (1973) has suggested a mechanism which might cause temperate and high-latitude spread F: the equilibrium in which the nighttime F region is supported by $\vec{\mathbf{E}} \times \vec{\mathbf{B}}$ drifts is unstable if, in addition to the supporting eastward field, a north-south electric field component exists.

7.3. The Distribution of Hydrogen Ions and the Upper Ionosphere

In the upper F region (i.e. significantly above the $F2$ peak), production and recombination rates are negligibly small, and the behavior of the ionospheric plasma is entirely controlled by transport processes.

7.3.1. AMBIPOLAR DIFFUSION IN THE GRAVITATIONAL FIELD

We have seen in Chapter I that in the case of neutral species, gravitational equilibrium implies an exponential decrease in density with altitude, the scale height for each species above the turbopause being inversely proportional to its mass. The case of electrons and ions is quite different, because of the polarization space charges which tend to prevent any charge separation: although the mass difference of the two species is extreme, their concentration must be equal at all points, so they must have the same scale height. Indeed when charge neutrality is violated, electrostatic forces and the corresponding energies are very large, and even a minute deviation from neutrality induces an electric field which wipes out this deviation. In studying the simultaneous diffusion, called 'ambipolar', of electrons and ions in the gravitational fields, we therefore assume that $n_e = n_i$ and that the tiny differences that in fact exists between n_i and n_e are felt as a polarization field $\vec{\mathbf{E}}_p$, which materializes the coupling. We also assume provisionally that the neutral atmosphere has zero bulk velocity, and neglect all other electric fields. Then, if $\vec{\mathbf{g}}$ is the gravity, the equations of transport for ions and electrons are, in the case where only one ion species is present (cf. Section 1.2)

$$\frac{1}{n_i}\vec{\nabla}(n_i kT_i) = +m_i\vec{\mathbf{g}} + e(\vec{\mathbf{E}}_p + \vec{\mathbf{V}}_i \times \vec{\mathbf{B}}) - m_i\nu_{in}\vec{\mathbf{V}}_i - m_i\nu_{ie}(\vec{\mathbf{V}}_i - \vec{\mathbf{V}}_e) \quad \text{(VII.12)}$$

$$\frac{1}{n_e}\vec{\nabla}(n_e kT_e) = +m_e\vec{\mathbf{g}} - e(\vec{\mathbf{E}}_p + \vec{\mathbf{V}}_e \times \vec{\mathbf{B}}) - m_e\nu_{en}\vec{\mathbf{V}}_e - m_e\nu_{ei}(\vec{\mathbf{V}}_e - \vec{\mathbf{V}}_i).$$

The term $m_i\nu_{ie}(\vec{\mathbf{V}}_i - \vec{\mathbf{V}}_e)$ is equal to $-m_e\nu_{ei}(\vec{\mathbf{V}}_e - \vec{\mathbf{V}}_i)$ since the momentum per cm^3 per sec given by the ions to the electrons is the same as that received by the electrons from the ions.

We already know that the terrestrial magnetic field considerably limits velocities in perpendicular directions, and that $V_\perp$ is much less than $V_{//}$. Therefore, we are mostly interested in the projection of Equation (VII.12) along the field direction, whose inclination with respect to the horizontal we call I:

$$\frac{1}{n_i}\nabla_{//}(n_i kT_i) = -m_i \sin I\, g + eE_p - m_i\nu_{in}V_{i//} - m_i\nu_{ie}(V_{i//} - V_{e//})$$

$$\frac{1}{n_e}\nabla_{//}(n_e kT_e) = -m_e \sin I\, g - eE_p - m_e\nu_{en}V_{e//} - m_e\nu_{ei}(V_{e//} - V_{i//}).$$

Taking the sum of these equations eliminates the polarization field, and also the opposing terms in ν_{ei} and ν_{ie}. There remains, since $n_e = n_i$,

$$\frac{1}{n_e}\nabla_{//}n_e k(T_e + T_i) = -(m_i + m_e)\, g \sin I - m_i\nu_{in}\, V_{i//} - m_e\nu_{en}V_{e//}.$$

The impossibility of accumulating charge in the magnetosphere and ionosphere implies that $V_{e_{//}}$ cannot be much greater than $V_{i_{//}}$. Since furthermore $m_e\nu_{en}$ is much smaller than $m_i\nu_{in}$ and m_i is much greater than m_e, the above equation simplifies into

$$\frac{1}{n_e}\nabla_{//}n_e k(T_e + T_i) = -m_i g \sin I - m_i\nu_{in}\, V_{i//}.$$

If the ionosphere is horizontally stratified, which is true at least to first approximation, then

$$\nabla_{//}n_e k(T_e + T_i) = \sin I\, \mathrm{d}/\mathrm{d}z\, n_e k(T_e + T_i)$$

z being the vertical coordinate. Thus

$$V_{i//} = (-\sin I/m_i\nu_{in})\left[\frac{1}{n_e}\mathrm{d}/\mathrm{d}z\left(n_e(T_e + T_i)\right) + m_i g\right]. \quad \text{(VII.13)}$$

If ν_{in} approaches zero, $V_{i_{//}}$ can remain finite only if the term in brackets vanishes. This defines a distribution of charged particles as a function of altitude that is known as diffusive equilibrium:

$$\frac{1}{n_e}\frac{d}{dz}[n_e k(T_e + T_i)] + m_i g = 0. \tag{VII.14}$$

If we compare this equation with the hydrostatic equilibrium of a neutral species of the same atomic mass, which is

$$\frac{1}{n_n} d(n_n k T_n)/dz + m_i g = 0. \tag{VII.15}$$

T_n and n_n being the neutrals temperature and numerical density, we see that ambipolar diffusive equilibrium is equivalent to that of a neutral species of the ion atomic mass, and temperature $T_e + T_i$. In particular, when at night $T_e = T_i = T_n$, independent of altitude in the regions under consideration, the scale height of O is kT/mg, and that of O^+ ions $2kT/mg$, twice as much.

We have established, under Section 1.4.2.3, that the coupling with neutral motions modifies the diffusion velocity of the ions, according to the relationship:

$$\vec{V}_i = \vec{V}_{io}/\sin^2 I.$$

We note that if V_{io} is given by Equation (VII.13), the equilibrium velocity V_i is:

$$V_i = -(m_i \nu_{in} \sin I)^{-1}\left[\frac{1}{n_e} d/dz\left\{n_e k(T_e + T_i)\right\} + m_i g\right].$$

Its vertical projection therefore becomes:

$$-(m_i \nu_{in})^{-1}\left[\frac{1}{n_e} d/dz\left\{n_e k(T_e + T_i)\right\} + m_i g\right].$$

i.e. the value that it would have in the absence of the magnetic field, or in the presence of a vertical magnetic field.

7.3.2. IONIC COMPOSITION AT THE BASE OF THE PLASMASPHERE

Equations (VII.12) were written down under the assumption that only one type of ion exists. There is no difficulty in complicating the system of equations by adding as many equations as there are ion species: (NO^+, O_2^+, O^+, He^+, H^+).

Here we shall only write the equilibrium equations that are obtained by letting the collision frequencies approach zero. We assume that all the ions have the same temperature T_i, the mass and density of an ion species j being m_j and n_j, and those of the electrons m_e and n_e. We can then write:

$$d/dz\,(n_j k T_j) = -n_j m_j g + n_j e E. \tag{VII.16a}$$

$$d/dz\,(n_e k T_e) = -n_e m_e g - n_e e E. \tag{VII.16b}$$

Through summation of these equations and the use of overall electrical neutrality $\Sigma n_j = n_e$ we obtain:

$$d/dz\,(n_e k(T_e + T_i)) = -\Sigma_j n_j m_j g - n_e m_e g.$$

where the term $n_e m_e g$ is negligible. Using the definition of mean mass

$$m^+ = \Sigma_j n_j m_j / \Sigma_j n_j = \Sigma_j n_j m_j / n_e$$

we can write the equation as

$$d/dz[n_e k(T_e + T_i)] = -n_e m^+ g$$

which is the same as the equation of diffusive equilibrium for a single species, except that m_i has been replaced by the mean mass m^+.

Under the simple hypotheses, valid at night, that $T_e = T_i = T_n$ and that $dT/dz = 0$, this equation becomes

$$dn_e/dz = -(m^+ g/2kT)n_e.$$

and Equation (VII.16b) becomes:

$$-1/2\, m^+ g n_e = -n_e eE - n_e m_e g.$$

Since $n_e m_e g$ is negligible, we have

$$eE = 1/2 m^+ g.$$

Putting this value into Equation (VII.16a), we obtain

$$d/dz(n_j kT) = -n_j(m_j - m^+/2)g,$$

or

$$dn_j/dz = -(m_j - m^+/2)\, g n_j/kT. \qquad \text{(VII.17)}$$

For a light ion such as H^+ in a plasma where O^+ ions predominate, $m_j - m^+/2 < 0$. Consequently the concentration of H^+ ions is an increasing function of altitude until the point where the mean atomic mass of the ions present decreases to the value 2. We can therefore speak of a layer of H^+ that 'floats' upon the layer of O^+, thanks to the field of ambipolar diffusion.

If we define $H = kT/m_p g$, where m_p is the proton mass, we can write Equation (VII.17) for the concentration $n(H^+)$ of H^+ ions, in the regions where O^+ ions are in the majority ($m^+ = 16$), as

$$dn(H^+)/dz = (7/H)n(H^+)$$

from which

$$n(H^+) = n(H^+)_{z_0}\, e^{7(z - z_0)/H}.$$

This equation is based on a diffusive equilibrium without chemical reactions. However we have seen that we must take into account the charge exchange reaction $H^+ + O \leftrightarrow H + O^+$ which can be very rapid and tends to establish at low altitudes a chemical equilibrium in which the concentrations are related by $n(H^+)n(O) = 9/8 n(H)n(O^+)$.

If we make the hypothesis that all temperatures are equal and independent of altitude, we can write:

$$n(H) = n(H)_{z_0} e^{-(z - z_0)/H}$$

$$n(O) = n(O)_{z_0} e^{-16(z - z_0)/H}.$$

Now we have seen that if O^+ ions predominate, they are distributed according to the law

$$n(O^+) = n(O^+)_{z_0} e^{-8(z - z_0)/H}.$$

Then we find for the concentration of H ions

$$n(H^+) = \frac{9}{8} \frac{n(H^+)_{z_0} e^{-(z - z_0)/H} n(O^+)_{z_0} e^{-8(z - z_0)/H}}{n(O)_{z_0} e^{-16(z - z_0)/H}}$$

so that

$$n(H^+) = n(H^+)_{z_0} e^{7(z - z_0)/H}.$$

This formula turns out to be identical with the one for diffusive equilibrium previously derived. Thus the distribution of H^+ with altitude is the same no matter whether diffusive equilibrium or chemical equilibrium prevails.

Nonetheless, it is necessary to distinguish between chemical and transport processes if one is interested in dynamic processes, such as the flux of protons that can leave the ionosphere during the day and fill the plasmaspheric tubes of force, thus forming a reservoir which can contribute to the maintenance of the ionosphere at night. The diffusion of protons is then limited by $H^+ - O^+$ Coulomb collisions, and the collision frequency $\nu_{H^+O^+}$ must be included in a system of equations that generalizes the system we have been considering. Hanson and Ortenburger (1961) demonstrated that these collisions are sufficiently effective to create a diffusive barrier that keeps the protonosphere from responding to the short term variations of the ionosphere, and which shows the accuracy of the image of a proton layer floating on the layer of O^+ ions without mixing with them.

7.3.3. THE POLAR WIND

The dynamic aspect we have mentioned is essential for the phenomenon named the 'polar wind' by Axford (1968) and analyzed theoretically by Banks and Holzer (1968, 1969 a,b). In the neighborhood of the poles, where the lines of force of the magnetic field open towards the tail of the magnetosphere and the low density regions of the interplanetary medium, the equilibrium state we have described no longer exists. The hydrodynamic equations also allow solutions in which the flux at large distance retains a finite value, a physically acceptable condition for the tubes of force in the polar regions.

In the lower part of the upper F region, O^+ ions dominate over the lighter ions: H^+ and He^+ ions are accelerated upward by the O^+ electrons ambipolar diffusion electric fields. Theoretical descriptions of the phenomenon involve either an hydrodynamic approach (Banks and Holzer 1968, 1969a,b) or a kinetic approach (Lemaire and Scherer, 1973). H^+ velocities are estimated to be around 1 km s^{-1} at 1000 km and reach 10 km s^{-1} in the 2000–4000 km altitude range. As a result of the high speed of escape of H^+, O^+ remains the dominant ion up to at least 4000 km and correlatively high-altitude electron concentrations are reduced below their mid-latitude level because of the smaller scale height associated with O^+ dominance (Figure VII.11).

Even outside the polar cap, geomagnetic tubes of force are being convected in such a way that some sort of polar wind also plays a role down the low L values

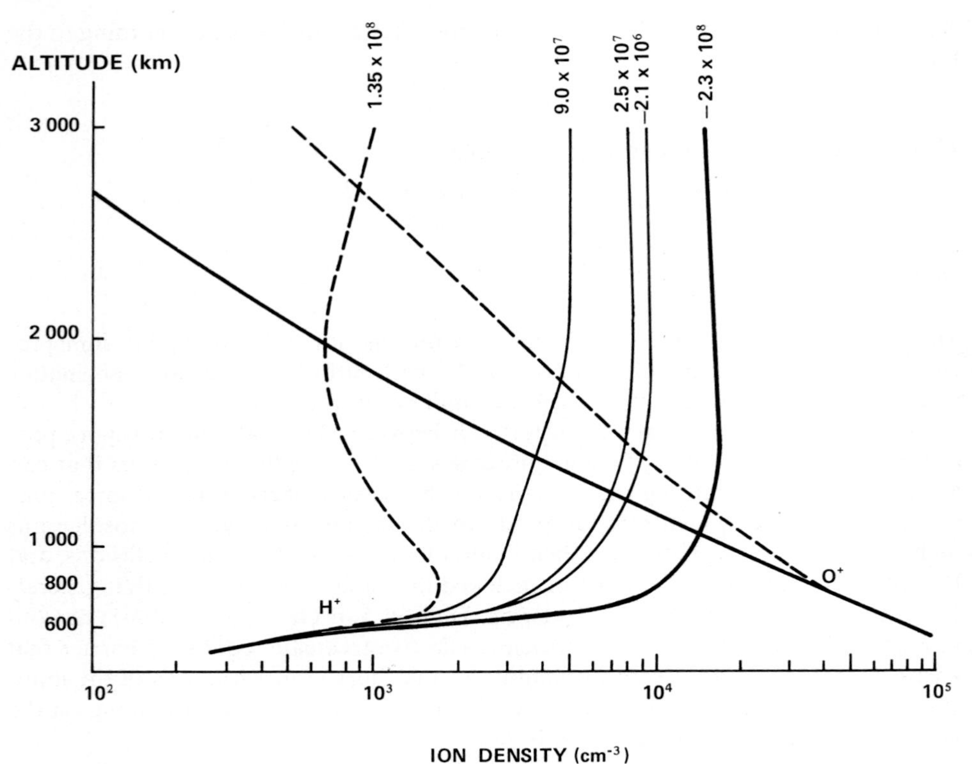

Fig. VII.11. The effect of vertical plasma flows on the composition of the upper ionosphere. Model parameters are the following: exospheric temperature is 1000°K; H^+ is in chemical equilibrium at 500 km, where O^+ has a concentration of 1.38×10^5 cm^{-3} and H 1.92×10^5 cm^{-3}. The H^+ concentration profiles are computed for the different vertical fluxes shown (in cm^{-3}), a negative sign indicating a downward flux. Zero flux corresponds to ambipolar diffusive equilibrium. The two profiles of O^+ concentration shown correspond to the two extreme profiles for H^+, the smaller values for O^+ being obtained for the higher H^+ concentration: the scale height of O^+ ions, when they become minor ions in a proton plasma, is halved. This is how the 'polar wind' explains the latitudinal distribution of topside ions shown for instance in Figure V.8 (Banks and Holzer, 1968).

corresponding to the plasmasphere. The relationship between the polar wind phenomenon and the plasmapause and plasmatrough formation has been reviewed by Blanc (1975) from which we borrow the following description. Over the polar cap, where field tubes are open, the polar wind prevails as a steady state, and is associated with plasma escape. Along closed convecting field tubes, high-speed replenishment prevails due to a very low plasma pressure along the tube, and should be followed by low-speed replenishment after about 1 day if the field tube were not opened again by convection before that delay. As a result, a permanent state of high-speed transport, associated with all features of plasma velocity and density distributions found for the polar wind, prevails along those convecting field lines. Along closed corotating field lines, tubes of the plasmasphere, as the velocity at the apex must be zero, diffusive equilibrium is the only steady flow solution. Apart from a possible influence of the 'convective interchange instability', recently investigated by Richmond (1973) and

Lemaire (1974), this state should prevail for those field lines during periods of long magnetic quiet. But under the effect of magnetic disturbances, the magnetic field geometry and the global instantaneous pattern of perpendicular plasma flow are modified, so that previously corotating field lines become convecting, and lose their plasma when reaching the magnetopause and drifting over the polar cap. Banks *et al.* (1971), using a theoretical polar wind model, studied the consequence of such an event on the dynamic behavior of the corotating field lines after the storm recovery. Essential features of their model are shown on Figure VII.12: between the new convection boundary, associated with quiet magnetic conditions, and the one which

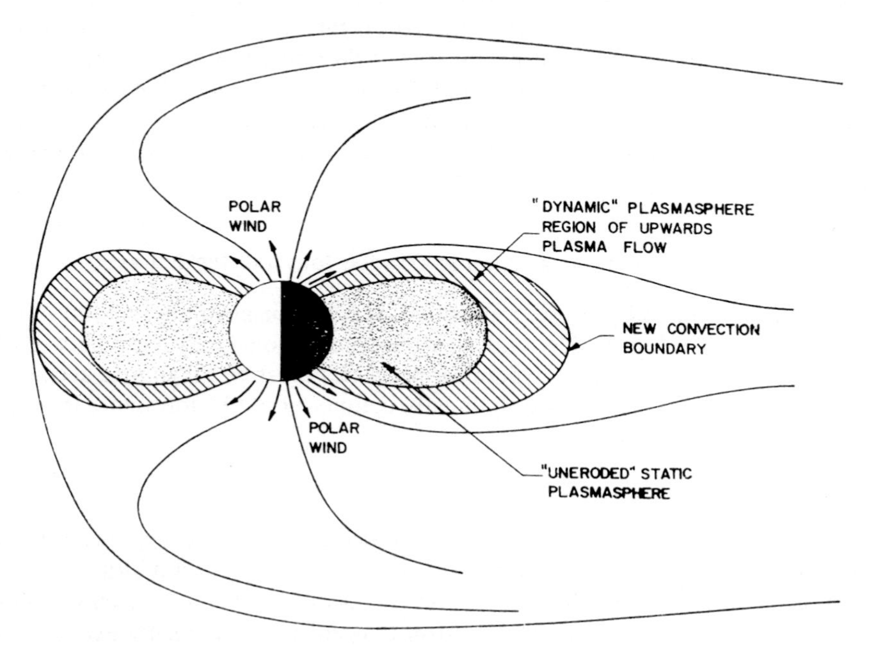

Fig. VII.12. Sketch of the refilling of the outer plasmasphere following a contraction of the plasmapause due to an enhanced convection in the magnetosphere.

prevailed during the storm, a 'dynamic' region of upward plasma flow is found, where field tubes are replenished by parallel transport from the ionosphere. Following Banks' hydrodynamic approach, this flow is first a high-speed one, connected to a region with a zero bulk motion around the apex through formation of a bow shock that progresses downstream and finally reaches the critical point of the flow. Then low-speed flow prevails, and finally the steady state of diffusive equilibrium is recovered. The total recovery time, increasing with increasing L shells, is about 4 to 5 days at the quiet time convection-corotation boundary. As a consequence of frequent occurrence of magnetic storms, as first suggested by Park (1970) from radio whistler studies, the outer part of the quiet time corotating region might be in a continual state of replenishment.

Thus, far from being restricted to the polar cap, static and dynamic features of plasma distribution associated with the polar wind should prevail over the whole convecting region, and even under the effect of magnetic activity, be a transient state in the outermost part of the corotating region. Plasma distribution associated with quasi-diffusive equilibrium would then only be found in an inner part of the corotating region, whose extension would depend on the past history of magnetic activity. The light ion trough (Chapter III, Figure III.4) is attributed to this effect. The long recovery times involved explain that one to one correlation is not always observed between the plasmatrough and the plasmapause locations.

Wherever the polar wind takes place, it constitutes a sink for the light neutral atmospheric constituents which keep being ionized by solar radiation and ion-molecule reactions. The flux of H that escapes in this way is estimated at some 10^9 $cm^{-2}\ s^{-1}$ and that of He at some $10^6\ cm^{-2}\ s^{-1}$. We note that the escape of He has the effect of balancing the Earth's global He budget (Axford, 1968) which, in its absence would bear a non-existent profit, since the thermal escape of He is negligible, while the production through radiative disintegration at ground level is estimated at $10^6\ cm^{-2}\ s^{-1}$.

7.4. The Behaviour of Atomic Oxygen Ions and the *F* Layer

The electron concentration in the ionosphere reaches an absolute maximum at an altitude of about 300 km, which, under extreme conditions, may vary between 200 and 450 km. Both chemical and dynamical processes must be taken into account for explaining the behavior of this region, mostly composed of O^+ ions, and, in the first place, the existence of this '*F* peak'.

7.4.1. VERTICAL STRUCTURE OF THE *F* REGION

Diffusive equilibrium, which prevails in the upper *F* region, implies (see Equation (III.14)) that the electron concentration decreases exponentially with altitude. On the other hand, we have seen in Chapter VI that chemical equilibrium, which prevails in the lower *F* region, implies that the electron concentration increases exponentially with altitude like that of O^+ ions. The existence and the variations of the maximum in electron concentration result directly from these two competitive processes, in the intermediate region where chemistry and transport are both effective.

7.4.1.1. *Formation of the F2 peak*

The general budget describing the electron concentration (cf. Chapter VI) is:

$$dn_e/dt = Q - \beta n_e - \frac{d}{dz}(n_e V_z) \qquad \text{(VII.18)}$$

under the assumption that the ionosphere is horizontally stratified, where, we recall, $Q(z)$ is the electron creation rate from photo or impact ionization, which is proportional to the concentration of the dominant neutral constituent, O, and $\beta(z)$ is the effective loss frequency of electrons, which is proportional to the concentration of the minor molecular neutrals N_2 and O_2 because electrons recombine with molecular ions NO^+ and O_2^+ created by the ion-molecule reaction of O^+ with N_2 and O_2. In the absence of winds and electric fields, the influence of which will be discussed later on, the velocity V_z is just the vertical component of the diffusion velocity. From Equation (VII.13), V_z is expressed as:

$$V_z = -(\sin^2 I/m_i \nu_{in})\left[\frac{1}{n_e}\frac{d}{d_z} n_e k(T_e + T_i) + m_i g\right]$$

which can be written

$$V_z = -D \sin^2 I \left[\frac{1}{n_e} dn_e/dz + \frac{1}{H_i}\right] \tag{VII.19}$$

where $D = k(T_e + T_i)/m_i \nu_i$ is the ambipolar diffusion coefficient and H_i the ambipolar scale height, defined by Equation (VII.14):

$$\frac{1}{H_i} = \frac{m_i g}{k(T_e + T_i)} + \frac{1}{T_e + T_i}\frac{d(T_e + T_i)}{dz}.$$

The steady-state equilibrium $dn_e/dt = 0$ corresponds to an electron concentration profile defined by the equation

$$Q - \beta n_e + \frac{d}{dz}\left[n_e D \sin^2 I\left(\frac{1}{n_e} dn_e/dz + \frac{1}{H_i}\right)\right] = 0. \tag{VII.20}$$

If H is the scale height of O, Q varies with height as $\exp(-z/H)$ and β as $\exp(-2z/H)$, the scale height of the molecular constituents (O_2 and N_2) being close to $H/2$. D varies essentially as $1/\nu_i$ i.e. as $\exp(z/H)$. Equation (VII.20) is thus consistent with both chemical equilibrium $Q = \beta n_e$ at low altitude and with diffusive equilibrium as defined by Equation (VII.14) at high altitude.

Figure VII.13d shows an observed profile of $\frac{d}{dz}(n_e V_z)$, which decreases exponentially above 300 km, with the same scale height as the concentration n(O). This result is consistent with the fact that recombination is negligible in this region and that production (proportional to n(O)) is just balanced by flux divergence. The production rate has been multiplied by an arbitrary constant factor to reach the agreement shown on Figure VII.13d. Below 300 km, the difference between production and flux divergence is a measure of the recombination rate which varies, as theoretically expected, with a scale height equal to that of N_2.

A numerical solution of Equation (VII.20) shows that the ionospheric peak occurs at an altitude where diffusion and chemical frequencies are equal, i.e. where β is of the order of $D \sin^2 I/H^2$ (Figure VII.14). This approximate relation in fact defines the altitude of the maximum in the layer with precision (Rishbeth and Garriott, 1969) because β and $1/D$ are both exponentially decreasing functions of altitude like neutral density. As for the value of the electron concentration at the maximum, during the day, it is nearly equal to the ratio of creation rate to chemical loss frequency at the altitude z_m of the maximum, Q_m/β_m. In other words, the value of n_e at the altitude z_m is that which would occur in the absence of diffusion. Similarly, at night, the decay should take place with dependence $\exp^{-(\beta_m t)}$.

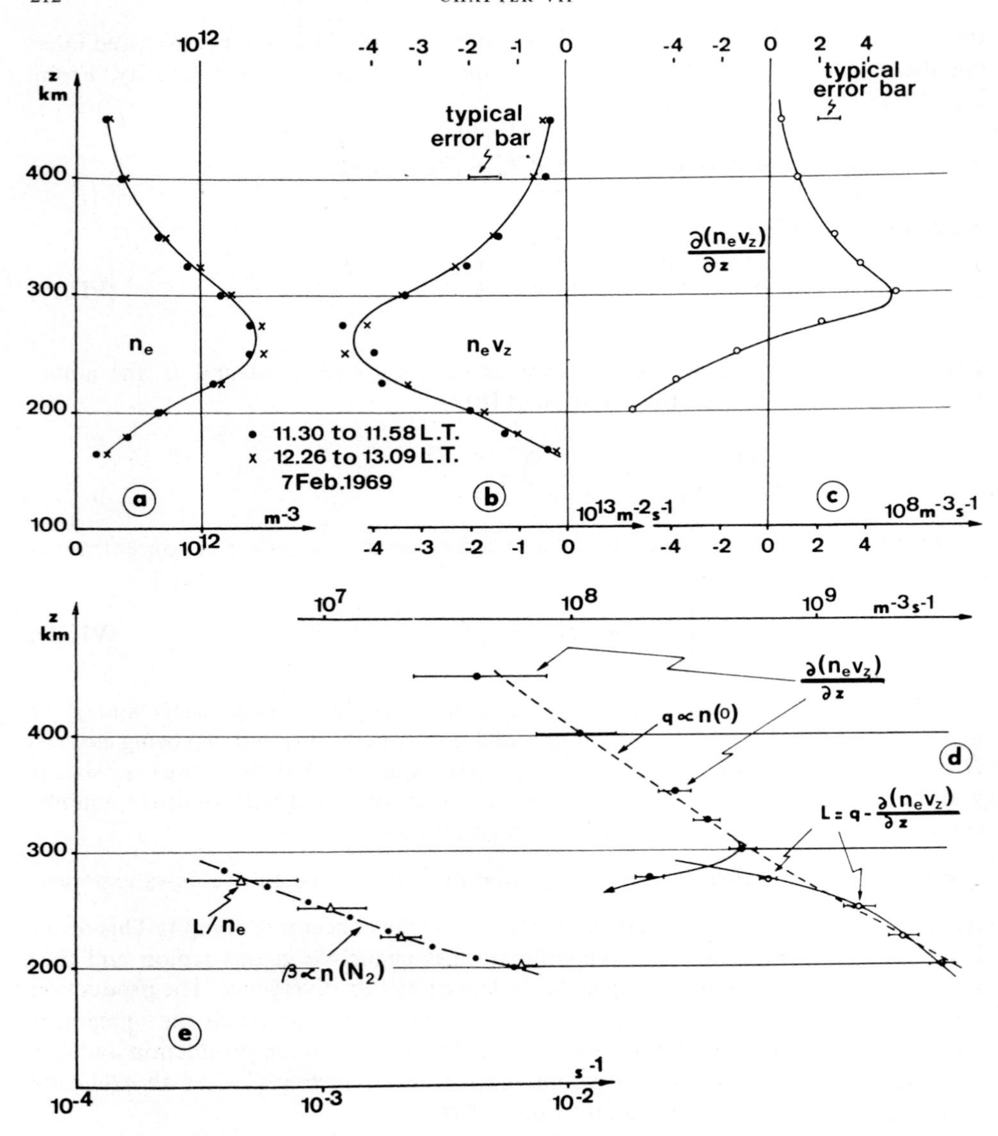

Fig. VII.13. The height profiles of the different terms in the continuity equation of electrons (Equation (VII. 18)), from incoherent scatter data (Vasseur, 1971) for electron concentration n_e and vertical drift v_z: (a), (b); (c) gives the experimental $d/dz\ (n_e\ v_z)$ linear scale; (d) full circles with error bars represent the same data as in (c), the horizontal scale being logarithmic. The dashed line is a fit for the production rate with the O scale height. Below 300 km, the difference between the extrapolated production rate and the experimental flux divergence gives the chemical loss rate $L = \beta n_e$ (open circles); (e) triangles with error bars represent the chemical loss frequency β derived from L by dividing with the experimental values for n_e. Dashed line is a fit with the N_2 scale height.

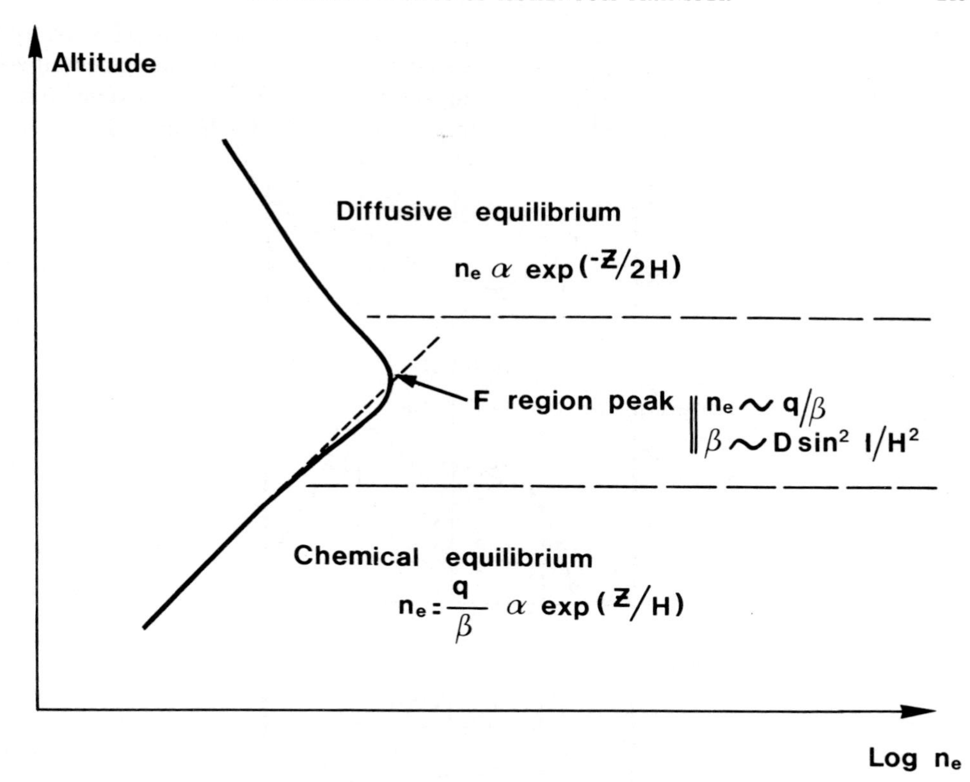

Fig. VII.14. The $F2$ peak (cf. text).

7.4.1.2. *Effects of Neutral Winds and Electric Fields*

An atmospheric wind or an electric field causes an incremental vertical velocity ΔV_z of the plasma, which must be added to the vertical component of the diffusion velocity.

The essential effect on plasma distribution is a change in the altitude of the maximum by a shift of the order of $\Delta V_z H^2 m_i \, v_i / k(T_e + T_i) = \Delta V_z H^2/D \sin^2 I$. As for the value of the concentration at the maximum, the rules edicted in the absence of drift are still valid, that is to say, it will vary as q_m/β_m, increasing as the peak region is lifted up, and decreasing as it dips down into the molecular neutral layer.

Electric fields are thought to have only a minor influence on the behavior of the $F2$ peak at mid-latitudes (Stubbe and Chandra, 1970; Bramley and Rüster, 1971).

Winds, i.e. large scale motions of the neutral air, can be estimated from horizontal pressure gradients deduced from static neutral atmosphere models (Fig. I.8). Poleward winds (daytime) reduce the altitude z_m of the peak and consequently decrease the maximum electron concentration. Equatorward winds (nighttime) increase z_m and n_{e_m}. Winds and electron concentration profiles have to be computed in a self-consistent procedure because of ion drag.

Taking winds into account can thus explain the diurnal variations (Rishbeth, 1972) (see Figure VII.15), and the seasonal variations of *foF*2, which are essentially due to the varying length of the day affecting the wind pattern, to the extent that thermospheric composition changes are themselves included (cf. Chapter VI).

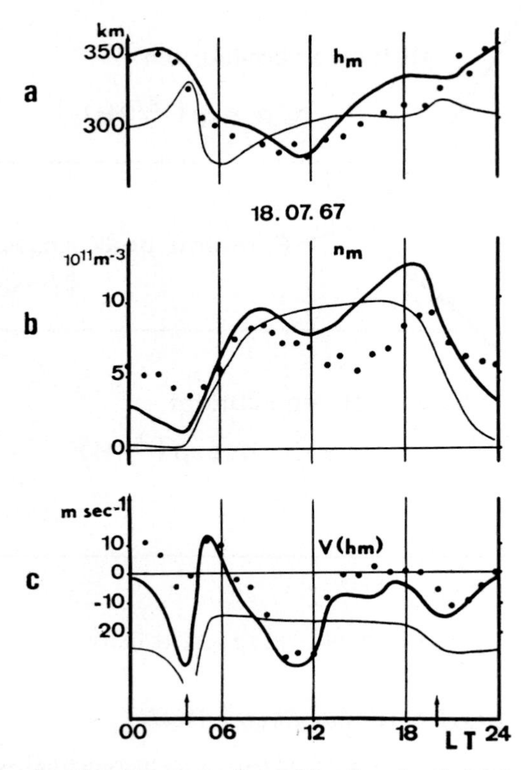

Fig. VII.15. Dots are incoherent scatter observations of the diurnal variation (on July 18, 1967) of height h_m, electron concentration n_m, and vertical ion drift velocity V at the $F2$ peak. Curves are model computations neglecting (thin lines) and including (thick lines) the effects of drifts. Arrows on the local time abscissa indicate sunrise and sunset at ground level (Vasseur, 1970).

In models of the thermosphere, the variations versus latitude and local time of pressure are identical at all longitudes. Therefore the pressure gradient is the same at all longitudes. Although the ion drag term depends somewhat on longitude through the direction of $\vec{\mathbf{B}}$, the resulting longitude variations of wind velocity are small. However the vertical ion drift velocity V_z can be written:

$$V_z = |U| \cos(\theta - \delta) \sin I \cos I,$$

where δ and I are the declination and the inclination of the Earth's magnetic field and θ the angle between $\vec{\mathbf{U}}$, assumed horizontal, and geographic North. Therefore the phase of the diurnal variation of V_z depends on δ and its amplitude on I, when the $\vec{\mathbf{U}}$ vector rotates throughout the day. In this way, wind effects can account for the dependence of the *foF*2 on declination, discovered by Eyfrig (1963).

7.4.1.3. *Atmospheric Waves and Travelling Ionospheric Disturbances*

Many observations, using ionosondes, backscattering and total electron content measurements, had shown that during magnetic storms, large disturbances in electron density travel with sonic velocity from the pole towards the equator. Several theoreticians suspected that these phenomena were in fact the manifestation of large-scale gravity waves generated in the auroral region (Hines, 1960). This view has been confirmed by incoherent scatter data on ion velocity and temperature which are strongly coupled to the neutral velocity and temperature. Figure VII.16 shows an example of such gravity waves observed at Saint-Santin (Testud, 1973). Similar observations have been performed at Arecibo (Harper, 1971).

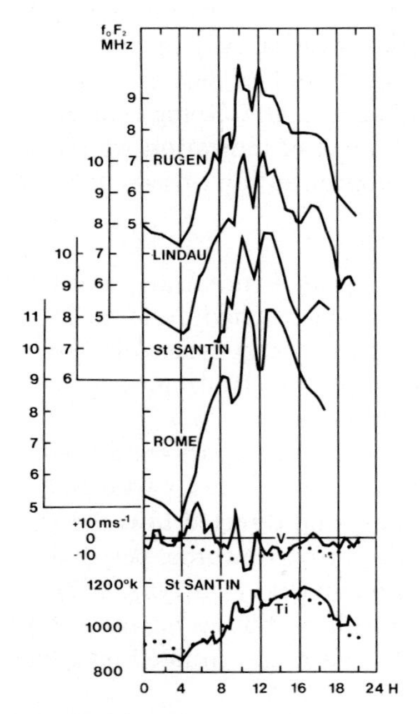

Fig. VII.16. A traveling ionospheric disturbance: four ionosondes from top to bottom in order of decreasing latitude, observe a similar oscillation of *F*2 critical frequency (peak electron concentration) with an increasing time delay. The incoherent scatter sounder in Saint Santin exhibits correlated oscillations of the ion vertical drift velocity and temperature (dots are a mean of undisturbed days observations).

Testud (1973) developed a calculation, including dissipative processes, to determine first the free wave spectrum launched by the auroral source, and then the propagation from the source region to the point of observation. According to this theory, the response of the atmosphere does not appear as a wave train but as a single pulse. This might seem to be in contradiction with the observations of Figure VII.16. However auroral magnetograms reveal that, in this particular occurrence, two events took place in the same longitudinal sector as Saint-Santin at 07.40 and 10.05 h. These events are linked with magnetic reversals of the interplanetary magnetic field observed by EXPLORER 33 and appear as a manifestation of a substorm growth phase. Thus the oscillations of Figure VII.16 can be interpreted as a succession of two positive impulses 70 and 65 min after the auroral events, a propagation time in good agreement with the theoretical value (75 min).

Larger period atmospheric waves have also been observed, and could be the result of excitation of global modes of oscillation (Volland and Mayr, 1972).

7.4.2. LATITUDINAL STRUCTURE OF THE *F* REGION

The geometry of the Earth's magnetic field superimposes latitudinal variations to the vertical structure discussed in the preceding section. Near the magnetic equator, the magnetic field lines are included in totality in the *F* layer and transport phenomena tend to move the plasma from the magnetic equator, where a minimum appears in the F_2 peak density, to two crests located 15° north and south: this is the equatorial anomaly. At higher geomagnetic latitudes (60°), the field lines are no longer always corotating with the Earth and may lose their plasma, causing the appearance of a plasma 'trough', most easily identified at night. Going to even higher latitudes, particle precipitation is an additional source of ionization which explains the abnormally high values of *F* region electron concentration observed in the auroral and polar regions.

7.4.2.1. *The Equatorial Anomaly*

The daytime value of the *F* peak concentration has a minimum at the magnetic equator in between two maxima situated at 15 to 20° North and South latitudes. This equatorial anomaly is particularly evident in the synoptic results from topside and bottomside sounding (cf. Figure III.4) as well as from optical airglow surveys (Appleton, 1946; Barbier and Glaume, 1962).

A probable explanation of the phenomenon, known as the 'equatorial fountain', involves the following motion of charged particles: the electric polarization field caused by the dynamo system and directed towards the East at the equator during the day causes the charges to move upwards, perpendicularly to the horizontal current and the terrestrial magnetic field (Martyn, 1947). The plasma thus transported to the upper equatorial ionosphere diffuses downward along the magnetic field lines and finally accumulates into two ridges situated on either side of the magnetic equator (Figure VII.17).

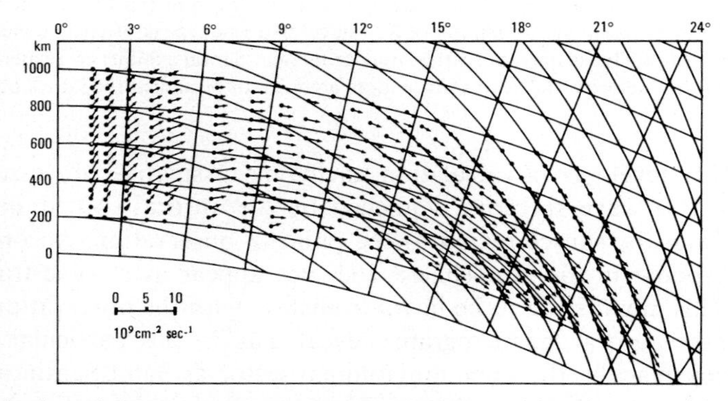

Fig. VII.17. The flux of electrons, in a magnetic meridian plane, driven by a zonal (eastward) electric field. Magnetic field lines shown are those which intersect the vertical at the equator every 200 km.

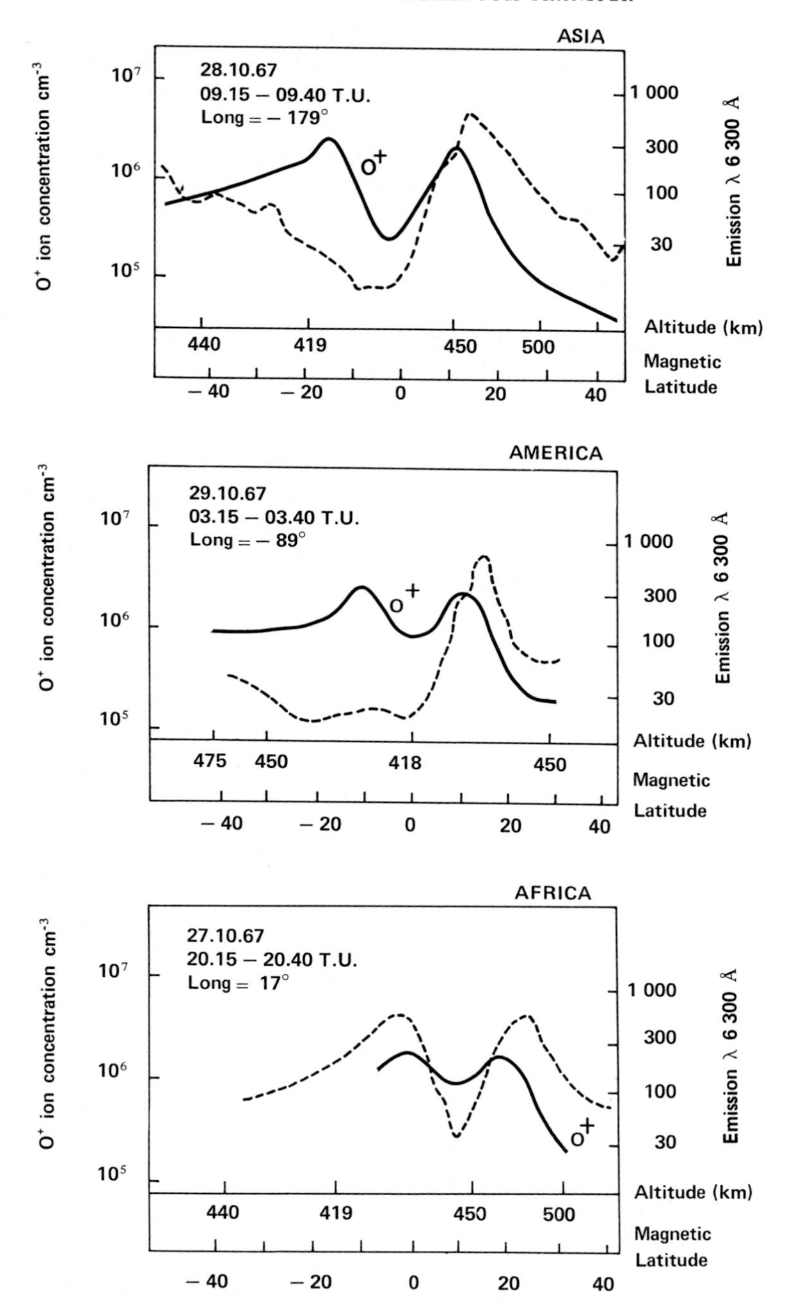

Fig. VII.18. Simultaneous observations by the satellite OGO 4 of the O^+ concentration (full line) and of the 6300 Å emission intensity. The two crests of the equatorial anomaly are clearly visible in O^+ concentration over three different continents. The 6300Å airglow due to dissociative electron recombination exhibits a similar pattern over Africa, but may also behave differently under the action of neutral winds (Asia and America).

The detection of meteoric ions, by satellite-borne mass spectrometers, at very high altitudes in low latitudes (Hanson and Sanatani, 1970) can only be explained by such a transport mechanism. As we shall see, meteoric ions produced near the equator cannot be gathered together and evacuated downward by wind shears (*Es*) as they are elsewhere; but on the other hand they are a tracer of the equatorial fountain.

Horizontal neutral winds affect the equatorial *F*2 layer. According to Bramley and Young (1968) a transequatorial wind, blowing in the North–South or South–North direction throughout the equatorial anomaly reduces both 'crests' of the anomaly. At the upwind crest, the height of the peak is increased. The corresponding decrease of β_m is overcompensated by the removal of ionization across the magnetic equator to the other hemisphere. At the downwind crest, the influx of ionization is outweighed by the effect of the reduction of the peak height, and the consequent increase of β_m caused by the downward drift. These changes of altitude of the peak strongly influence the geographical distribution of the 6300Å airglow due to dissociative electron recombination. The effect of winds thus explains that these intertropical arcs of the red line are not always coincident with the crests (Figure VII.18) (Thuillier, 1973) although the first observations had suggested a close correlation between the two phenomena. Further refinements in the explanation of the behavior of the equatorial ionosphere would require more detailed knowledge of the distribution of atmospheric pressure and electric fields.

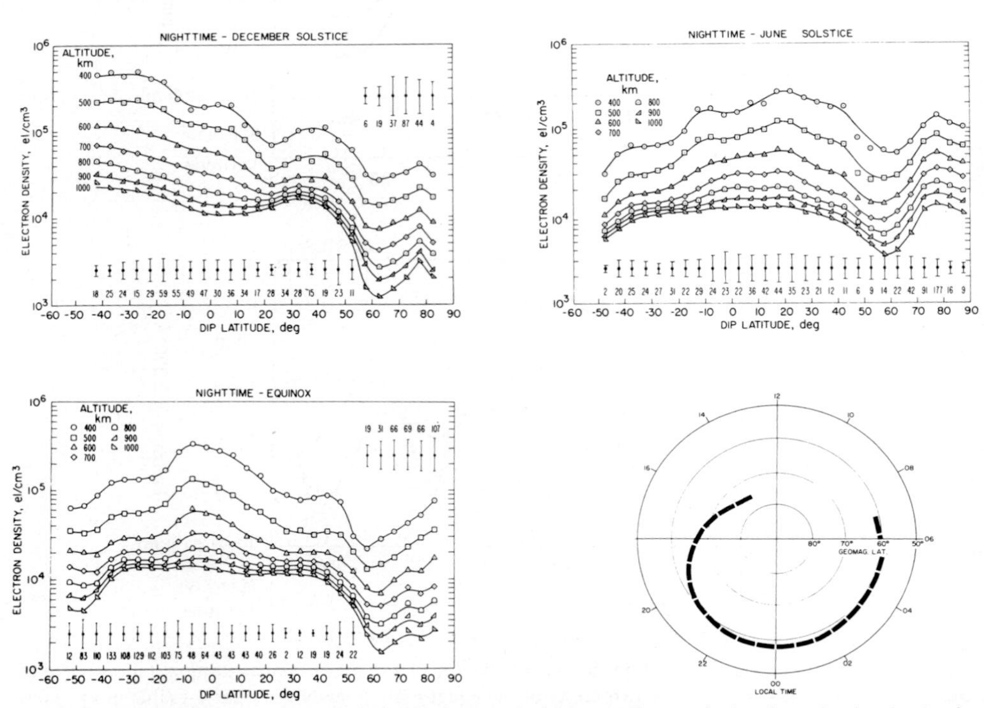

Fig. VII.19. Iso-density contours deduced from top-side sounding and showing clearly the 'main trough' during nighttime. The average location of this trough is indicated on the hemispherical map (lower right panel).

7.4.2.2. *The Mid-Latitude Trough*

Analysis of ground-based ionosonde data by Reber and Ellis (1956) revealed a nighttime minimum of the $F2$ peak concentration between 50° and 60° magnetic latitude. Muldrew (1965) found a corresponding dip in electron concentration of the topside F region. This minimum, referred to as 'the main trough', which exists in the upper ionosphere up to 1000 km, is most easily identified at night. Figure VII.19 illustrates the average location of the trough over Eastern Canada and the North Atlantic as a function of local time.

The average location of the trough and its variation with the index of magnetic activity K_p are found to be similar to those of the plasmapause as seen with satellites and whistler observations. The existence of the trough is evidently related to the flows of plasma discussed in Section 3.3, which take place along the convecting field lines. The time constants necessary for replenishment of the flux tubes are large enough to explain local discrepancies between the locations of the trough and the plasmapause.

7.4.2.3. *The Auroral and Polar F Layer*

Poleward of the trough, the electron concentration is frequently exhibiting a sharp increase (Figure VII.19). On occasions it can be quite irregular, with a ratio of maximum to minimum as large as 25 to 1. Such secondary peaks and troughs have a north–south extent of a few hundred kilometers.

Sometimes, at night, a strong E layer is observed, probably due to low energy electron precipitations. This 'night E' may have a peak value exceeding the peak value of F layer. Sometimes a valley is observed between the E and the F layers, but on occasions, the E layer becomes thick enough to fill in the valley so that the electron concentration decreases monotically versus altitude above the peak of the E layer.

It has been postulated that the ionization of the E layer by soft particles might be a significant source for the F layer, where the plasma is moved upward by some transport process, such as the expansion of the neutral atmosphere heated by precipitations (Bates *et al.*, 1973a). Such an explanation is consistent with the observation that the energetic electron boundary appears somewhat poleward of the positions reported for the plasma trough.

7.4.3. MAINTENANCE OF THE NIGHTTIME $F2$ LAYER

The maximum concentration of the $F2$ layer never decreases below about 10^5 cm^{-3}, even during long winter nights. An explanation of this phenomenon must take into account several features. Existing neutral winds directed towards the equator cause the charged particles to move upwards, consequently diminishing the factor β. The diffusive descent from higher levels of charges accumulated during the day in the magnetic tube of force provides an ionization input in to the F region.

Parks and Banks (1974) have shown that a vertical motion of ionization (caused either by a neutral wind or an electric field) actually induces a change in the flux coming from the magnetosphere. As a consequence, the true effect of an induced motion is quite different from what was predicted by earlier computations assuming

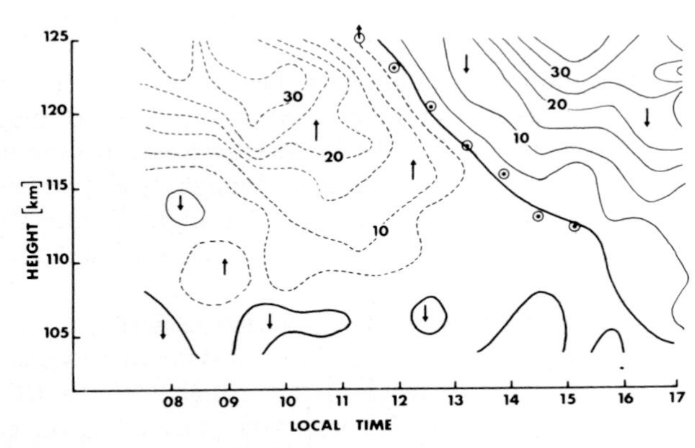

Fig. VII.20. The circles indicate the location of a sporadic E layer which follows the node of the vertical ion drift velocity: the thick line corresponds to the convergence level separating the upwards ionic velocity contours (dashed lines) from the downwards velocity contours (thin lines), labeled in ms^{-1}, deduced from incoherent scatter observations at Arecibo (Harper *et al.*, 1975).

a constant downward flux. In particular, if an upward motion decreases β, it simultaneously decreases the downward input flux of ionization and the peak density of the layer n_{e_m} remains nearly constant, in contradiction with the earlier conventional results predicting a significant increase of n_{e_m}. The height of the layer however increases more or less as previously predicted. As a consequence, equatorward neutral winds or eastward electric fields cause no major change in the equilibrium nighttime peak density, but help in explaining the maintenance of the $F2$ layer by reducing the necessary downward flux from the magnetospheric tubes. Still, this flux must be consistent with the maximum daytime upward flux allowed by the diffusive barrier described in paragraph 2.2 of the present chapter.

The question of maintenance of the nighttime F layer has been discussed at length in many papers. Some authors have concluded to the necessity of a nocturnal source of ionization, such as soft electrons and protons of 100 to 1000 eV energy. The present relatively poor state of knowledge of neutral winds and electric fields does not allow ruling out completely such a possibility. However, the maintenance of the layer is no longer considered as a mysterious phenomenon, and is mainly attributed to dynamic effects which reduce the recombination rate to a value small enough to be balanced by a very small downward flow of ionization from the magnetosphere, compatible with the electron content of the magnetospheric flux tubes.

7.5. The Formation of Sporadic *E* Layers at Temperate Latitudes

As already noted, whenever the lifetime of the ions is sufficiently long, the distribution of charged particles is affected by their motions, which is to say that the divergence of the particle flux is no longer negligible in the continuity equation. This is the case when atomic ions are involved, in the F layer and upper ionosphere; but as seen above this is also true for the small number of metallic ions of meteoric origin in the E layer, and to lesser extent for the molecular ions (Walker and McElroy 1966).

The study of short term motions (period about an hour) of the neutral atmosphere at 100 km shows the existence of well-defined wind 'shears'. This is due to the vertical 'corkscrew' propagation of gravity waves and tides, in which the wind velocity rotates with altitude and the phase of the entire system has a downward motion corresponding to the upward transport of energy. Thus at two points differing in altitude by several kilometers, one can have winds in opposite directions which cancel each other at an intermediate point ('node') that itself moves downward.

The results of Section 1.3.1 show that if ν_i is not much greater than ω_{Bi}, a component of wind directed eastward causes an ascending component of ion motion, while a component of wind directed westward causes a downward vertical drift of the ions, because of the north–south horizontal component of the magnetic field. This being the case, if the neutral wind shifts from a west–east to an east–west direction as the altitude increases, there is accumulation of charged particles in the neighborhood of the node (the electrons, less subject to collisions with the neutrals, are constrained by polarization fields to follow the ions along the lines of force). For a shear in the opposite sense, there is a contrary tendency for charged particles to be forced away from the node, in one direction or the other. The altitudes of the wind nodes and of the layers of convergence or divergence of the plasma are offset in the presence of an electric field (Axford and Cunnold, 1966). Figure VII.20 shows a downward motion of an *Es* layer, following the node of the ion convergence level.

The mechanism of stratification of ionization by wind shears becomes less and less effective as the altitude decreases and the collision frequency ν_i increases. It also becomes ineffective near the poles, where the horizontal components of the magnetic field become very small. But it is also ineffective near the magnetic equator, where the vertical component of the field vanishes, because there the electrons cannot drift vertically at all, in the absence of an electric field perpendicular to the magnetic field; in that case as mentioned above, the whole plasma is subject to the equatorial fountain mechanism.

Therefore, it is only between 100 and 140 km, at temperate latitudes, that ions are subject to convergence stratification by wind shears. This explanation of the sporadic *E* layers at temperate latitudes was developed by Whitehead (1961, 1967) and Cuchet according to an hypothesis put forward by Dungey (1959), after a correlation between the map of the frequency of *Es* occurrence and that of strength of the horizontal component of the magnetic field had been noticed.

The recombination of molecular ions with electrons is so rapid that their spatial distribution cannot be much affected by this, except when their lifetime is increased, perhaps at night, as was mentioned in Chapter VI. This is not so however for the atomic ions of meteoric origin which, for reasons we have discussed, have very long lifetimes. These ions are thus gathered into thin layers, whose thickness is without doubt limited only by diffusion. Such thin layers of meteoric ions, 'trapped' at the nodes of the wind and 'drowned' in the thick swarm of molecular ions in photochemical equilibrium, always exist, as has been shown in measurements by mass spectrometry (Narcisi *et al.* 1967a,b). The sporadic *E* layers, on the other hand, as their name implies, only exist in a random manner and for limited times, of the order of an hour. Their partial transparency to radiofrequency waves, their virtual height (independent of the probing frequency), as well as measurements *in situ* (Figure V.2), show that they are thin and abrupt and most often have a downward movement that slows down, and vanishes near 90 km. The layers of meteoric ions stratified by the wind must have all of these characteristics. The appearance of *Es* layers therefore would correspond only to those occasions when the density of the patchy meteoric ions becomes greater than the equilibrium density of O_2^+ and NO^+ ions. The elec-

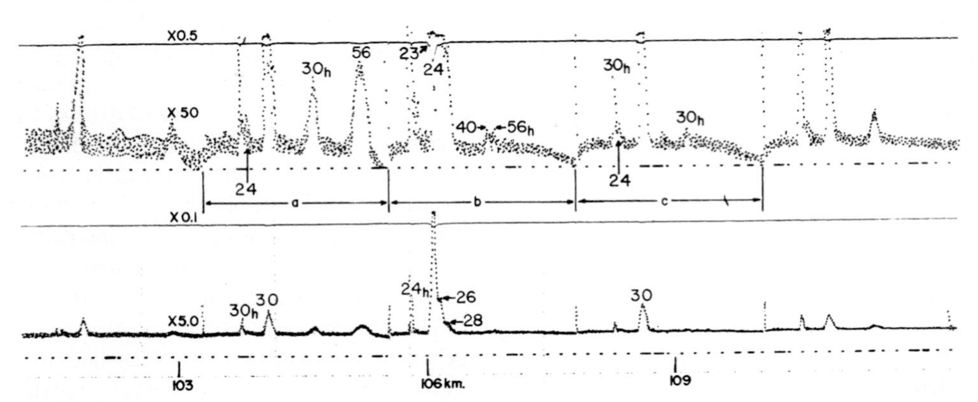

Fig. VII.21. Output current of the mass spectrometer for three successive mass scannings: a, b, c, respectively, centered around 103, 106 and 109 km altitude.

tron layer accompanying the meteoric ions then 'pierces' through the profile of electron density of the normal *E* layer, while otherwise it has no influence on this profile.

It should be noted that this mechanism, because of the downward propagation of the phase of the waves that trap the long lived ions at the wind nodes, scavenges the *E* region of its meteoric ions, which could otherwise accumulate there (Chimonas and Axford, 1968). They are evacuated towards the *D* region where they are neutralized, as explained in Chapter VI, perhaps serving there as precursors in the formation of agglomerate hydrates (Burke, 1972b).

The accuracy of these views has been confirmed through measurements by mass spectrometry from rocket probes in the presence of *Es*. In fact Figure VII.21 gives the only example we have so far of the mass spectrometric measurement of the ion composition of a strong 'blanketing' sporadic *E* layer (Young *et al.*, 1967). Despite the poor altitude resolution it can be seen from the three successive ion spectra a, b and c centered at 103, 106 and 109 km respectively, that the large peak in ion density, corresponding to the large peak in electron (10^6 cm^{-3}) density simultaneously observed on an ionosonde from the ground near 105 km, is entirely composed of mass 56 (Fe), 24 and 26 (Mg isotopes), and 28 (Si). Indeed, mass 30 (NO^+) which dominates the ion composition above and below with a concentration of 10^4 is virtually absent within the layer as it recombines with the electrons brought along with the atomic ions by the wind shear mechanism.

On the equatorial belt as predicted by the theory, the layering of metallic ions is observed to be inhibited (Aikin and Goldberg, 1973) but they are subject to the equatorial fountain mechanism as mentioned above.

While the diurnal and seasonal variations in the frequency of appearance of *Es* have not yet been clearly explained, they are very surely correlated in a complex way with variations in the arrival of meteors, in photoionization of metallic atoms, and in the excitation and propagation of the atmospheric oscillations that produce favorable wind shears.

At the end of a long winter night, the paraboloidal antenna of one of the stations of the French multistatic incoherent scatter facility is pointed to intersect at 300 km altitude the beam of the transmitting antenna 60 miles away, and receives the weak scatter from thermal fluctuations in the *F* layer. While this region of the Earth will remain in the dark for another hour, the temperature of the electron gas in the ionosphere is observed to have already started its diurnal climb well over neutral and ion temperatures, as the summer sun has arisen over the conjugate atmosphere in South Africa, filling the 10,000 km long geomagnetic arch to France with hot photoelectrons. (Courtesy of P. Bauer, Centre de Recherches en Physique de l'Environnement, Issy les Moulineaux)

CHAPTER VIII

THERMODYNAMICS OF ENERGY BALANCE

In this chapter we account for the energy of thermal agitation of the electrically charged species in the ionosphere. Such a study is necessary because, as we mentioned in Part 1, solar radiation prevents the ionospheric plasma from reaching thermodynamic equilibrium. We shall begin by recalling the nature and essential characteristics of the relevant processes of energy exchange. Next, we shall interpret the behavior of the electron temperature and ion temperature in the low- and mid-latitude regions, and shall try to show how the study of the energy balance of the charged particles contributes to the general improvement of our understanding of the photochemical ionosphere. Finally we will tackle the magnetospheric control of ionospheric temperatures and the dissipation of auroral energy inputs.

8.1. Energy Budget of the Charged Particles

8.1.1. GENERAL OUTLINE

At middle and low latitudes, the major heat input to the ionosphere is through the solar XUV flux; at the initial stage of photoionization of neutral molecules, solar radiation injects energy into the population of charged particles through the agency of the supra-thermal kinetic energy of the primary photoelectrons. This energy is dissipated in the manner outlined in Figure VIII.1 (Dalgarno *et al.*, 1963; Geisler and Bowhill, 1965a,b; Banks, 1967a).

Let us consider a slab in the ionosphere, sufficiently narrow in altitude for it to be assumed homogeneous. In the presence of solar radiation, photoelectrons are being continually created, and their mean energy, of the order of 10 eV, is much higher

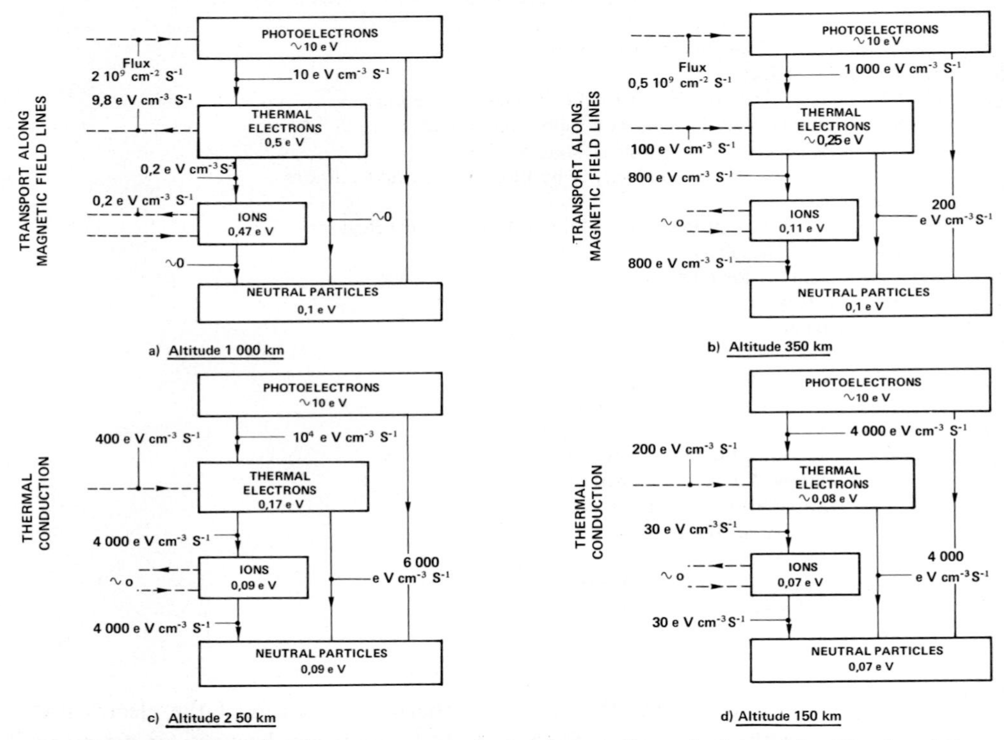

Fig. VIII.1. Sketch of the daytime energy exchanges between ionospheric particles. The dotted lines indicate an energy exchange between the considered slab and the nearby regions by transport phenomena along the magnetic field lines (non-local heating by photoelectrons and thermal conduction of the electron and ion gases). The full lines indicate the energy exchange between various species inside the slab. The mean energy decreases when going downward in the figure from photoelectrons to thermal electrons, ions and eventually neutral particles. Actual values of the mean energy and of the energy exchange largely vary around the typical values indicated on the figure depending upon the electron concentration and upon the solar zenith angle which is important mainly for the altitude 150 km.

than that of the molecules of the neutral atmosphere, whose temperature is about 1000°K, some 0.1 eV. If the mean free path of these photoelectrons is large compared with the thickness of the slab, we must consider the possibility, that most of them have, of leaving the region, following the lines of force of the terrestrial magnetic field, which is the axis of their helicoïdal motion. Similarly, photoelectrons created above and below the slab can penetrate into it.

In the course of their interactions with the other particles of the medium, these photoelectrons lose their kinetic energy and end up assimilated into the population of thermal electrons. They give up energy first of all to the neutral particles, mostly through inelastic collisions. Since more important processes contribute to the energy balance of the neutral particles (cf. Chapter I), which are always much more numerous than the charged particles in the ionosphere, we can consider the neutral gas as a reservoir of very high capacity, able to absorb this imput of energy without appreciable change in its temperature. The electrons also give up their energy to the other charged particles in the medium through elastic collisions. Because of the difference in masses, interactions with other electrons are much more efficient for energy transfer than those with ions. Therefore, the ensemble of electrons receives an energy input, and their kinetic temperature increases until loss mechanisms can compensate. There are three types of losses: collisions with neutral particles, collisions with ions, and transport to adjacent regions through thermal conduction along the magnetic lines of force. Thermal conduction might be a loss mechanism, but it might also result in a gain of energy, depending on the divergence of the conductive flux, i.e. the sign and the value of the gradient of electron temperature at the boundaries of the slab.

The energy that ions receive from electrons tends to increase the ion temperature, while the loss mechanisms that limit this increase are again collisions with neutral particles and, possibly, conduction of heat through the ion gas.

Typical temperatures of the various species, and of energy exchange rates, are given in Figure VIII.4 for mid-latitude regions. In the high-latitude regions, precipitation of energetic particles and Joule heating dominates the energy input due to the solar E.U.V. flux.

We now turn to the details of each of the processes that affect these thermal balances.

8.1.2. ENERGY SPECTRUM OF THE PHOTOELECTRONS

In Chapter VI, we examined the production of photoelectrons and ions through the action of solar radiation upon the neutral particles. Figure VI.6 shows the variation with altitude of the number of photoelectrons created per cm^3 per second for the Sun at zenith. Now we must be concerned not only with the number of photoelectrons created, but also with their distribution in energy.

The kinetic energy of a photoelectron is the difference between the energy of the incident photon and the ionization energy. The latter depends on the state of internal excitation of the neutral particle before ionization, and the state of excitation of the ion created. It is necessary, therefore, to consider all possible product ions, and we cannot be content here with an overall ionization cross section for each constituent. When, for each wavelength of the solar radiation, we have determined the cross section of each photoionization reaction, we must still introduce into the calculation the characteristics of the neutral atmosphere as a function of altitude as well as a satisfactory description of the solar radiation.

An example of such a calculation, whose complexity is evident, is the work of Tohmatsu *et al.* (1965). Of course, the results obtained depend on the assumptions made about the neutral atmosphere and solar radiation. Nonetheless, we note the

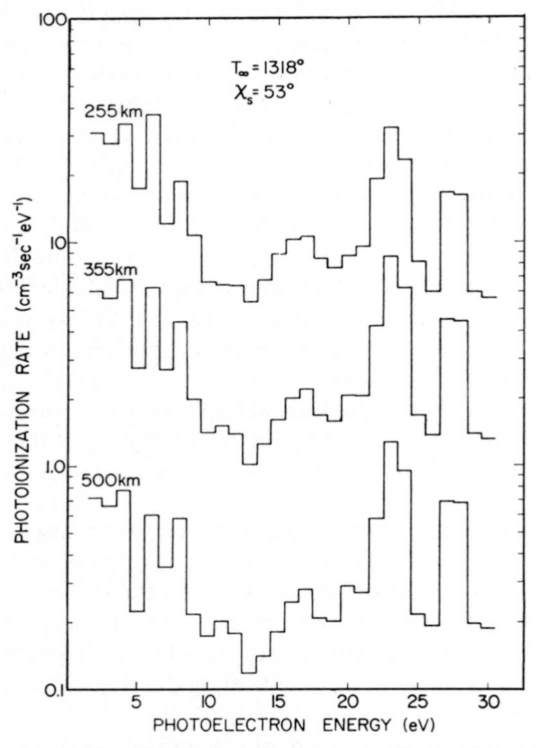

Fig. VIII. 2. Energy spectrum of the primary photoelectrons according to Cicerone and Bowhill (1971) for a neutral atmosphere model referred by the value of $T_\infty = 1318°$ K and a solar zenith angle $\chi_s = 53°$.

simple result obtained by Tohmatsu and his colleagues: the number of photons as a function of their energy E is approximately described by a curve of the form e^{-E/E_0}, where E_0 is the mean energy, a function of altitude which, however, remains around 8 eV from 200 to 300 km. More recent results, such as those of Nagy *et al.* (1969) and Cicerone and Bowhill (1971) bring out a more complex structure of the spectrum (Figure VIII.2) with a strong secondary maximum between 18 and 22 eV, which results essentially from taking into account a more precise structure of the solar spectrum (Hinterreger, 1970).

8.1.3. COULOMB INTERACTIONS

Neglecting for a while the problem of thermal conductivity, Coulomb collisions appear at two stages: first during the interaction of energetic photoelectrons with the population of thermalized electrons, and second during the exchange between the latter population and the ion population.

Energy losses in collisions of this type can be studied theoretically. For two particles with given velocities and impact parameter, the elementary laws of mechanics allow the calculation of the angle ψ through which the collision causes the relative velocity of the two particles to turn. Starting from this result, we need only integrate over the possible impact parameters and over one of the particle velocities in order to have the energy exchange between a photoelectron and the ambient electrons, which have a Maxwellian velocity distribution. A similar computation but with a double integration in velocity space, results in the energy exchanged per cm^3 per sec between any two populations of charged particles.

The difficulty that we mentioned in Chapter VII concerning the validity of the concept of binary interaction also enters here. Therefore the parameter $\ln \Lambda$, which results from the approximations described in Chapter VII, appears again. Recall that in the ionosphere $\ln \Lambda$ varies between 13 and 15.

8.1.3.1. *Energy exchange between a photoelectron and thermal electrons*

In the case where the energy E of the photoelectron is large compared with the thermal energy of ambient electrons, the loss of energy per unit of path length is given by the very simple formula:

$$\frac{\mathrm{d}E}{\mathrm{d}l} = 1.95 \times 10^{-12} \frac{n_e}{E}\left(1 + \frac{\ln (E/kT_e)}{\ln \Lambda}\right) \tag{VIII.1}$$

where the lengths are in cm and the energies in eV. Of course this formula is no longer valid when E approaches zero.

We mentioned, in the study of incoherent scattering (Chapter IV), that photoelectrons can create collective oscillations of the electron gas. This is an energy loss that is neglected in the preceding calculation; it can be shown that this loss amounts to 10 to 15% of the one that we have taken into account by limiting ourselves to binary collisions.

8.1.3.2. *Electron-Ion Collisions*

When the calculation of the energy exchange between two populations of particles with masses m_1 and m_2 is carried out, it is noticed that if the temperatures T_1 and T_2 of the two populations are of the same order of magnitude, the energy exchange L_{12} depends on m_1 and m_2 by a factor of proportionality $(m_1 m_2)^{1/2}/(m_1 + m_2)^{3/2}$. Application of this to the cases $m_1 = m_2 = m_e$ (electron mass), $m_1 = m_2 = M$ (ion mass), and $m_1 = m_e$, $m_2 = M$, shows that energy exchange between electrons is more efficient than energy exchange between ions, which in turn is more effective than exchange between electrons and ions. These inequalities are the reason why one can speak of separate Maxwellian distributions for electrons and ions, without the corresponding temperatures being at all identical.

In the case of collisions between electrons and atomic oxygen ions O^+, the value of the energy exchange L_{ei} is given by

$$L_{ei} = 4.8 \times 10^{-7} n_e^2 T_e^{-3/2} (T_e - T_i) \text{ eV cm}^{-3} \text{ s}^{-1} \tag{VIII.2}$$

where the temperatures are in °K. The considerations of the preceding paragraph show that for another type of ion with atomic mass M, the above result should be multiplied by $16/M$.

A notable characteristic of L_{ei} is that its variation with T_e is not uniform; it has a maximum $L_{ei\,max}$ at $T_e = 3T_i$, and approaches 0 when T_e approaches infinity. This can cause a runaway effect. If the electron gas is subject to a power input greater than $L_{ei\,max}$, its temperature increases beyond $3T_i$ and the larger its temperature increases, the less efficient electron-ion collisions for extracting energy. The electron temperature therefore increases until other processes (collisions with neutrals, conduction) can bring about the losses necessary for equilibrium. Actually, it is found

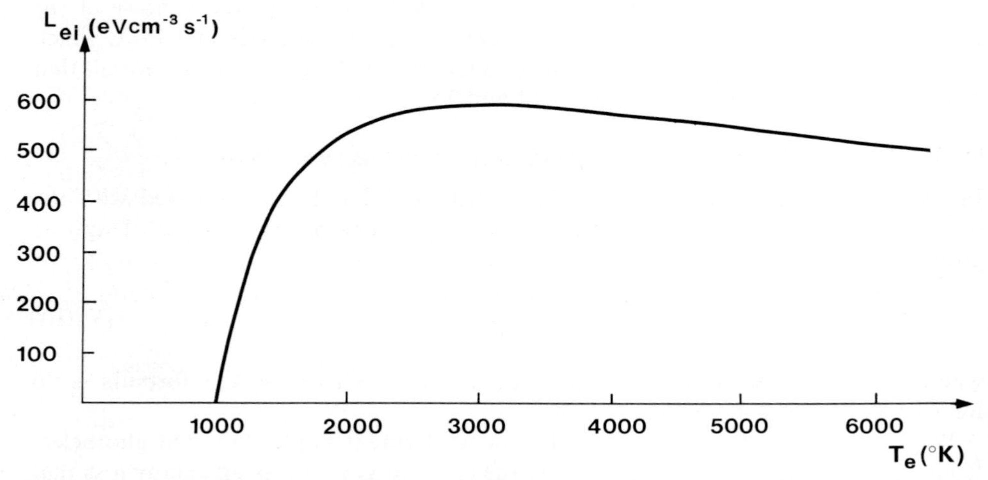

Fig. VIII.3. Energy exchange rate L_{ei} between the electron gas with a temperature T_e (in abcissa) and the ion gas with a temperature of 1000° K. The electron concentration n_e is assumed to the 3.1×10^5 cm^{-3}; L_{ei} is proportional to n_e^2.

that in the ionosphere the condition $T_e > 3T_i$ is practically never fulfilled. On the other hand, as soon as $T_e - T_i$ is of the order of T_i, L_{ei} is not much less than $L_{ei\,max}$, and varies only slightly with $T_e - T_i$ (Figure VIII.3).

8.1.4. CHARGED-NEUTRALS INTERACTIONS

Unlike Coulomb collisions, collisions between charged and neutral particles cannot be studied in a satisfactory way by entirely theoretical approaches. The corresponding cross sections are determined by a synthesis of theoretical work and experimental data obtained in the laboratory.

8.1.4.1. *Collisions between Photoelectrons and Neutral Particles*

Exchange of energy between a photoelectron and neutral particles results essentially from inelastic collisions in the course of which the internal energy of the neutral molecule is increased, perhaps by ionization, perhaps by a transition between atomic states, perhaps by excitation of rotational or vibrational states. The study of all such interactions has given rise to an abundant literature. Green and Stolarsky (1972) give good estimates of the resulting cross sections.

Elastic collisions result in very little energy loss by photoelectrons. Their importance comes essentially from deflections limiting the possibility of moving along the magnetic field lines. Elastic collisions thus intervene most effectively to inhibit the importance of 'non local' heating by photoelectrons created far away from the region studied. The corresponding cross sections for O and O_2 have been given by Sunshine *et al.* (1967) and those for N_2 by Fisk (1936).

8.1.4.2. *Collisions Between Thermal Charged Particles and Neutral Particles*

8.1.4.2.1. *Electron-Neutral Collisions.* Banks (1966a) showed that in elastic collisions the exchange of energy per unit volume, unit time, is expressed in the form:

$$\frac{du_1}{dt} = -3n_1 \frac{m_1}{m_1 + m_2} k(T_1 - T_2)\nu_{12} \qquad \text{(VIII.3)}$$

where ν_{12} is the frequency, defined in Chapter VII, of collisions between particles of types 1 and 2, with densities n_1 and n_2, masses m_1 and m_2 and temperatures T_1 and T_2. This formula applies also to electron-ion collisions and formula (VIII.2) is alternatively deduced from Equations (VIII.3) and (VII.3).

Inelastic collisions are the most efficient from the point of view of energy transfer. The important cases are excitation of the fine structure of the 3P level of O (Dalgarno and Degges, 1968), and excitation of rotational (Dalgarno, 1969) and vibrational (Stubbe, 1971) states of O_2 and N_2.

The following table gives the heat transfer rates which correspond to the main collision processes between thermal electrons and neutral particles. The densities are expressed in cm^{-3}, the temperatures in degrees Kelvin and the heat transfer rates in eV cm^{-3} s^{-1}. When no simple analytical expression is available, the reader is referred to an appropriate paper.

Elastic collisions with	O	$5.3\ 10^{-19} \sqrt{T_e}\,(1 + 5.7\ 10^{-4}\, T_e)\, n_o n_e (T_e - T_n)$
	O_2	$1.21\ 10^{-18} \sqrt{T_e}\,(1 + 0.036 \sqrt{T_e})\, n_{O_2} n_e (T_e - T_n)$
	N_2	$1.77\ 10^{-19}\, T_e (1 - 1.21\ 10^{-4}\, T_e) n_{N_2} n_e (T_e - T_n)$
Excitation of rotation	O_2	$6.9\ 10^{-14} \sqrt{T_e}\, n_{O_2} n_e (T_e - T_n)$
	N_2	$2.9\ 10^{-14} \sqrt{T_e}\, n_{N_2} n_e (T_e - T_n)$
Excitation of vibration	O_2	Lane and Dalgarno (1969)
	N_2	Rees *et al.* (1967)
Fine structure	O	$3.3\ 10^{-15} [1 - 1.3(T_n/1000 - 1)(2 - T_n/1000)]\, n_o n_e (T_e - T_n)$.

Such collisions with neutral particles are the dominant cooling process of electrons in the *E* and *F* region (see Figure VIII.1).

8.1.4.2.2. *Ion-neutral collisions.* The relation VIII.3 can also be used to determine the transfer of energy by elastic collisions between ions and neutral particles (Banks,

1966b). Even though they are inelastic, resonant charge-exchange reactions between neutral and ionized particles of the same chemical nature can be approached in a very similar way, if it is borne in mind that such collisions conserve the kinetic energy and momentum of the particles concerned. Banks (1966b) gives the corresponding cross sections. The rate of the important energy exchange between O^+ ions and O atoms is given by:

$$L_{in} = 2.1\ 10^{-15}\,(T_i + T_n)^{1/2}\,n_o n_e\,(T_i - T_n)\,\text{eV cm}^{-3}\,\text{s}^{-1} \qquad \text{(VIII.4)}$$

if n_o and n_e are expressed in cm^{-3} and T in degrees Kelvin.

8.1.5. THERMAL CONDUCTIVITY

8.1.5.1. *Generalities on the Thermal Conductivity of a Gas*

We repeat here the simple, but physically meaningful way in which Chapman and Cowling (1952) presented the phenomenon of thermal conduction. Let us take a gas at rest whose temperature T is a function of the altitude variable z. We consider the flux of heat across a unit area on a plane z = constant, e.g. the plane $z = 0$. The number of particles per unit time crossing this area in a downward (or upward) direction is $1/2n\langle|V_z|\rangle$, where n is the particles concentration and $\langle|V_z|\rangle$ the mean of the absolute value of the component of their velocity parallel to the axis Oz. The particles crossing the plane $z = 0$ in the upward direction have on the average a kinetic energy not equal to $E(0)$, the value at $z = 0$, but $E(-l)$, the value on the plane $z = -l$, where l is a distance of the order of magnitude of the mean free path. Likewise, the particles that cross the plane downwards have a mean kinetic energy $E(+l)$. The net ascending flux of energy is thus:

$$\varphi = (1/2n\,\langle|V_z|\rangle\,E(-l) - 1/2n\,\langle|V_z|\rangle\,E(+l).$$

If $E(z)$ varies only slightly over distance this can be written as

$$\varphi = -n\,\langle|V_z|\rangle\,l/\mathrm{d}E/\mathrm{d}z.$$

Aside from a factor C, the heat capacity, the mean energy E is equal to the temperature T. Therefore

$$\varphi = -C\,n\,\langle|V_z|\rangle\,l\,\mathrm{d}T/\mathrm{d}z. \qquad \text{(VIII.5)}$$

This simple approach can be used to obtain qualitative results about the conductivity of the electrons and ions. First of all, the mean free path l is inversely proportional to the concentration n, so that the conductivity is independent of n, at least to a first approximation. Also the conductivity is proportional to $\langle|V_z|\rangle$, and therefore the electron conductivity is much greater than that of the ions, since the mean free paths are nearly identical.

8.1.5.2. *Thermal Conductivity of a Fully Ionized Gas*

Spitzer and Härm (1953) resolved numerically the Fokker-Planck equation describing the evolution of small deviations from a Maxwellian distribution in an inhomogeneous gas, which can be used to describe conductivity. On the right hand side of the Fokker-Planck equation, there appear terms representing diffusion in velocity space due to Coulomb collisions. These terms are calculated in a way similar to the energy losses in paragraph 1.3 and it is therefore natural to see $\ln \Lambda$ appear in the final result.

It is shown in this manner that a temperature gradient causes a heat flux $\vec{\varphi} = -K_1 \vec{\nabla} T$ in the electron gas, but that it also generates a current $\vec{\mathbf{j}} = \alpha \vec{\nabla} T$. At equilibrium state this current will be cancelled by the current due to a polarization electric field, but this field in turn generates a heat flux. The calculations of Spitzer and Härm determine both the current $\vec{\mathbf{j}}$ and heat flux $\vec{\varphi}$ that result from a temperature gradient $\vec{\nabla} T$ and an electric field $\vec{\mathbf{E}}$;

$$\vec{\mathbf{j}} = \sigma \vec{\mathbf{E}} + \alpha \vec{\nabla} T$$
$$\vec{\varphi} = -\beta \vec{\mathbf{E}} - \mathrm{K}_1 \vec{\nabla} T.$$

The polarization field that cancels $\vec{\mathbf{j}}$ is $\vec{\mathbf{E}} = -\alpha/\sigma \vec{\nabla} T$, so that $\vec{\varphi} = -K \vec{\nabla} T$, with

$$K = K_1 - \alpha\beta/\sigma.$$

The correction to K found in this way is not negligible; for the usual case of a singly ionized gas, $K/K_1 = 0.42$.

For typical ionospheric conditions ($\ln \Lambda = 13$ to 15), the formula for the electron thermal conductivity is:

$$K = 7.7 \times 10^5 T_e^{5/2} \text{ eV cm}^{-1} \text{ s}^{-1} \text{ °K}^{-1}. \tag{VIII.6}$$

Similar considerations lead to the following formula (Banks, 1967) for the value of the conductivity of protons:

$$K_{H^+} = 0.46 \times 10^5 T_i^{5/2}. \tag{VIII.7}$$

This conductivity plays an important role in the upper part of the ionosphere.

8.1.5.3. *Influence of the Terrestrial Magnetic Field*

Since a magnetic field has no influence on the motion of charged particles moving parallel to it, conductivity in the direction of the field is the same as in its absence. On the other hand, the motion of electrons and ions perpendicular to the magnetic field is severely restricted as long as the gyrofrequencies are greater than the collision frequencies.

Thus, above the dynamo region, thermal conductivity in the directions perpendicular to the magnetic field is practically zero. An elementary argument leads to the conclusion that the heat flux across 1 cm² of a horizontal surface will be given by

$$\varphi = -K \sin^2 I \, \mathrm{d}T_e/\mathrm{d}z, \tag{VIII.8}$$

where I denotes the inclination of the magnetic field to the horizontal plane z = constant, K being the conductivity in the absence of the magnetic field.

8.1.5.4. *Decrease of the Conductivity by Electron-Neutral Collisions*

Banks (1966c) showed that below 250 km collisions of electrons with neutral particles decrease their mean free path, and consequently their thermal conductivity, in comparison with the values for a completely ionized gas.

8.1.6. JOULE EFFECT

Under the action of an electric field, charged particles drift relative to one another and relative to neutral particles. Collisions between species limit the drift velocities as described in Chapter VII; moreover they convert some of the drift energy into heat. The rate of frictional heating of species 1 due to collision with species 2 is:

$$Q_{12} = \nu_{12} \frac{m_1 m_2}{m_1 + m_2} n_1 (v_1 - v_2)^2 \tag{VIII.9}$$

where v_1 and v_2 are the drift velocities (Morse, 1963; Tanenbaum, 1965).

The drift velocities v_1 and v_2 can be computed from the results of Chapter VII. They are proportional to the applied electric field E, and the corresponding heat input Q varies therefore as E^2.

The large values of the parallel conductivity prevent the existence of electric fields along the magnetic field (with the exception of phenomena of anomalous resistivity due to turbulence or wave-particle interactions, which are expected to take place in the magnetosphere rather than in the ionosphere). Above the dynamo region, electrons and ions drift with the same velocity, under the action of a perpendicular electric field. Assuming an electric field of 10 mV m^{-1}, the drift velocity is 200 m s^{-1}; the heat input to the electrons is 1,6 10^{-12} W m^{-2} or 10 eV cm^{-3} s^{-1} at an altitude of 200 km and 0,6 10^{-12} W m^{-2} or 4 eV cm^{-3} s^{-1} at an altitude of 300 km; these figures have been computed from Equation (VIII.9) by using typical collision frequency and electron concentration values. Equation (VIII.9) shows that the heat input to ions is greater than that to electrons by a factor of the order $m_i \nu_{in}/2m_e \nu_{en}$. The typical values of heat input to ions for similar conditions are 1,6 10^{-9} W m^{-2} or 10^4 eV cm^{-3} s^{-1} at an altitude of 200 km and 0,5 10^{-9} W m^{-2} or 3300 eV cm^{-3} s^{-1}. Equation (VIII.9) implies also that neutral particles receive roughly the same heat input per unit volume (not per particle) as the ions, their mass being roughly equal. Electric fields of 10 mV m^{-1} are typical in the auroral zone, and electric fields of 30 mV m^{-1} corresponding to heat inputs 10 times greater than the values above have been observed by incoherent scattering at Chatanika (Banks *et al.*, 1974).

An alternative approach is to use the concept of conductivity to evaluate the total heat input. The Hall current flowing perpendicular to E does not correspond to any energy dissipation and only the Pedersen conductivity is to be taken into account; the total heat input is

$$Q = \sigma_1 E^2.$$

The two approaches are equivalent. This is easily shown for the friction between ions and neutrals which is the dominant interaction. From Equation (VIII.9), the heat input to ions and to neutrals are respectively

$$\begin{aligned} Q_{in} &= \nu_{in} \frac{m_i m_n}{m_i + m_n} n_i v_i^2 \\ Q_{ni} &= \nu_{ni} \frac{m_i m_n}{m_i + m_n} n_n v_i^2. \end{aligned} \tag{VIII.10}$$

Computing $Q = Q_{in} + Q_{ni}$ and using the equality (see Chapter VIII.1.1) $m_i n_i \nu_{in} = m_n n_n \nu_{ni}$, one obtains

$$Q = n_i m_i \nu_{in} v_i^2.$$

Then $$v_i = E/B$$

and $$Q = \frac{n_i m_i \nu_{in}}{B^2} E^2. \qquad \text{(VIII.11)}$$

Neglecting the electrons, Equations (VIII.8) and (VIII.9) show that, in the region we have considered so far, where $\nu_{in} \ll \omega_{Bi}$, $\dfrac{n_i m_i \nu_{in}}{B^2} = \sigma_1$. We recover the value found by the second approach $Q = \sigma_1 E^2$.

8.2. Electron and Ion Temperatures in the Mid and Low Latitude Regions

We shall now inquire as to how the various processes described allow to interpret the measurements of electron temperature in middle latitudes. After a brief study of the *D* and *E* regions, we shall devote some space to the *F* region, where thermodynamic balance and coupling with the magnetosphere offer interesting problems.

8.2.1. EXPERIMENTAL RESULTS

8.2.1.1. *D and E Regions*

According to measurements made by incoherent scatter sounders the electron temperature is equal to that of the ions below 120 km altitude, and their common value is in agreement with the currently accepted values for the temperature of the neutral atmosphere. Experiments based on Langmuir probes have yielded much higher electron temperatures, but it seems probable that they involve errors tied in with the operation of such probes (see Chapter V).

The existence of thermodynamic equilibrium below 120 km is in agreement with the theoretical analysis of the electron energy balance that we have presented. Indeed, neutral particles are so numerous that electron-neutral collisions are very efficient, and a much higher energy input than that from photoelectrons would be necessary for an appreciable temperature difference between electrons and neutrals to be sustained. This might be possible in the high latitude regions where intense precipitation of energetic particles could provide a sufficient energy input. The fact is not yet established.

Conflicting results concerning enhancements of T_e in sporadic *E* layers have been reported. Due to the uncertainty in Langmuir probe measurements, the existence of such phenomena remains controversial.

8.2.1.2. *F Region*

All measurements agree that in the *F* region, during the day, the electron temperature is greater than that of the other particles. Nevertheless, in the past, when a direct comparison between incoherent scattering and Langmuir probes was made, the latter quite frequently indicated much larger electron temperatures, by a factor that could

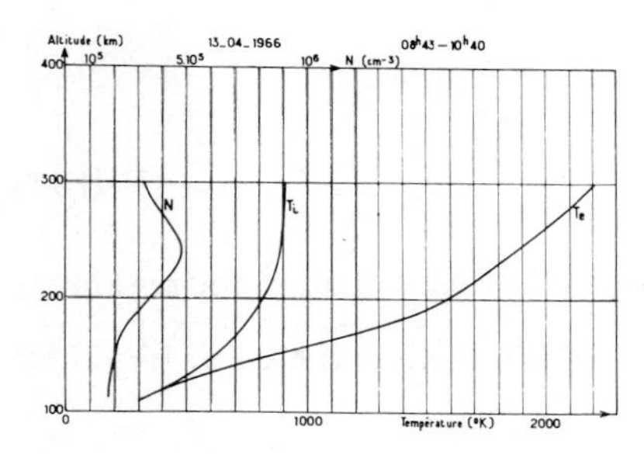

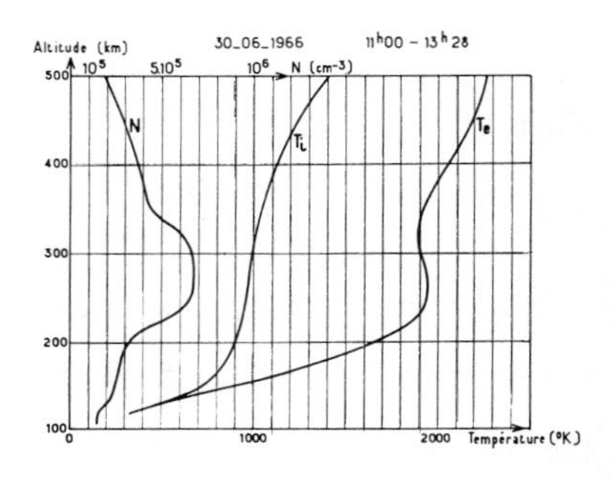

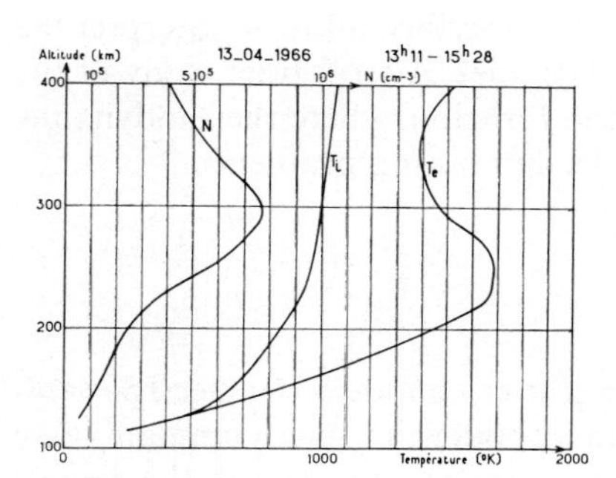

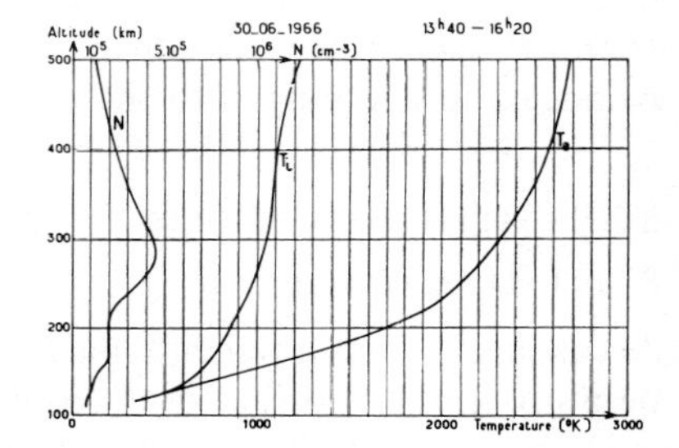

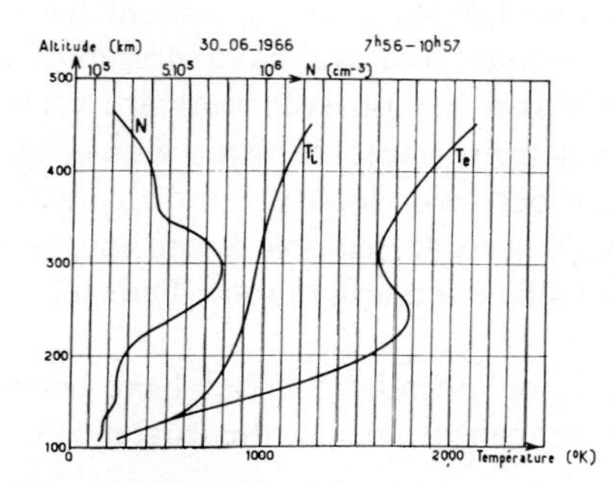

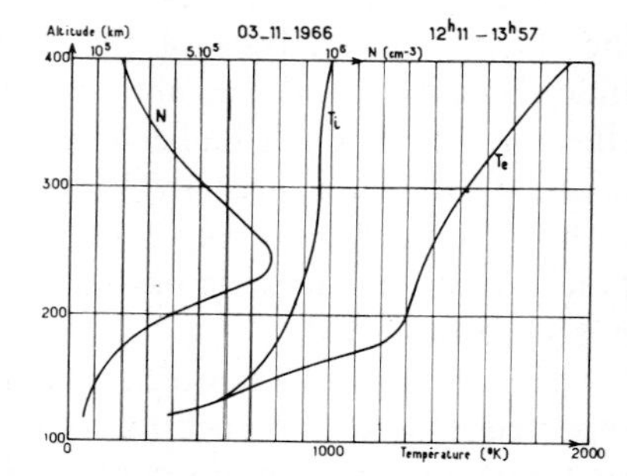

Fig. VIII.4. Typical electron and ion temperature profiles (lower scale) observed over Saint-Santin (France) by incoherent scattering. At a fixed altitude, $T_e - T_i$ is smaller when the electron concentration (upper scale) is higher. Generally speaking, $T_e - T_i$ increases with altitude, although large values of electron concentration around 300 km altitude can cause the appearance of a secondary minimum.

reach 1.7 (Hanson *et al.*, 1969; Brace *et al.*, 1969). However, the agreement between the two techniques has been improved, thanks to the use of better Langmuir probes, and can now be excellent (Taylor and Wrenn, 1970).

Figure VIII.4 shows some examples of electron temperature profiles measured by incoherent scattering. We see that the shape of this profile varies considerably, from a curve that increases monotonically with altitude to one that has relative maxima and minima. This latter profile is always seen when the peak electron concentration lies around 300 km and reaches a high value, greater than 6×10^5 cm^{-3}.

This variation in profile shape is only an instance of the anti-correlation that one always obtains between electron temperature and electron concentration, the one increasing when the other decreases at a given altitude. It has been shown that there even exists, at a given place and altitude, a strict relation between $T_e - T_i$ and n_e (Lejeune and Waldteufel, 1970), as shown on Figure VIII.5.

It is clear from Figure VI.6 that the production of photoelectrons is nearly independent of the solar zenith angle χ, provided this angle is smaller than a maximum value χ_M which decreases when altitude decreases. On the contrary, for $\chi > \chi_M$, the production rapidly decreases with increasing χ. This explains why the data on Figure VIII.5 have been selected with a χ value less than 65°. Moreover, generally speaking, the electron temperature is more sensitive to the value of the solar zenith angle when the altitude decreases.

Actually, what is physically significant is not the temperature of the electrons, but rather their energy balance, which depends also upon their concentration. Therefore, since losses through collisions increase with the latter, the existence of an anti-correlation between T_e and n_e seems natural enough. We should therefore wonder whether the results of simultaneous observations of electron concentration, and electron and ion temperatures are compatible with the electron energy balance.

8.2.2. ENERGY BALANCE OF THE ELECTRONS

Referring to the results in Section 1, the energy balance of the electrons can be summarized by the equation:

$$C\, \partial T_e/\partial t = Q - L_{ei} - L_{en}\, \partial/\partial z(K\, \partial T_e/\partial z), \tag{VIII.12}$$

where C is the heat capacity of the electron gas, Q the energy input to the electrons, L_{ei} and L_{en} the energy losses through collisions with ions and neutrals respectively, and the derivative $\partial/\partial z(K \partial T_e/\partial z)$ expresses the gain or loss of energy through thermal conduction.

The term $C \partial T_e/\partial t$ is always negligible even at sunrise and sunset (Da Rosa, 1966). The rate of increase of electron temperature at sunrise is about 0.2°K s^{-1}, some 12° min^{-1} and the corresponding value of $C \partial T_e/\partial t$ is 2 eV cm^{-3} s^{-1} for an electron density $n_e = 10^5$ cm^{-3}. This value is reached by L_{ei} when the difference between the electron and ion temperatures $T_e - T_i$ is 10°K, i.e. in less than a minute.

For $n_e = 10^6$ cm^{-3}, $C \partial T_e/\partial t$ would be 30 eV cm^{-3} s^{-1}, but, since L_{ei} varies as n_e^2, it would be sufficient that $T_e - T_i = 1$°K for L_{ei} to reach that value.

Attempts have sometimes been made, starting from a situation where $T_n = T_e = T_i$ and $\partial T_e/\partial t = 0$, to use Equation (VIII.12), reduced to $C \partial T_e/\partial t = Q$, in order to interpret the observed $\partial T_e/\partial t$ in terms of Q. Such a procedure is incorrect because

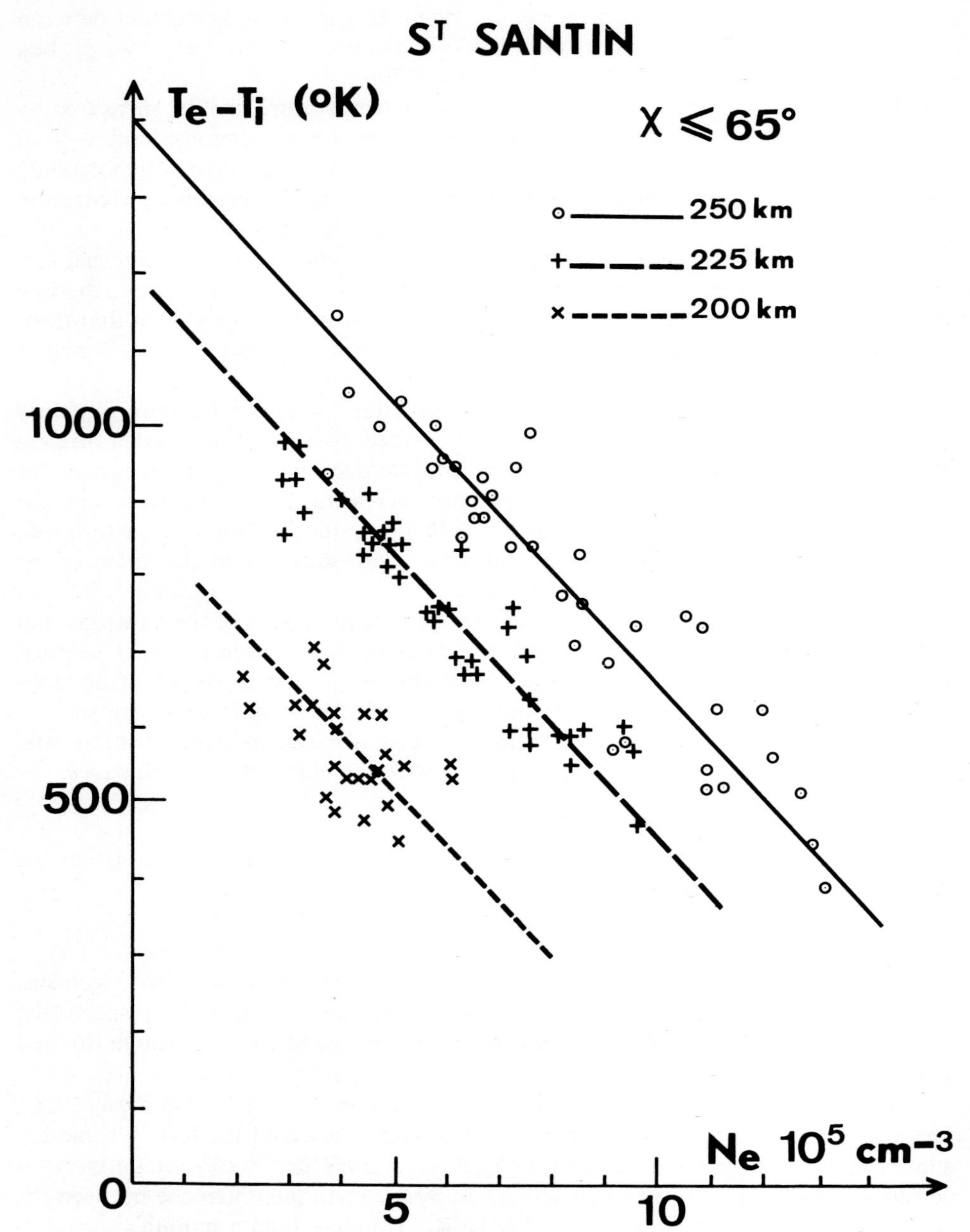

Fig. VIII.5. Values of $T_e - T_i$ as a function of the electron concentration n_e for all the data obtained over Saint-Santin during the year 1967, when the solar zenithal angle was smaller than 65°. The accuracy on $T_e - T_i$ is about 25°K; the accuracy on n_e is 2%. At a fixed altitude, all the data points are aligned on a straight line with a mean square scattering of about 60°K.

once the value of $\partial T_e/\partial t$ is measurable, the loss terms have already become preponderant. Therefore, the equilibrium equation

$$Q = L_{ei} + L_{en} - \partial/\partial z(K\partial T_e/\partial z) \quad \text{(VIII.13)}$$

is the correct one to use, even when T_e changes with time.

If simultaneous measurements of the profiles of electron concentration, and electron and ion temperatures, such as those obtained in incoherent scattering experiments, are available, all the elements necessary for the calculation of the terms L_{ei} and $\partial/\partial z(K \partial T_e/\partial z)$ are on hand. Moreover, we shall see below that the ion temperature T_i is identical to the neutral temperature T_n in the lower ionosphere, where the term L_{en} is important. Therefore, we can assume $T_n = T_i$. Furthermore, there exist models of the atmosphere that give the densities of the different species with good accuracy. Therefore, we can also calculate the term L_{en}, and thus all of the right hand side of the equation is known.

As for the left hand side, we have already examined the problem of the energy spectrum of the primary photoelectrons. We have also reviewed several essential results on the collisions that thermalize these photoelectrons. A first problem is that the collisions with neutrals are inelastic and the loss of energy does not occur continuously but by discrete steps. An additional difficulty in the calculation of the energy given up to the ambient electrons is the phenomenon sometimes called non-local heating. As we have mentioned, the photoelectrons can move along the lines of force of the terrestrial magnetic field and when the mean free path is large compared with the scale height, as is the case in the upper ionosphere, a photoelectron undergoes a large displacement before losing all its energy. A computer calculation can take this into account, the only difficulty being how to handle the changes in direction that the velocity undergoes during collisions (Banks and Nagy, 1970; Cicerone and Bowhill, 1970, 1971). This is where elastic collisions, which practically do not change the energy of the photoelectrons, nonetheless play a role, because they reduce their excursion in altitude.

Figure VIII.6 is an example of comparison between the two sides of the above equation. The left hand side Q_c was calculated, taking into account, besides the measured electron concentration, models for the neutral atmosphere and solar radiation. The right hand side Q_m was determined from the measured temperatures and electron concentration. These two quantities vary with altitude in very similar ways and this similarity is observed by all authors (Lejeune and Petit, 1969; Swartz and Nisbet, 1973; Lejeune, 1973). The limitations and significance of such comparisons are tied in with the uncertainties in the quantities that enter into Q_m and Q_c and are not directly accessible to measurement: the photoionization cross section of O is only known to within a factor of 2. The cross section of electron-neutral collisions is also subject to large errors, and it was only in 1968 that the existence of the most effective collision process, that which involves the fine structure of O, was discovered by Dalgarno and Degges (1968). Also, the models of the neutral atmosphere that must be used still give a very imperfect description of the seasonal variations in composition, particularly for the concentration of O_2. Finally, as we saw in Chapter VI, the solar flux of ionizing radiations of various wavelengths is not so well known. In this regard, it should be noted that in Figure VIII.6 Q_c had to be multiplied by a

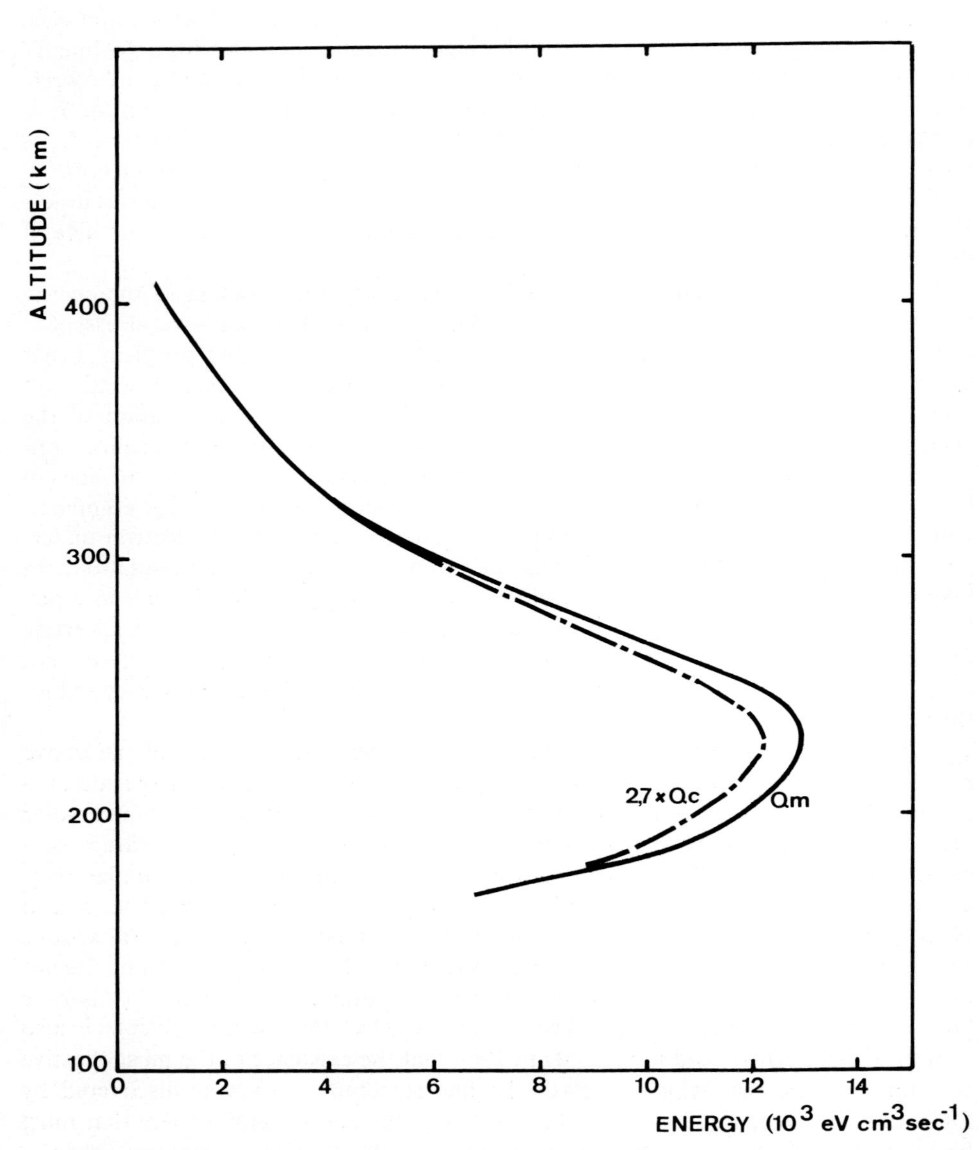

Fig. VIII.6. Energy balance of the electrons. Q_c represents the heat input to the thermal electrons from photoelectrons. Q_m represents the other terms of the energy balance: loss to ions, to neutral particles, thermal conduction which can be deduced from temperature measurements. Note that the horizontal scale is linear to appreciate the good quality of the fit.

factor of 2.7 in order to obtain agreement in absolute value with Q_m. Such a multiplication can be interpreted as implying the existence of a solar flux 2.7 times greater than had been assumed at first. All the comparisons we have already referred to, lead to similar conclusions, and certain comparative studies of the calculated and observed values of the electron concentration and ionic composition lead to analogous results (Swartz and Nisbet, 1973; Scialom, 1974).

We see that the study of the energy balance of the electrons in the ionosphere at temperate latitudes brings into question the value of the flux of solar ionizing radiation, the direct measurements of which are few, often fragmentary, and sometimes contradictory (Chapter VI).

8.2.3. ENERGY BALANCE OF THE IONS

An equation similar to Equation (VIII.13) describes the energy balance of ions. The energy input comes from electron-ion collisions, in mid-latitude regions, and is equal to L_{ei}, which represents a loss to the electrons, per unit volume, unit time. Therefore, we have

$$L_{ei} = L_{in} - \partial/\partial z(K_i \partial T_i/\partial z). \qquad \text{(VIII.14)}$$

In the lower atmosphere, because of the closeness in mass of the ions and neutral particles, and the large concentration of the latter, L_{in} has the form $\alpha(T_i - T_n)$, with α very large. Below 250 km, we must have $T_i = T_n$ within a few degrees, and measurements of T_i indeed give results in agreement with the generally accepted values of T_n.

On the other hand, within the magnetosphere, the concentration of neutral particles approaches zero, and the term L_{in} becomes zero no matter what T_i is. Thus we have (Figure VIII.1a)

$$L_{ei} = -\mathrm{d}/\mathrm{d}z(K_i \mathrm{d}T_i/\mathrm{d}z). \qquad \text{(VIII.15)}$$

Since K_i is much less than K_e, this equation implies (Banks, 1967a,b; Bauer, *et al.*, 1970a) that T_i is very close to T_e, $T_e - T_i$ being of the order of 100°K.

Thus the ion temperature should begin to increase at an altitude of 250 km and progressively approaches the electron temperature. This has been confirmed by experimental data (Evans, 1967).

With the help of the energy balance of the ions, measurements of ion temperature can be used to study the neutral thermosphere, whose temperature is otherwise difficult to measure directly.

8.2.3.1. *Measurement of Exospheric Temperature*

We have seen that at low altitude T_i is almost equal to T_n. Equation (VIII.14) can be used to calculate the small correction $T_i - T_n$, and furthermore Equation (VIII.14) reduces to

$$L_{ei} = L_{in} \qquad \text{(VIII.16)}$$

because the rate of conduction is negligible. To use this equation, n_e, T_e and T_i must be known, as is the case with incoherent scattering experiments. A model can then be used for the density of the neutral atmosphere, so that finally $T_i - T_n$, and hence T_n, can be deduced from Equation (VIII.16).

The results obtained can be compared with those deduced indirectly from the atmospheric drag of satellite orbits. The general agreement is good, but systematic differences appear, the discussion of which is outside the framework of this book. Let us recall, however, that atmospheric drag is a function of the density of the neutral atmosphere, and requires considerable modelling to produce exospheric temperatures. This increases interest in the derivation of neutral atmospheric density (Bauer *et al.*, 1970b) which we now describe.

8.2.3.2. *Measurement of Thermospheric Density*

If simultaneous or almost simultaneous measurements at several altitudes are available, Equation VIII.16 can be used to deduce not only T_n, but also the concentration of O, which is the major constituent at the altitudes considered, 300 km and higher. The basic idea is that the measurements below 300 km altitude permit a determination of the neutral temperature, which can be assumed to remain constant above. Thermal balance of ions in the upper F region can thus be used to deduce the O density. The actual procedure is as follows:

The temperature of the neutral atmosphere can be represented by an expression of the form:

$$T_n(z) = T_\infty - (T_\infty - T(z_0))e^{-S(z-z_0)}$$

which depends on the three parameters T_∞, $T(z_0)$ and S. On the other hand, O is in diffusive equilibrium, which implies (cf. Chapter I) that its density n(O) at an altitude z is known as a function of its density at a reference altitude z_0, provided the tem-

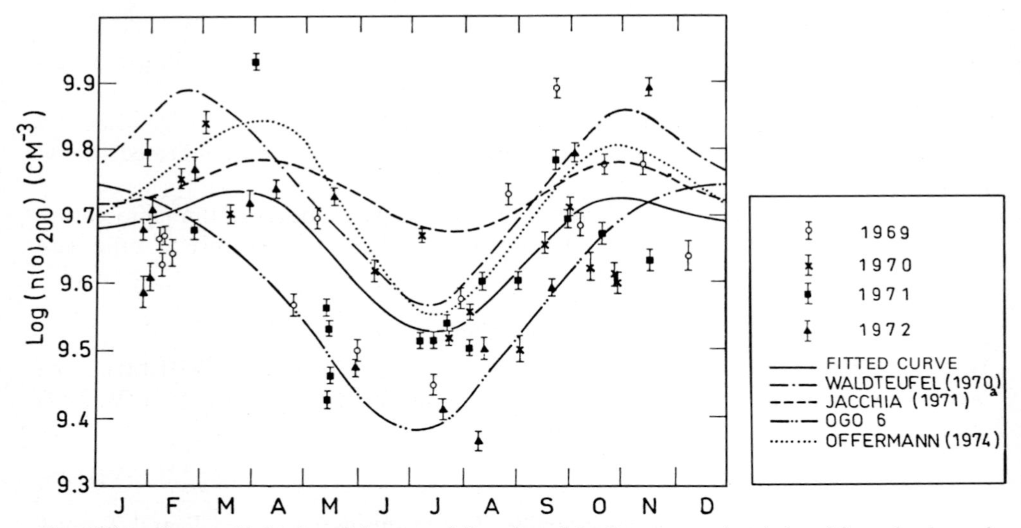

Fig. VIII.7. Seasonal behavior of the mean daily value of the O concentration deduced from the thermal budget of the ions (Alcayde, 1975). The full line indicates a least square fitting of the data by a semi-annual and annual variation superimposed to a constant value. Curves published by other authors using various techniques are given for comparison.

perature profile between z_0 and z is known. So, temperature and density are known at all altitudes, provided the unknown values of T_∞, S, $T(z_0)$ and $n(z_0)$ are determined. It suffices to have measurements from four different altitudes to be able to write four independent equations that correspond at each altitude to the energy balance described in Equation (VIII.16). If there are measurements available from more than four altitudes, the accuracy in the determination of T_∞, S, $T(z_0)$ and $n(z_0)$ can be improved by use of least square type methods. Examples of the results obtained are given in Figure VIII.7.

As in the case of the electrons, study of the energy balance of the ions thus allows the charged particles to be used to improve our knowledge of more general aspects of the terrestrial environment. This is why the study of the ionosphere through incoherent scattering has turned out to be, in an unforeseen way, a powerful means of investigation of the upper neutral atmosphere.

8.3. Geomagnetic Control of Ionospheric Temperature

8.3.1. THE GRADIENTS IN ELECTRON TEMPERATURE AT HIGH ALTITUDE

A general phenomenon at temperate latitudes is involved here. Such a temperature gradient implies a heat flux of the order of 10^9 to 10^{11} eV cm^{-2} s^{-1} continually issuing from the magnetosphere, and there must be a source that yields this same energy to the electrons in the magnetosphere. The photoelectrons created in the ionosphere that escape into the magnetosphere following magnetic field lines give part of their energy to the ambient electrons. We shall see that, because of the small number of neutral particles in the magnetosphere, the ions there are practically at the same temperature as the electrons, and the only way the charged particles can lose their energy is thermal conduction towards the ionosphere (Figure VIII.1a). In order for this conduction to take place, the temperature of the magnetosphere must reach, at the apex of the line of force, a value of the order of 3000 to 5000°K, in agreement with the measurements of Brace (1969) and Santani and Hanson (1970). The influx of photoelectrons seems to be large enough (Bauer *et al.*, 1970a) to explain the observed heat fluxes.

No gradient in electron temperature appears in measurements made in the neighbourhood of the magnetic equator (Farley *et al.*, 1967). Quite on the contrary, the electron temperature (Figure VIII.8) approaches that of the ions. This different behaviour is not surprising if it is borne in mind that the horizontal magnetic field inhibits all vertical conduction and also prevents any escape of the photoelectrons towards the magnetosphere.

8.3.2. THE NIGHTTIME *F* REGION TEMPERATURES

During the night there is no longer any significant production of photoelectrons and therefore electrons, ions and neutral particles must all tend towards the same temperature. We have mentioned that in the ionosphere the time constants for cooling are very small. It is another matter in the magnetosphere, where local loss mechanisms do not exist; the magnetospheric energy reservoir, heated during the day by

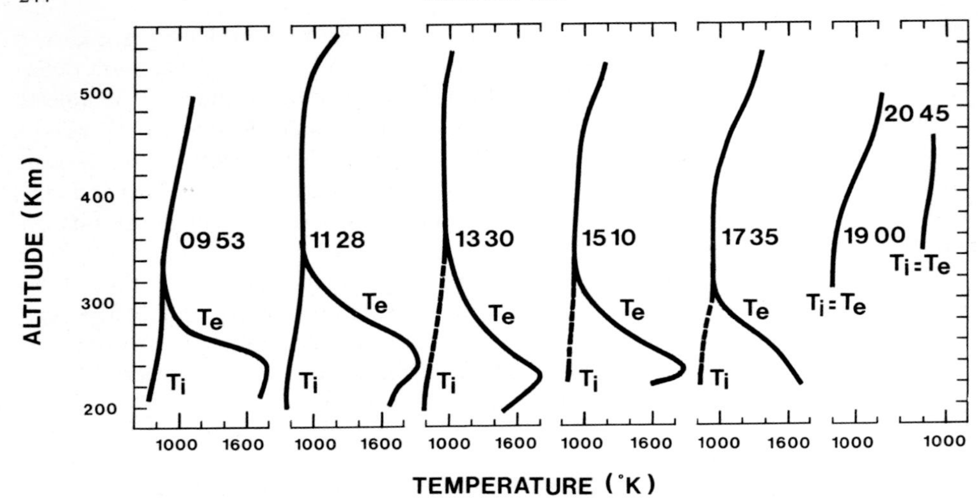

Fig. VIII.8. Typical electron and ion temperature profiles observed at the geomagnetic equator (Jicamarca incoherent scatter facility). Note the difference of behavior with the mid-latitude results (Figure VIII.4) in the upper F region: the horizontal magnetic field inhibits any vertical transport, reducing T_e at high altitudes.

photoelectrons, is thus progressively emptied into the ionosphere (Nagy *et al.*, 1968). The duration of this transient period depends on the length of the line of force, which is probably the reason why one observes at Nançay (France, $L = 1.8$) that the temperatures of electrons and ions are roughly equal during most of the night, while at Millstone Hill (near Boston, U.S.A., $L = 3.1$) Evans (1967) observed that the ratio T_e/T_i always remains greater than 1 (1.1 to 1.2).

Moreover, at Nançay, the nighttime electron temperature is slightly, but systematically, smaller than the ion temperature, the averaged $T_i - T_e$ value being 15° to 20°. This feature has been interpreted by Mazaudier (1974). The ions are strongly coupled to the neutral particles and their temperature keeps in the F region a small, but significant vertical gradient, as the neutral temperature does. The electron temperature would sustain a similar gradient, if it were exactly the same as the ion temperature. However, the thermal conductivity of the electrons is very large and such a gradient implies a significant downward flux of thermal energy. The electrons are cooled in this way below the ion temperature; equilibrium is achieved when the subsequent energy transfer via collisions compensate for the thermal flux. This process explains quantitatively the observed positive $T_i - T_e$ values.

Finally, we shall mention a spectacular effect due to photoelectrons that escape into the magnetosphere. During winter nights, as soon as the Sun rises at the magnetically conjugate point, the electron temperature increases (Carlson, 1966). This is caused by the photoelectrons created in the conjugate ionosphere, which have a double effect: firstly, after having traversed all the magnetosphere they deposit their energy locally; secondly, they heat the magnetosphere and thus cause a descending heat flux. Figure VIII.9, which shows the simultaneous increase of the electron temperature and the intensity of the line at 6300Å, is an argument in favor of this

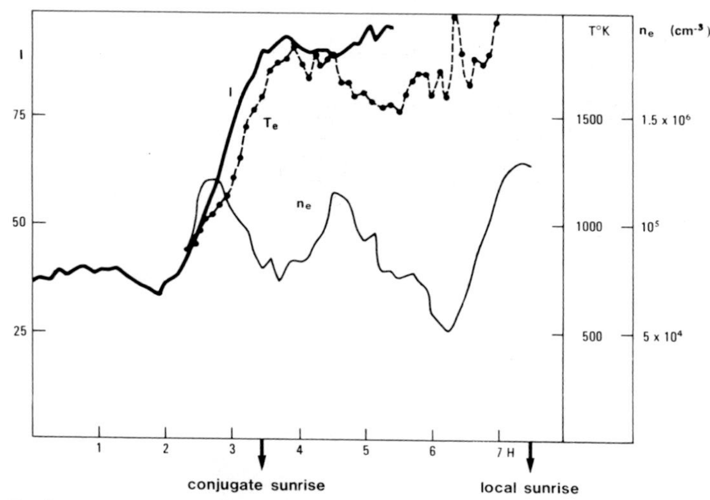

Fig. VIII.9. During a winter night, when the solar zenith angle at the conjugate point becomes smaller than 100° (about 1 h before sunrise at ground), photoelectrons created at the conjugate point and travelling along the magnetic field lines cause a simultaneous increase of both the intensity of the 6300Å O line (*I* expressed in Rayleighs) and the electron temperature, while electron concentration is essentially unaffected. (Duboin *et al.*, 1968)

interpretation, since the photoelectrons can be the cause of both observed phenomena. The increase starts about 1 h before conjugate sunrise and stops shortly after local sunrise, both the intensity of the optical line and the electron temperature saturating at the high values they have reached. The beginning of the phenomenon is consistent with the fact that production of photoelectron starts for solar zenith angles χ of 105 to 100°. Saturation coincides with the time when photoelectrons production becomes constant ($\chi = 90$ to 85°; see Figure VI.6) in the upper *F* conjugate region from which photoelectrons can escape. Plasma lines enhancement (Chapter IV, Section 4.2.2.) are another experimental evidence of the phenomenon. To avoid confusion, let us specify that the conjugate photoelectrons are important from an energetic point of view, but that their contribution to ionization remains small. Nonetheless, certain authors (Shawhan *et al.*, 1970), think that their role may not always be completely negligible.

8.3.3. JOULE HEATING IN THE HIGH LATITUDES

In the auroral zone, the solar E.U.V. flux is no longer the major heat source. As mentioned in Section 1.6, electric fields provide a significant energy input to ions and neutrals. On the other hand, electrons are heated by elastic scattering of the secondary electrons produced by the primary energetic particles bombardment.

8.3.3.1. *Exospheric Temperature*

Blamont and Luton (1972) have deduced the temperature of the neutral components of the atmosphere from the profile of the 6300Å O line in the dayglow observed from the OGO-6 satellite. Exospheric temperature is high in the polar region even during

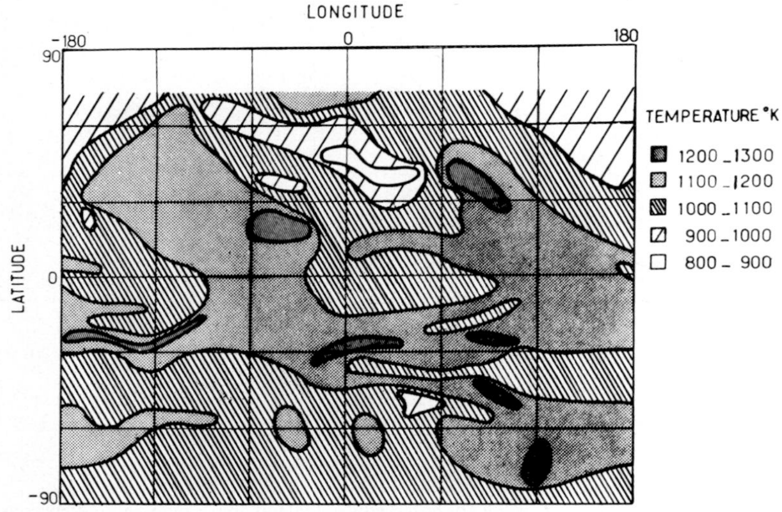

Fig. VIII.10. Neutral temperature map deduced from the profile of the 6300Å O line measured on OGO 6 at 260 km altitude for a quiet day ($K_p \sim 0$): September 27, 1969.

quiet days (Figure VIII.10). Geomagnetic activity affects mainly the polar regions where the temperature increases by several hundred degrees and changes very rapidly (Figure VIII.11).

Even in the lower latitude regions, an influence of geomagnetic activity on the neutral temperature has been evidenced from satellite-drag data and from incoherent scatter results. Jacchia (1965), using the first approach, found that the exospheric

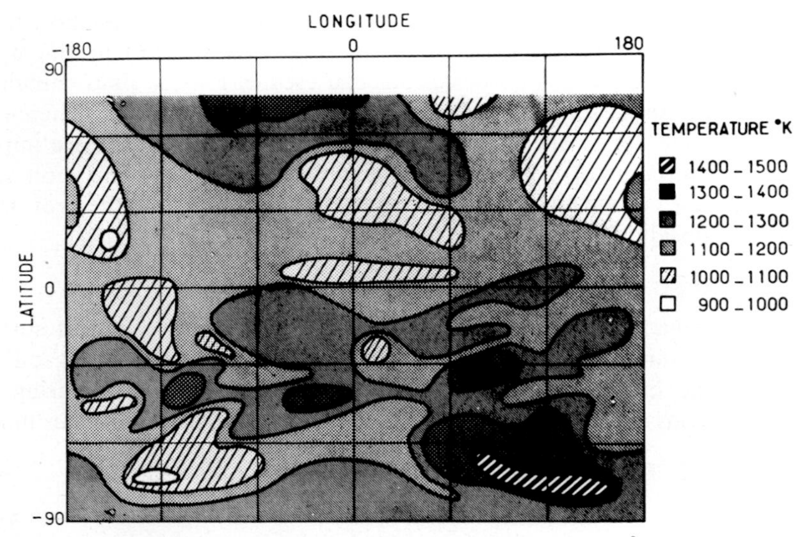

Fig. VIII.11. Neutral temperature map deduced from the profile of the 6300Å O line measured on OGO 6 at 260 km altitude during a geomagnetic storm ($K_p \sim 5$): September 28, 1969.

temperature varies as $28K_p$, where K_p is the conventional magnetic index 6 h before the temperature measurement. A similar linear relationship with K_p has been observed at Saint-Santin by Waldteufel (1970) and at Arecibo by Waldteufel and Cogger (1971); the main difference is with the value of the coefficient which is evaluated to be 16 at Saint-Santin and 12 at Arecibo.

Auroral heating of the neutral atmosphere is clearly an important phenomenon affecting the whole dynamics of the thermosphere by inducing pressure gradients and subsequent winds. Furthermore, the expansion of the neutral atmosphere when heated might drive the ionospheric plasma in the F region upward along magnetic field lines (Bates *et al.*, 1973a). Ionization produced by auroral precipitations and transported upward by this process could cause the auroral F layer.

8.3.3.2. *Ion Temperature*

Rees and Walker (1968) have shown that an enhanced ion temperature results as a consequence of a substantial electric field normal to the magnetic field. Such Joule heating of F region ions has been actually detected by Bates *et al.* (1973b): their

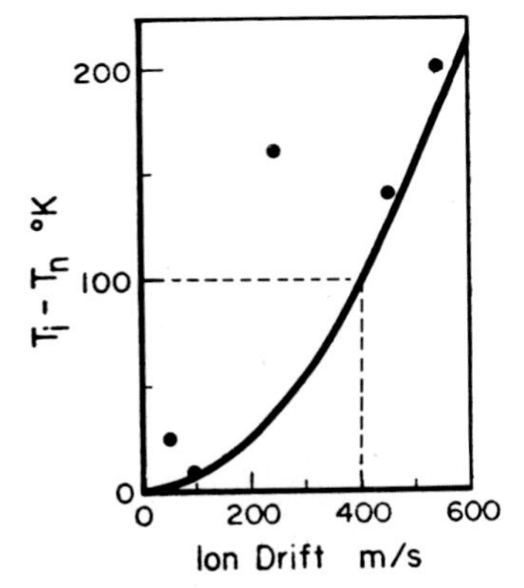

Fig. VIII.12. Dots are the observed difference between ion (T_i) and neutral (T_n) temperatures versus westward ion drift velocity. The continuous curve is the theoretical estimation deduced from Equation (VIII.18).

incoherent scatter observations show (Figure VIII.12) ion temperatures correlated with local magnetic activity and with the westward ion drift velocity. These data can be explained neither by an increase of the neutral temperature nor by an electron temperature enhancement increasing the heat transfer from the electrons to the ions. It is therefore concluded that the horizontal electric field responsible for the ion drift also elevates the ion temperature.

Assuming that the energy loss of the ions is mainly due to the elastic collisions with neutrals responsible for frictional heating, a simple relationship can be established between the temperature difference of ions and neutrals and their relative velocity. This result is derived by considering neutrals as an infinite heat reservoir, unaffected by Joule heating. If the heat input per unit volume is the same for ions and neutrals, the heat input per particle is much smaller for the denser neutral gas.

Equating the heat input derived from Equation (VIII.9) to the heat loss derived from Equation (VIII.3) (see paragraph 1.6)

$$\nu_{\text{in}} \frac{m_{\text{i}} m_{\text{n}}}{m_{\text{i}} + m_{\text{n}}} n_{\text{i}} (v_{\text{i}} - v_{\text{n}})^2 = 3n_{\text{i}} \frac{m_{\text{i}}}{m_{\text{i}} + m_{\text{n}}} \nu_{\text{in}} k (T_{\text{i}} - T_{\text{n}}) \qquad \text{(VIII.17)}$$

leads to the simple result

$$3k(T_{\text{i}} - T_{\text{n}}) = m_{\text{n}} (v_{\text{i}} - v_{\text{n}})^2. \qquad \text{(VIII.18)}$$

The temperature difference depends neither on particle concentration nor on collision frequencies. It varies as the square of the relative velocity and is 100°K for a relative velocity of 400 m s^{-1}. This is consistent with the data (Figure VIII.12).

8.3.4. ELECTRON TEMPERATURE IN THE AURORA

Many papers have been devoted to quantitative discussions of the thermal budget of the electrons at mid-latitude, taking into account the actually observed temperatures. As far as high latitude phenomena are concerned, the present state of the art is much less developed.

On the one hand, theoretical calculations of electron temperature have been performed, taking into account a heat input due either to likely electric fields or to electron precipitation (Rees and Walker, 1968). Electron precipitation is found to be the major heat source, except for extreme values of the electric field (50 mV m^{-1}) for

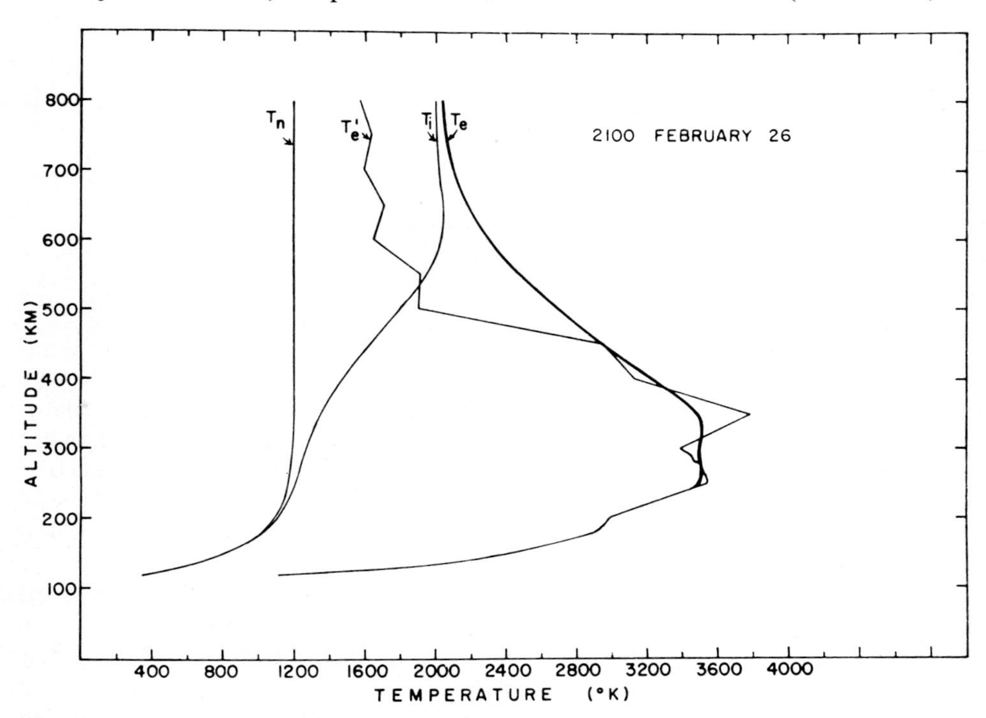

Fig. VIII.13. Altitude profiles of electron temperature T_{e}, ion temperature T_{i} and neutral gas temperature T_{n}, inferred from $\lambda = 3914$Å volume emission rates. T_{e}' is the electron temperature calculated without thermal conduction (the zigzags result from computational approximations).

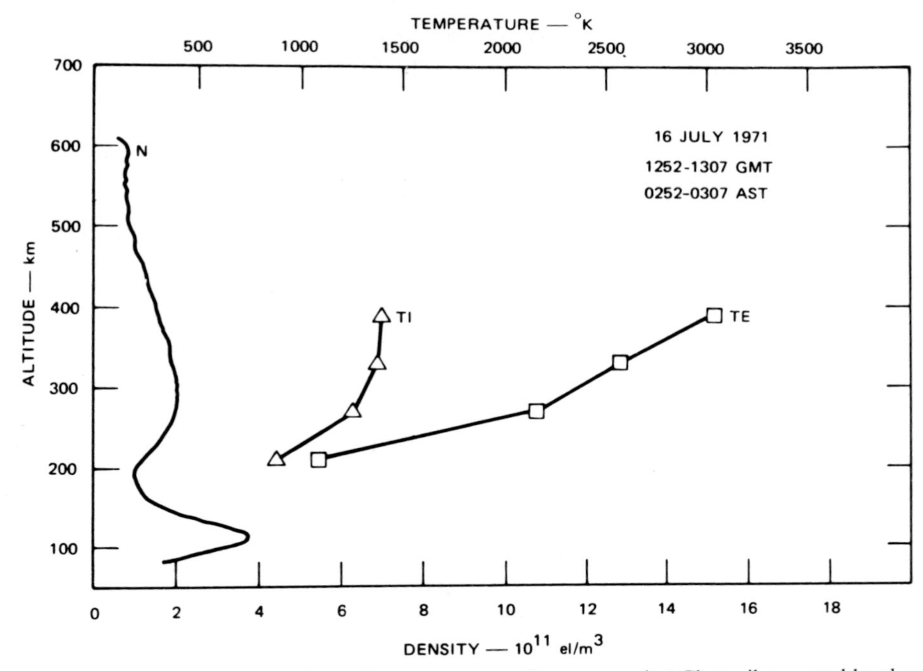

Fig. VIII.14. Representative density and temperature profiles measured at Chatanika around local magnetic midnight for a solar zenith angle of 90°. A very prominent auroral *E* layer can be seen centered at about 120 km altitude.

which Joule heating becomes dominant at low altitudes and important at high altitudes. Using height profiles of 3914Å radiation (N_2 ionization) volume emission rates, Walker and Rees (1968) have computed the temperature of the electron gas (Figure VIII.13) heated by secondary electrons resulting from the precipitated 10 keV electrons, taking into account conduction and the various losses described in Section 1 (except the one involving the fine structure level of the ground state of O).

On the other hand, electron temperature enhancements clearly related to auroral oval phenomena have been observed (Figure VIII.14) by the incoherent scatter facility of Chatanika (Watt, 1973). However, no attempt has been made up to now to study the electron energy budget by using temperature measurements and heat input estimations, deduced independently from simultaneous electron precipitation and electric field measurements. A basic difficulty is the rapid variations in space and time of auroral phenomena. Another problem of minor importance lies with the neutral atmosphere which might differ significantly from conventional models.

8.3.5. SUBAURORAL ARCS

The red arcs (SAR-arcs) which occur at subauroral latitudes were first observed by Barbier in 1956 in southern France. The SAR-arcs are subvisual and require sensitive instruments to be detected. They are stable homogeneous bands of the 6300Å emission line of O, roughly aligned in an east–west direction. The arcs observed by Barbier were unusual because they occurred at mid-latitudes, well equatorward of the auroral zone. They are generally separated by several degrees of geomagnetic latitude from the aurora. They are several hundred kilometers wide in the meridional direction. They extend from 300 to 700 km in altitude, and in longitude around the nightside of the globe and even possibly around the Earth.

The SAR-arcs appear from near the equatorward edge of the ionospheric trough (Figure VII.19). The electron temperature within the SAR-arcs is considerably enhanced compared with the neighbouring regions. According to Cole (1965), this increases the number of electrons in the high energy tail of the Maxwellian distribution which excites O to the ^{1}D level, and causes the optical phenomenon. Such an electron temperature enhancement is attributed to thermal conduction transferring energy from the magnetosphere down to the ionosphere, where electron concentration is small enough (10^4 to 10^5 cm^{-3}) to reduce the cooling by ions in the upper F region.

A theoretical modelling (Rees and Roble, 1975) predicts that a 20 kR SAR-arc corresponds to a topside heat flow of 2.5 10^{11} eV cm^{-2} s^{-1}. The altitude of the peak emission varies from 360 km to 380 km, in good agreement with the observations by Tohmatsu and Roach (1962).

The source of the magnetospheric heating which is the prime cause of the phenomenon is still under discussion. A good candidate is Landau damping of ion cyclotron waves generated by an anisotropic pitch angle distribution of trapped (ring current) protons in the outer plasmasphere.

REFERENCES

Adams, G. W. and Megill, L. R.: 1967, *Planetary Space Sci.* **15**, 1111.
Akasofu, S. I.: 1964, *Planetary Space Sci.* **12**, 273.
Akasofu, S. I. and Meng, C. I.: 1969, *J. Geophys. Res.* **74**, 293.
Akasofu, S. I., Meng, C. I. and Kimball, D. S.: 1966, *J. Atmospheric Terrest. Phys.* **28**, 489.
Aikin, A. C. and Goldberg, R. A.: 1973, *J. Geophys. Res.* **78**, 1229.
Alcayde, D.: 1975, *Structure et thermodynamique de la thermosphère à moyenne latitude,* Doctorat d'Etat, Fac. des Sciences de Toulouse.
Alcayde, D., Bauer, P. and Fontanari, J.: 1974, *J. Geophys. Res.* **79**, 629.
Alfvén, H.: 1950, *Cosmical Electrodynamics*, Clarendon Press, Oxford, England.
Allen, C. W.: 1948, *Terr. Magn. Atm. Elect.* **53**, 433.
Amayenc, P.: 1975, *Mouvements thermosphériques de grande échelle. Etude par diffusion incohérente à moyenne latitude.* Doctorat d'Etat, Université de Paris VI.
Ananthakrishnan, S. and Ramanathan, K. R.: 1969, *Nature* **223**, 488.
Appleton, E. V.: 1925, *Proc. Phys. Soc. London* **37**, 16 D.
Appleton, E. V.: 1946, *Nature* **157**, 691.
Appleton, E. V. and Barnett, M. A. F.: 1925a, *Nature* **115**, 333.
Appleton, E. V. and Barnett, M. A. F.: 1925b, *Proc. R. Sci.* **A109**, 621.
Appleton, E. V.: 1925, *Proc. Phys. Soc. London* **37**, 16 D.
Appleton, E. V. and Piggott, W. R.: 1948, *Proc. URSI* **7**, 320.
Appleton, E. V. and Piggott, W. R.: 1954, *J. Atmospheric Terrest. Phys.* **5**, 141.
Aubry, M. P., Russel, C. T. and Kivelson, M. G.: 1970, *J. Geophys. Res.* **75**, 7018.
Auer, G., Cap, F., Floriani, D. and Cratzl, H.: 1970, Institute for theoretical physics, University of Innsbruck, Austria, Scientific report n° 75.
Axford, W. I.: 1968, *J. Geophys. Res.* **73**, 6855.
Axford, W. I. and Cunnold, D. M.: 1966, *Radio Sci.* **1**, 191.
Axford, W. I., Cunnold, D. M. and Gleeson, L. J.: 1966, *Planetary Space Sci.* **14**, 909.
Axford, I. W. and Hines, C. O.: 1961, *Can. J. Phys.* **39**, 1433.
Baker, W. G.: 1953, *Phil. Trans. Roy. Soc. London A* **246**, 281.
Baker, W. G. and Martyn, D. F.: 1953, *Phil. Trans. Roy. Soc. London A* **246**, 295.
Bailey, D. K., Bateman, R., Berkner, L. V., Booker, H. G., Montgomery, G. F., Purcell, E. M., Salisbury, W. W. and Wiemer, J. B.: 1952, *Phys. Rev.* **86**, 141.
Banks, P. M.: 1966a, *Planetary Space Sci.* **14**, 1085.
Banks, P. M.: 1966b, *Planetary Space Sci.* **14**, 1105.
Banks, P. M.: 1966c, *Ann. Geophys.* **22**, 577.
Banks, P. M.: 1967a, *J. Geophys. Res.* **72**, 3365.
Banks, P. M.: 1967b, *Planetary Space Sci.* **15**, 77.
Banks, P. M. and Holtzer, T. E.: 1968, *J. Geophys. Res.* **73**, 6846.
Banks, P. M. and Holtzer, T. E.: 1969a, *J. Geophys. Res.* **74**, 6304.
Banks, P. M. and Holtzer, T. E.: 1969b, *J. Geophys. Res.* **74**, 6317.
Banks, P. M. and Kockarts, G.: 1973, *Aeronomy*, Academic Press, New York and London.
Banks, P. M. and Nagy, A. F.: 1970, *J. Geophys. Res.* **75**, 1902.
Banks, P. M., Nagy, A. F. and Axford, W. I.: 1971, *Planetary Space Sci.* **19**, 1053.
Banks, P. M., Rino, C. L. and Wickwar, V. B.: 1974, *J. Geophys. Res.* **79**, 187.
Barbier, D. and Glaume, J.: 1962, *Planetary Space Sci.* **9**, 133.

Bardsley, J. N.: 1970, *Phys. Rev. A* **2**, 1359.
Barrington, R. E. and Thrane, E. V.: 1962, *J. Atmospheric Terrest. Phys.* **24**, 31.
Bates, H. F., Belon, A. E. and Hunsucker, R. D.: 1973a, *J. Geophys. Res.* **78**, 648.
Bates, H. F., Belon, A. E., Watkins, B., Hunsucker, R. D. and Baron, M. J.: 1973b, *J. Geophys. Res.* **78**, 1723.
Bauer, P., Lejeune, G. and Petit, M.: 1970a, *Planetary Space Sci.* **18**, 1447.
Bauer, P., Waldteufel, P. and Alcayde, D.: 1970b, *J. Geophys. Res.* **75**, 4825.
Baum, W. A., Johnson, F. S., Oberly, J. J., Rockwood, C. C., Strain, C. V. and Tousey, R.: 1946, *Phys. Rev.* **70**, 781.
Beckmann, P. and Spizzichino, A.: 1963, *The Scattering of Electromagnetic Waves From Rough Surfaces*, Pergamon Press, Oxford, London, New York, Paris.
Belrose, J. S.: 1970, *J. Atmospheric Terrest. Phys.* **32**, 567.
Berger, M. J., Seltzer, S. M. and Maeda, K.: 1974, *J. Atmospheric Terrest. Phys.* **36**, 591.
Bernard, R.: 1974, *Radio Sci.* **9**, 295.
Berthelier, J. J. and Godard, R.: 1972, *Space Res.* **12**, 1357.
Beynon, W. J. G. and Davies, K.: 1954, *J. Atmospheric Terrest. Phys.* **5**, 273.
Biondi, M. A.: 1969, *Can. J. Chem.* **47**, 1711.
Birkeland, K. : 1908–1913, *The Norwegian Aurora Polaris Expedition* 1902–1903. Vol. I. Sections 1 and 2, H. Ascheharg and Co., Christiania.
Blamont, J. E. and Barat, J.: 1967, *Ann. Géophys.* **23**, 173.
Blamont, J. E. and Barat, J.: 1968, *Ann. Géophys.* **24**, 375.
Blamont, J. E. and De Jaeger, C.: 1961, *Ann. Géophys.* **17**, 134.
Blamont, J. E. and Luton, J. M.: 1972, *J. Geophys. Res.* **77**, 3534.
Blanc, M.: 1975, *Field Aligned Plasma Flow and the Polar Wind, a Review*, Paper presented at the EISCAT Summer School June 5–13, Tromso, Norway.
Booker, H. G.: 1956, *J. Atmospheric Terrest. Phys.* **8**, 204.
Booker, H. G.: 1958, *P.I.R.E.* **46**, 298.
Booker, H. G.: 1974, *J. Atmospheric Terrest. Phys.* **36**, 2113.
Bossolasco, M. and Elena, A.: 1963, *Compt. Rend. Acad. Sci. Paris*, **256**, 4491.
Bourdeau, R. E.: 1961, *Space Res.* **2**, 554.
Bowen, P. J., Boyd, R. L. F., Henderson, C. L. and Willmore, A. P.: 1964, *Proc. Roy. Soc.* (*London*) *A* **281**, 514.
Bowhill, S. A.: 1958, *J. Atmospheric Terrest. Phys.* **13**, 175.
Bowles, K. L.: 1958, *Phys. Rev. Letters* **1**, 454.
Bowles, K. L.: 1961, *J. Res. Natl. Bur. Std.* **65D**, 1.
Boyd, R. L. F. and Raitt, W. J.: 1965, *Space Res.* **5**, 207.
Brace, L. H.: 1969, *Goddard Space Flight Center* Report No. X 261–69–158.
Brace, L. H., Carigan, G. R. and Findlay, J. A.: 1971, *Space Res.* **11**, 1079.
Brace, L. H., Carlson, H. C. and Mahajan, K. K.: 1969, *J. Geophys. Res.* **74**, 1883.
Bracewell, R. N., Budden, K. G., Ratcliffe, J. A., Straker, J. W., and Weekes, K.: 1951, *Proc. Inst. Elec. Engrs. London* **98**, 221.
Bramley, E. N. and Rüster, R.: 1971, *J. Atmospheric Terrest. Phys.* **33**, 269.
Bramley, E. N. and Young, M.: 1968, *J. Atmospheric Terrest. Phys.* **30**, 99.
Bricard, J. and Kastler, A.: 1944, *Ann. Géophys.* **1**, 53.
Brice, N. M.: 1967, *J. Geophys. Res.* **72**, 5193.
Brown, R. R., Barcus, J. R. and Parsons, N. R.: 1965, *J. Geophys. Res.* **70**, 2579.
Budden, K. G.: 1961, *Radio Waves in the Ionosphere*, Cambridge University Press, Cambridge.
Buneman, O.: 1961, *J. Geophys. Res.* **66**, 1978.
Burke, R. R.: 1972a, in C. F. Sechrist and M. A. Geller (eds.) *COSPAR Symposium on D- and E-Region Chemistry*, Aeronomy Report no. 48, Aeronomy Laboratory, University of Illinois, Urbana, Illinois, USA, p. 259.
Burke, R. R.: 1972b, *J. Geophys. Res.* **77**, 1998.
Calvert, W. and Schmid, C. W.: 1964, *J. Geophys. Res.* **69**, 1839.
Carlson, H. C., Jr.: 1966, *J. Geophys. Res.* **71**, 195.
Carlson, H. C. and Sayers, J.: 1970, *J. Geophys. Res.* **75**, 4883.

Carpenter, D. L.: 1966, *J. Geophys. Res.* **71**, 693.
Carru, H., Petit, M. and Waldteufel, P.: 1967, *Compt. Rend. Acad. Sci. Paris* **264**, 560.
Cauffman, D. P. and Gurnett, D. A.: 1972, *Space Sci. Rev.* **13**, 369.
Chalmers, A. J.: 1957, *Atmospheric Electricity*, Pergamon Press, Oxford.
Chanin, L. M., Phelps, A. V. and Biondi, M. A.: 1959, *Phys. Rev. Letters* **2**, 344.
Chapman, S.: 1931, *Proc. Phys. Soc. London,* **43**, 26 and 483.
Chapman, S.: 1968, *Ann. Géophys.* **24**, 497.
Chapman, S. and Cowling, T. G.: 1952, *The Mathematical Theory of Non-Uniform Gases*, Cambridge Univ. Press, New York and London.
Chapman, S. and Lindzen, R. S.: 1970, *Atmospheric Tides*, D. Reidel Publishing Co., Dordrecht, Holland.
Chappell, C. R.: 1972, in B. M. McCormac (ed.) *Earth's Magnetospheric Processes*, D. Reidel Publishing Company, Dordrecht, Holland, p. 280.
Chesnut, W. G., Hodges, J. C. and Leadabrand, R. L.: 1968, *Auroral Backscatter Wavelength Dependence Studies*, Final Report, Project 5535, Stanford Res. Inst., Menlo Park, California.
Chimonas, G. and Axford, W. I.: 1968, *J. Geophys. Res.* **73**, 111.
Chimonas, G. and Hines, C. O.: 1970, *J. Geophys. Res.* **75**, 875.
Cicerone, R. J. and Bowhill, S. A.: 1970, *Aeron. Report 39*, Aeron. Lab., University of Illinois, Urbana.
Cicerone, R. J. and Bowhill, S. A.: 1970, *Radio Sci.* **5**, 49.
Cicerone, R. J. and Bowhill, S. A.: 1971, *J. Geophys. Res.* **76**, 8299.
Clark, D. H., Raitt, N. J. and Willmore, A. P.: 1972, *J. Atmospheric Terrest. Phys.* **34**, 1865.
Cole, K. D.: 1965, *J. Geophys. Res.* **70**, 1689.
Colegrove, F. D., Hanson, W. B. and Johnson, F. S.: 1965, *J. Geophys. Res.* **70**, 4931.
Cornelius, D. W., Joyner, K. H., Dyson, P. L. and Butcher, E. C.: 1975, *J. Atmospheric Terrest. Phys.* **37**, 769.
Coroniti, F. V. and Kennel, C. F.: 1973, *J. Geophys. Res.* **78**, 2837.
Cox, L. P. and Evans, J. V.: 1970, *J. Geophys. Res.* **75**, 6271.
Cuchet, L.: 1965, *Ann. Géophys.* **21**, 477.
Cuchet, L.: 1968, *La théorie de la formation de la Couche Ionosphérique E Sporadique dans les Régions Tempérées*, Doctorat d'Etat, Faculté des Sciences, Paris.
Culhane, J. L., Willmore, A. P., Pounds, K. A. and Sanford, P. W.: 1964, *Space Res*, **4**, 741.
Dalgarno, A., McElroy, M. B. and Moffett, J. R.: 1963, *Planetary Space Sci.* **11**, 463.
Dalgarno, A.: 1969, *Can. J. Chem.* **47**, 1723.
Dalgarno, A. and Degges, T. C.: 1968, *Planetary Space Sci.* **16**, 125.
Dalgarno, A., Henry, R. J. W. and Stewart, A. L.: 1964, *Planetary Space Sci.* **12**, 235.
Da Rosa, A. V.: 1966, *J. Geophys. Res.* **71**, 4107.
Davies, K.: 1969, *Ionospheric Radio Waves*, Blaisdell Publishing Company.
De Forest, L.: 1913, *Proc. Inst. Radio Engrs.* **1**, 13.
Denny, B. W. and Bowhill, S. A.: 1973, *Partial Reflection Studies of D Region Winter Variability*, Aeronomy Report n° 56, University of Illinois, Urbana, U.S.A.
Derblom, H. and Ladell, L.: 1973, *J. Atmospheric Terrest Phys.* **35**, 2123.
Dickinson, R. E.: 1972, *Space Res.* **12**, 1015.
Donahue, T. M.: 1968, *Science* **159**, 489.
Donahue, T. M., Zipf, E. C. and Parkinson, T. D.: 1970, *Planetary Space Sci.* **18**, 171.
Dougherty, J. P. and Farley, D. T.: 1960, *Proc. Roy. Soc.*, A **259**, 79.
Doupnik, J. R. and Schmerling, E. R.: 1965, *J. Atmospheric Terrest. Phys.* **27**, 917.
Duboin, M. L., Lejeune, G., Petit, M. and Weill, G.: 1968, *J. Atmospheric Terrest. Phys.* **30**, 299.
Dungey, J. W.: 1959, *J. Geophys. Res.* **64**, 2122.
Dungey, J. W.: 1962, in De Witt, Hiéblot and Lebeau (eds.) *Geophysics, The Earth's Environment*, Gordon and Breach, New York, p. 503.
Drukarev, G.: 1946, *J. Phys.* (*USSR*) **10**, 81.
Eccles, W. H.: 1912, *Proc. Roy. Soc. A* **87**, 79.
Edwards, P. J., Burtt, G. J. and Knox, F.: 1969, *Nature* **222**, 1053.
Egeland, A.: 1971, *Interference of Radio Aurora on Incoherent Scatter Signals at High Latitude*, The Norwegian Institute of Cosmic Physics, Report n° 713.

Elford, W. G. and Roper, R. G.: 1967, *Space Res.* **7**, 42.
Evans, J. V.: 1965, *J. Geophys. Res.* **70**, 733.
Evans, J. V.: 1967, *Planetary Space Sci.* **15**, 1387.
Evans, J. V.: 1972, *J. Geophys. Res.* **77**, 2341.
Evans, J. V.: 1973, *J. Atmospheric Terrest. Phys.* **35**, 393.
Evans, J. V.: 1974, *J. Atmospheric Terrest. Phys.* **36**, 2183.
Evans, J. V. and Cox, L. P.: 1970, *J. Geophys. Res.* **75**, 159.
Eyfrig, R. W.: 1963, *Ann. Géophys.* **19**, 102.
Fabry, Ch.: 1928, *Compt. Rend. Acad. Sci.* Paris **187**, 777.
Farley, D. T.: 1963, *J. Geophys. Res.* **68**, 6083.
Farley, D. T. and Balsley, B. B.: 1973, *J. Geophys. Res.* **78**, 227.
Farley, D. T.: 1974, *Rev. Geophys.* **12**, 285.
Farley, D. T., Balsley, B. B., Woodman, R. F. and McClure, J. T.: 1970, *J. Geophys. Res.* **75**, 7199.
Farley, D. T., Dougherty, J. P. and Baron, D. W.: 1961, *Proc. Roy. Soc. London A***263**. 238.
Farley, D. T. and Fejer, B. G.: 1975, *J. Geophys. Res.* **80**, 3087.
Farley, D. T., McClure, J. P., Sterling, D. L. and Green, J. L.: 1967, *J. Geophys. Res.* **72**, 5837.
Fedor, L. S.: 1967, *J. Geophys. Res.* **72**, 5401.
Fejer, J. A.: 1955, *J. Atmospheric Terrest. Phys.* **7**, 322.
Fejer, J. A.: 1960, *Can. J. Phys.* **38**, 1114.
Fejer, J. A.: 1963, *J. Geophys. Res.* **68**, 2147.
Fejer, J. A.: 1967, *Can. J. Phys.* **39**, 716.
Feldstein, R. and Graff, P.: 1972, *J. Geophys. Res.* **77**, 1896.
Ferguson, E. E.: 1974, *Rev. Geophys.* **12**, 703.
Ferguson, E. E. and Fehsenfeld, F. C.: 1968, *J. Geophys. Res.* **73**, 6215.
Ferguson, E. E., Fehsenfeld, F. C. and Schmeltekopf, A. L.: 1969, *in* D. R. Bates and I. Esterman (eds.), *Advances in Atomic and Molecular Physics*, Academic Press, New York and London, Vol. 5, p. 1.
Findlay, J. A. and Brace, L. H.: 1969, *Proc. I.E.E.E.* **57**, 1054.
Fisk, J. B.: 1936, *Phys. Rev.* **43**, 167.
Fjeldbo, G., Eshelman, V. R., Garriott, O. K. and Smith, F. L.: 1965, *J. Geophys. Res.* **70**, 3701.
Fleury, L. and Taieb, C.: 1971, *J. Atmospheric Terrest. Phys.* **33**, 909.
Folkestad, K.: 1972, *J. Atmospheric Terrest. Phys.* **34**, 1935.
Fontanari, J. and Alcayde, D.: 1974, *Radio Sci.* **9**, 275.
Francis, S. H. and Perkins, F. W.: 1975, *J. Geophys. Res.* **80**, 3111.
Francis, W. E. and Bradbury, J. N.: 1975, Report no. D 409 467, Lockheed, Palo Alto Res. Lab.
Frank, L. A. and Gurnett, D. A.: 1971, *J. Geophys. Res.* **76**, 6829.
Friedman, H., Lichtmann, S. W. and Byram, E. T.: 1951, *Phys. Rev.* **83**, 1025.
Fritz, H.: 1874, *Petermann's Geograph. Mitt.* **20**, 347.
Fuller, L. F.: 1915, *Trans. Am. Inst. Elec. Engrs.* **34**, 809.
Gadsden, M.: 1967, *Planetary Space Sci.* **15**, 893.
Gadsden, M.: 1970, *Ann. Géophys.* **26**, 141.
Gardner, F. F. and Pawsley, J. L.: 1953, *J. Atmospheric Terrest. Phys.* **3**, 321.
Garriott, O. K., Da Rosa, A. V., Davis, M. J. and Villard, O. G.: 1967, *J. Geophys. Res.* **72**, 6099.
Garriott, O. K., Da Rosa, A. V. and Ross, W. J.: 1970, *J. Atmospheric Terrest. Phys.* **32**, 705.
Geisler, J. E.: 1966, *J. Atmospheric Terrest. Phys.* **28**, 703.
Geisler, J. E.: 1967, *J. Atmospheric Terrest. Phys.* **29**, 1469.
Geisler, J. E. and Bowhill, S. A.: 1965a, *J. Atmospheric Terrest. Phys.* **27**, 457.
Geisler, J. E. and Bowhill, S. A.: 1965b, *J. Atmospheric Terrest. Phys.* **27**, 1119.
Giraud, A.: 1966, *Compt. Rend. Acad. Sci. Paris* **262**, 566.
Goldberg, R. A. and Aikin, A. C.: 1971, *J. Geophys. Res.* **76**, 8352.
Gordon, W. E.: 1958, *Proc. I.R.E.* **46**, 1824.
Gray, R. W. and Farley, D. T.: 1973, *Radio Sci.* **8**, 123.
Green, J. S. A.: 1946, *Amalgamated Wireless of Australasia. Tech. Rev.* **7**, 177.
Green, A. F. and Stolarsky, R. S.: 1972, *J. Atmospheric Terrest. Phys.* **34**, 1703.
Gurnett, D. A. and Frank, L. A.: 1973, *J. Geophys. Res.* **78**, 145.
Hagfors, T.: 1961, *J. Geophys. Res.* **66**, 1699.

Hagg, E. L.: 1967, *Can. J. Phys.* **45**, 27.
Hanson, W. B., Brace, L. H., Dyson, P. L. and McClure, J. P.: 1969, *J. Geophys. Res.* **74**, 400.
Hanson, W. B., McClure, J. P. and Sterling, D. L.: 1973, *J. Geophys. Res.* **78**, 2353.
Hanson, W. B. and Ortenburger, I. B.: 1961, *J. Geophys. Res.* **66**, 1425.
Hanson, W. B. and Sanatani, S.: 1970, *J. Geophys. Res.* **75**, 5503.
Hanson, W. B. and Sanatani, S.: 1971, *J. Geophys. Res.* **76**, 7761.
Hanson, W. B., Zuccaro, D. R., Lippincott, C. R. and Sanatani, S.: 1973, *Radio Sci.* **8**, 333.
Harper, R. M.: 1971, Ph.D. Thesis, Rice University, Houston.
Harper, R. M., Wand, R. H., and Whitehead, J. D.: 1975, *Radio Sci.* **10**, 357.
Harris, I. and Priester, W.: 1962, *J. Atmospheric Sci.* **19**, 286.
Hartree, D. R.: 1929, *Proc. Camb. Phil. Soc.* **25**, 97.
Hedin, A. E., Mayr, H. G., Reber, C. A., Carignan, G. R. and Spencer, N. W.: 1973, *Space Res.* **13**, 315.
Heikkila, W. J.: 1972, in E. R. Dyer (ed.), *Critical Problems of Magnetospheric Physics*, IUCSTP Secretariat, Washington, D.C., p. 67.
Heppner, J. P.: 1972, in E. R. Dyer (ed.), *Critical Problems of Magnetospheric Physics*, IUCSTP Secretariat, Washington, D.C., p. 107.
Herman, J. R.: 1966, *Rev. Geophys.* **4**, 255.
Hesstvedt, E.: 1968, *Geophys. Publik. New.* **27**, 4.
Hines, C. O.: 1951, *J. Geophys. Res.* **56**, 63, 197, 207, 535.
Hines, C. O.: 1960, *Can. J. Phys.* **38**, 1441.
Hines, C. O.: 1963, *Quart. J. Roy. Meteorol. Soc.* **89**, 1.
Hines, C. O.: 1968, *J. Atmospheric Terrest. Phys.* **30**, 837.
Hines, C. O.: 1972, *Phil. Trans. Royal Soc. Lond.* A**271**, 457.
Hinterreger, H. E.: 1965, *Space Sci. Rev.* **4**, 461.
Hinterreger, H. E.: 1970, *Ann. Géophys.* **26**, 547.
Hinterreger, H. E.: 1976, *J. Atmospheric Terrest. Phys.* **38**, 791.
Hirao, K. and Oyama, K.: 1973, *Space Res.* **13**, 489.
Hoffman, J. H., Hanson, W. B., Lippincott, C. R. and Ferguson, E. E.: 1973, *Radio Sci.* **8**, 315.
Holmes, J. C., Johnson, C. Y. and Young, J. M.: 1965, *Space Res.* **5**, 756.
Hoult, D. P.: 1965, *J. Geophys. Res.* **70**, 3183.
Howard, H. J., Tyler, G. L., Fjeldbo, G., Kliore, A. J., Levy, G. S., Brunn, D. L., Dickinson, R., Edelson, R. E., Martin, W. L., Postal, R. B., Seidel, B., Sesplaukis, T. T., Shirley, D. L., Stelzried, C. T., Sweetnam, D. N., Zygielbaum, A. I., Esposito, P. B., Anderson, J. D., Shapiro, I. I., and Reasenberg, R. D.: 1974, *Science* **183**, 1297.
Hudson, M. K. and Kennel, C. K.: 1975, *J. Geophys. Res.* **80**, 4581.
Hudson, R. D.: 1971, *Rev. Geophys.* **9**, 305.
Hunt, B. G.: 1973, *J. Atmospheric Terrest. Phys.* **35**, 1755.
Hunten, D. M.: 1967, *Space Sci. Rev.* **6**, 493.
Hyman, E., Strickland, D. J., Julienne, P. S. and Strobel, D. S.: 1976, *J. Geophys. Res.* **81**, 4765.
Istomin, V. G. and Pokhunkov, A. A.: 1963, *Space Res.* **3**, 117.
Jacchia, L. G.: 1965, *Smithsonian Contrib. Astrophys.* **8**, 215.
Jackson, J. E. 1969, *Proc. I.E.E.E.* **57**, 977.
Jaggi, R. K. and Wolf, R. A.: 1973, *J. Geophys. Res.* **78**, 2852.
Johnson, F. S.: 1956, *J. Geophys. Res.* **61**, 71.
Johnson, F. S. and Gottlieb, B.: 1969, *Space Res.* **9**, 442.
Jones, R. A. and Rees, M. H.: 1973, *Planetary Space Sci.* **21**, 537.
Justus, C. G.: 1966, *J. Geophys. Res.* **71**, 3767.
Kato, S.: 1973, *J. Geophys. Res.* **78**, 757.
Keating, G. M., Mullins, J. A. and Prior, E. J.: 1970, *Space Res.* **10**, 439.
Keating, G. M. and Prior, E. J.: 1968, *Space Res.* **8**, 982.
Kelley, M. C. and Mozer, F. S.: 1973, *J. Geophys. Res.* **78**, 2214.
Keneshea, T. J., Narcisi, R. S. and Swider, W.: 1970, *J. Geophys. Res.* **75**, 845.
Klemperer, W. K.: 1963, *J. Geophys. Res.* **68**, 3191.
Kockarts, G.: 1963, *Acad. Roy. Belg. Bull. Cl. Sci.* **49**, 1135.
Kohl, H.: 1965, in *Kleinheubacher Ber.*, Fernmeldetechnisches Zentralamt, Darmstadt, Germany.

Kohl, H. and King, J. W.: 1967, *J. Atmospheric Terrest. Phys.* **29**, 1045.
Krankowsky, D., Arnold, F., Wieder, H., Kissel, J. and Zähringer, J.: 1972, *Radio Sci.* **7**, 93.
Krassovsky, V. I.: 1960, *Space Res.* **1**, 90.
Krautkramer, J.: 1950, *Arch. Elect. Ubertragung* **4**, 133.
Kupperian, J. E. Jr., Byram, E. T., Chubb, T. A. and Friedman, H.: 1959, *Planetary Space Sci.* **1**, 3.
Landmark, B. and Lied, F.: 1961, *J. Atmospheric Terrest. Phys.* **23**, 92.
Lane, N. F. and Dalgarno, A.: 1969, *J. Geophys. Res.* **74**, 3011.
Larmor, J.: 1924a, *Nature* **114**, 650.
Larmor, J.: 1924b, *Phil. Mag.* **48**, 1025.
Lebedinets, V. N., Manochina, A. V. and Shushkova, V. B.: 1973, *Planetary Space Sci.* **21**, 1317.
Lejeune, G.: 1973, Photoelectrons et bilan énergétique des electrons dans l'ionosphère diurne, Doctorat d'Etat, Université de Paris VI.
Lejeune, G. and Petit, M.: 1969, *Planet. Space Sci.* **17**, 1763.
Lejeune, G. and Waldteufel, P. W.: 1970, *Ann. Géophys.* **26**, 223.
Lemaire, J.: 1974, *Planetary Space Sci.* **22**, 757.
Lemaire, J. and Scherer, M.: 1973, *Rev. Geophys.* **11**, 427.
Leu, M. T., Biondi, M. A. and Johnson, R.: 1973, *Phys. Rev. A.* **7**, 292.
Lindemann, F. A.: 1919, *Phil. Mag.* **38**, 669.
Lindemann, F. A. and Dobson, G. M. B.: 1923, *Proc. Roy. Soc. London A* **102**, 411.
Liu, V. C.: 1969, *Space Res.* **9**, 304.
Loomis, E.: 1860, *Ann. J. Sci. Arts* (2) **30**, 89.
Manson, J. E.: 1977, in O. R. White (ed.) *The Physical Output of the Sun and its Variation*, Colorado Associated University Press, Boulder, Colorado, U.S.A., p. 261..
Mariner Stanford Group: 1967, *Science* **158**, 1678.
Maris, H. B.: 1928, *Terr. Mag. and Atmos. Elec.* **33**, 293.
Marubashi, K.: 1970, *Rep. Ionos. Space Res. Jap.* **24**, 322.
Martyn, D. F.: 1947, *Proc. Roy. Soc. London* A **189**, 241.
Matsushita, S.: 1965, *J. Geophys. Res.* **70**, 4395.
Mazaudier, C.: 1974, *Thèse 3ème cycle*, Université de Paris VI.
McAfee, J. R.: 1968, *J. Geophys. Res.* **73**, 5577.
McAfee, J. R.: 1969, *J. Geophys. Res.* **74**, 802.
McDiamid, I. B. and Burrows, J. R.: 1969, *J. Geophys. Res.* **74**, 6239.
McElhinny, M. W. and Merrill, R. T.: 1975, *Rev. Geophys.* **13**, 687.
McIlwain, C. E.: 1961, *J. Geophys. Res.* **66**, 3681.
McIlwain, C. E.: 1972 in B. M. McCormac (ed.) *Earth's Magnetospheric Processes*, D. Reidel Publishing Company, Dordrecht, Holland, p. 268.
McKinley, D. W. R.: 1961, *Meteor Science and Engineering*, McGraw Hill, New York, p. 67.
Mead, G. D.: 1964, *J. Geophys. Res.* **69**, 1181.
Mechtly, E. A., Bowhill, S. A., Smith, L. G. and Knoebel, H. W.: 1967, *J. Geophys. Res.* **72**, 5239.
Mechtly, E. A. and Smith, L. G.: 1968, *J. Atmospheric Terrest. Phys.* **30**, 363.
Mehr, F. J. and Biondi, M. A.: 1969, *Phys. Rev.* **181**, 264.
Meier, R. R.: 1969, *J. Geophys. Res.* **74**, 3561.
Mitra, S. N.: 1949, *Proc. Inst. Elec. Engrs.* London **96**, 441.
Möller, H. G. and Tauriainen, A.: 1975, *J. Atmospheric Terrest. Phys.* **37**, 161.
Morse, T. F.: 1963, *Phys. Fluids* **6**, 1420.
Moseley, J. T., Olson, R. E. and Peterson, J. R.: 1975, in M. R. C. McDowell and E. W. McDaniel (eds.) *Base Studies in Atomic Physics*, North Holland Publishing Company, Amsterdam, **5**, p. 1.
Muldrew, D. B.: 1965, *J. Geophys. Res.* **70**, 2635.
Muldrew, D. B.: 1972, *Radio Sci.* **7**, 779.
Nagy, A. F., Bauer, P. and Fontheim, E. G.: 1968, *J. Geophys. Res.* **73**, 6259.
Nagy, A. F., Brace, L. H., Carignan, G. R. and Kanal, M.: 1963, *J. Geophys. Res.* **68**, 6401.
Nagy, A. F., Fontheim, E. G., Stolarski, R. S. and Beutler, A. E.: 1969, *J. Geophys. Res.* **74**, 4667.
Narcisi, R. S.: 1966, *Ann. Géophys.* **22**, 224.
Narcisi, R. S.: 1973, in B. M. McCormac (ed.) *Physics and Chemistry of Upper Atmospheres*, D. Reidel Publishing Company, Dordrecht, Holland, p. 171.

Narcisi, R. S., Bailey, A. D. and Della Luca, L.: 1967a, *Space Res.* **7**, 123.
Narcisi, R. S., Bailey, A. D. and Della Luca, L.: 1967b, *Space Res.* **7**, 446.
Narcisi, R. S., Bailey, A. D., Wlodyka, L. E. and Philbrick, C. R.: 1972, *J. Atmospheric Terrest. Phys.* **34**, 647.
Narcisi, R. S., Sherman, C., Wlodyka, L. E., and Ulwick, J. C.: 1974, *J. Geophys. Res.* **79**, 2843.
Narcisi, R. S. and Swider, W.: 1976, *J. Geophys. Res.* **81**, 4770.
Nichols, H. W. and Schelleng, J. C.: 1925a, *Nature* **115**, 334.
Nichols, H. W. and Schelleng, J. C.: 1925b, *Bell Syst. Tech. J.* **4**, 215.
Nicolet, M.: 1945, *Inst. Roy. Met. Belg. Mem.* **19**, 74.
Nishida, A.: 1968, *J. Geophys. Res.* **73**, 1795.
Northrop, T. G. and Teller, E.: 1960, *Phys. Rev.* **117**, 215.
Norton, R. B., Van Zandt, T. E. and Denison, J. S.: 1963, in *Proceedings of the International Conference on the Ionosphere*, Chapman and Hall Ltd, London.
Ogawa, T. and Shimazaki, T.: 1975, *J. Geophys. Res.* **80**, 3945.
Olson, W. P. and Pfister, K. A.: 1974, *J. Geophys. Res.* **79**, 3739.
Oshio, M., Maeda, R. and Sakgami, H.: 1966, *J. Radio. Res. Lab. Japan* **13**, 245.
Oyama, K. I.: 1976, *Planetary Space Sci.* **24**, 183.
Park, C. G.: 1970, *J. Geophys. Res.* **75**, 4249.
Park, C. G. and Banks, P. M.: 1974, *J. Geophys. Res.* **79**, 4661.
Parthasarathy, R., Lerfald, G. M. and Little, C. G.: 1963, *J. Geophys. Res.* **68**, 358.
Paulikas, G. A.: 1975, *Rev. Geophys.* **13**, 709.
Payzant, J. D., Cunningham, A. J. and Kebarle, P.: 1973, *J. Chem. Phys.* **59**, 5615.
Perkins, F.: 1973, *J. Geophys. Res.* **78**, 218.
Perkins, F. and Salpeter, E. E.: 1965, *Phys. Rev.* **139**, A 55.
Peterson, J. R.: 1976, *J. Geophys. Res.* **81**, 1433.
Petit, M.: 1968, *Ann. Géophys.* **24**, 1.
Phelps, A. V. and Pack, J. L.: 1959, *Phys. Rev. Letters* **3**, 340.
Pilkington, G. R., Münch, J. W., Braun, H. J. and Möller, H. G.: 1975, *J. Atmospheric Terrest. Phys.* **37**, 337.
Pines, D. and Bohm, D.: 1952, *Phys. Rev.* **85**, 338.
Prakash, S., Subbaraya, B. H., and Gupta, S. P.: 1971, *J. Atmospheric Terrest. Phys.* **33**, 129.
Prölss, G. W., von Zahn, U. and Raitt, W. J.: 1975, *J. Geophys. Res.* **80**, 3715.
Raitt, W. J., von Zahn, U. and Cristophersen, P.: 1975, *J. Geophys. Res.* **80**, 2277.
Ratcliffe, J. A.: 1959, *The Magneto-Ionic Theory*, Cambridge University Press, London and New York.
Ratcliffe, J. A.: 1974, *J. Atmospheric Terrest. Phys.* **36**, No. 12.
Rawer, K. and Suchy, K.: 1967, in S. Flügge (ed.) *Handbuch der Physik, Geophysik III/2*, Springer-Verlag, Heidelberg, New York, p. 1.
Reagan, J. B. and Watt, T. M.: 1976, *J. Geophys. Res.* **81**, 4579.
Reber, C. A., Harpold, D. N., Horowitz, R. and Hedin, A. E.: 1971, *J. Geophys. Res.* **76**, 1845.
Reber, G. and Ellis, G. R.: 1956, *J. Geophys. Res.* **61**, 1.
Rees, M. H.: 1975 in B. McCormac (ed.), *Atmosphere of the Earth and Planets*, D. Reidel Publishing Co., Dordrecht, Holland, p. 323.
Rees, M. H. and Roble, R. G.: 1975, *Rev. Geophys.* **13**, 201.
Rees, M. H. and Walker, J. C. G.: 1968, *Ann. Géophys.* **24**, 193.
Rees, M. H., Walker, J. C. G. and Dalgarno, A.: 1967, *Planetary Space Sci.* **15**, 1097.
Reid, G. C.: 1970, *J. Geophys. Res.* **75**, 2551.
Revah, I.: 1969, *Ann. Géophys.* **25**, 1.
Richmond, A. D.: 1973, *Radio Sci.* **8**, 1019.
Richmond, A. D., Matsushita, S., and Tarpley, J. D.: 1976, *J. Geophys. Res.* **81**, 547.
Rishbeth, H.: 1972, *J. Atmospheric Terrest. Phys.* **34**, 1.
Rishbeth, H. and Garriott, O. K.: 1969, *Introduction to Ionospheric Physics*, Academic Press, New York and London.
Rishbeth, H. and Setty, C. S. G. K.: 1961, *J. Atmospheric Terrest. Phys.* **20**, 263.
Roederer, J. G.: 1970, *Dynamics of Geomagnetically Trapped Radiation*, Springer-Verlag, Berlin, Heidelberg, New York.

Roemer, M.: 1971, in F. Verniani (ed.) *Physics of the Upper Atmosphere*, Editrice compositori, Bologna, p. 229.
Roper, R. G. and Elford, W. G.: 1963, *Nature* **197**, 963.
Rosenbluth, M. N. and Rostoker, N.: 1962, *Phys. Fluids* **5**, 776.
Rowe, J. N.: 1970, *Ionosphere Research Laboratory Report No. 346*, The Pennsylvania State University.
Rowe, J. N., Ferraro, A. J., Lee, H. S. and Mitra, A. P.: 1969, *J. Atmospheric Terrest. Phys.* **31**, 1077.
Rowe, J. N., Mitra, A. P., Ferrado, A. J. and Lee, H. S.: 1974, *J. Atmospheric Terrest. Phys.* **36**, 755.
Rush, D. W., Torr, D. G., Hays, P. B. and Torr, M. R.: 1976, *Geophys. Res. Letters* **3**, 537.
Rutherford, J. A., Mathis, R. F., Turner, B. R. and Vroom, D. A., 1971, *J. Chem. Phys.* **55**, 3785.
Sabine, E.: 1852, *Phil. Trans. Roy. Soc. London A* **142**, 103.
Sagalyn, R. C. and Smiddy, N.: 1964, *J. Geophys. Res.* **69**, 1809.
Salpeter, E. E.: 1961, *J. Geophys. Res.* **66**, 982.
Sanatani, S. and Hanson, W. B.: 1970, *J. Geophys. Res.* **75**, 769.
Sayers, J.: 1970, *J. Atmospheric Terrest. Phys.* **32**, 663.
Schunk, R. W., Raitt, W. J. and Banks, P. M.: 1975, *J. Geophys. Res.* **80**, 3121.
Schuster, A.: 1911, *Proc. Roy. Soc. London A* **85**, 44.
Schwabe, H.: 1844, *Astron. Nachr.* **21**, 233.
Scialom, G.: 1974, *Radio Sci.* **9**, 253.
Sen, H. K. and Wyller, A. A.: 1960, *J. Geophys. Res.* **65**, 3931.
Shapley, A. H. and Beynon, W. J. G.: 1965, *Nature* **206**, 1242.
Shawhan, S. D., Block, L. P. and Fälthammar, C. G.: 1970, *J. Atmospheric Terrest. Phys.* **32**, 1885.
Shimazaki, T.: 1962, *J. Geophys. Res.* **67**, 4617.
Silsbee, H. C. and Vestine, E. H.: 1942, *Terr. Magn.* **47**, 195.
Simon, A.: 1963, *Phys. Fluids* **6**, 382.
Skinner, N. R., Kelleher, R. F., Hacking, J. B. and Banson, C. W.: 1971, *Nature Phys. Sci.* **232**, 19.
Smith, D.: 1972, *Planetary Space Sci.* **20**, 1717.
Smith, L. G.: 1970, *J. Atmospheric Terrest. Phys.* **32**, 1247.
Smith, F. L. III and Smith, C.: 1972, *J. Geophys. Res.* **77**, 3592.
Sonin, A. A.: 1967, *J. Geophys. Res.* **72**, 4547.
Spencer, N. W., Brace, L. H., Carignan, G. R., Taeusch, D. R., and Niemann, H.: 1965, *J. Geophys. Res.* **70**, 2665.
Spitzer, L., Jr. and Härm, R.: 1953, *Phys. Rev.* **89**, 977.
Spizzichino, A. and Revah, I.: 1968, *Space Res.* **8**, 679.
Spreiter, J. R., Alksne, A. Y. and Summers, A. L.: 1968, in R. L. Carovillono, J. F. McClay and H. R. Radoski (eds.) *Physics of the Magnetosphere*, D. Reidel Publishing Company, Dordrecht, Holland, p. 301.
Spreiter, J. R. and Briggs, B. R.: 1962, *J. Geophys. Res.* **67**, 37.
Stix, T. H.: 1962, *The Theory of Plasma Waves*, McGraw-Hill Book Company Inc., New York, San Francisco, Toronto.
Stolarski, R. S. and Johnson, N. P.: 1972, *J. Atmospheric Terrest. Phys.* **34**, 1691.
Storey, L. R. O.: 1952, An investigation of whistling atmospherics. Ph.D Thesis, University of Cambridge, Great Britain.
Störmer, C.: 1917, *Terrest. Magn. Atm. Elect.* **22**, 23.
Strobel, D. F., Young, T. R., Meier, R. R., Coffey, T. P. and Ali, A. W.: 1974, *J. Geophys. Res.* **79**, 3171.
Stubbe, P.: 1971, *J. Sci. Industrial Res.* **30**, 379.
Stubbe, P. and Chandra, S.: 1970, *Planetary Space Sci.* **32**, 1909.
Sudan, R. N., Akinrimisi, J. and Farley, D. T.: 1973, *J. Geophys. Res.* **78**, 240.
Sunshine, G., Aubrey, B. B. and Bederson, B.: 1967, *Phys. Rev.* **154**, 1.
Swartz, W. E. and Nisbet, J. S.: 1973, *J. Geophys. Res.* **78**, 5640.
Swider, W.: 1964, *Planetary Space Sci.* **12**, 761.
Swider, W.: 1965, *J. Geophys. Res.* **70**, 4859.
Swider, W. and Narcisi, R. S.: 1970, *Planetary Space Sci.* **18**, 379.
Tanenbaum, B. S.: 1965, *Phys. Fluids* **8**, 683.
Tarpley, J. D.: 1970a, *Planetary Space Sci.* **18**, 1075.
Tarpley, J. D.: 1970b, *Planetary Space Sci.* **18**, 1091.

Taylor, G. N. and Wrenn, G. L.: 1970, *Planetary Space Sci.* **18**, 1663.
Taylor, H. A., Jr.: 1972, *Planetary Space Sci.* **20**, 1593.
Testud, J.: 1973, Thèse d'Etat, University of Paris, Paris.
Testud, J., Amayenc, P. and Blanc, M.: 1975, *J. Atmospheric Terrest. Phys.* **37**, 989.
Thomas, L.: 1961, *J. Atmospheric Terrest. Phys.* **23**, 101.
Thomas, L.: 1976, *J. Atmospheric Terrest. Phys.* **38**, 1345.
Thuillier, G.: 1973, *Explication de l'émission tropicale à 6300 Å de l'oxygène atomique: Mesures et Théorie*, Doctorat d'Etat, Université de Paris VI.
Thuillier, G. and Blamont, J. E.: 1973, in B. McCormac, (ed.) *Physics and Chemistry of Upper Atmospheres*, Reidel Publishing Co., Dordrecht, Holland, p. 219.
Timleck, P. L. and Nelms, G. L.: 1969, *Proc. I.E.E.E.* **57**, 1164.
Tohmatsu, T., Ogawa, T. and Tsuruta, H.: 1965, *Rep. Ionos. Space Res. Jap.* **19**, 482.
Tohmatsu, T. and Roach, F. E.: 1962, *J. Geophys. Res.* **67**, 1817.
Tohmatsu, T. and Wakai, N.: 1970, *Ann. Géophys.* **26**, 209.
Timothy, J. G.: 1977, in O. R. White (ed.) *The Physical Output of the Sun and its Variation*, Colorado Associated University Press, Boulder, Colorado, USA, p. 237.
Tremellen, K. W. and Cox, J. W.: 1947, *Proc. Inst. Elec. Engrs. London* **94**, 200.
Tsunoda, R. T.: 1976, *J. Geophys. Res.* **81**, 425.
Turco, R. P. and Sechrist, C. F.: 1970, *Aeronomy Report No. 41*, University of Illinois.
Turco, R. P. and Sechrist, C. F.: 1972, *Radio Science*, **7**, 703.
Tuve, M. A. and Breit, G.: 1925, *J. Terr. Mag.* **30**, 15.
Unwin, R. S. and Baggaley, W. J.: 1972, *Ann. Géophys.* **28**, 111.
Vallance Jones, A.: 1974 *The Aurora*, Geophysics and Astrophysics Monographs **9**, D. Reidel Publishing Co., Dordrecht, Holland.
Vasseur, G.: 1970, *J. Atmospheric Terrest. Phys.* **32**, 775.
Vasseur, G.: 1971, *La Dynamique de l'Atmosphère Neutre et Ionisée au-dessus de 200 km. Etude par Diffusion Incohérente*, Doctorat d'Etat, Université de Paris VI.
Vasyliunas, V. M.: 1972, in B. M. McCormac (ed.) *Earth's Magnetospheric Processes*, D. Reidel Publishing Company, Dordrecht, Holland, p. 29.
Vogan, E. L. and Campbell, L. L.: 1957, *Can. J. Phys.* **35**, 1176.
Volland, H.: 1969a *Planetary Space Sci.* **17**, 1581.
Volland, H.: 1969b *Planetary Space Sci.* **17**, 1709.
Volland, H. and Mayr, H. G.: 1972, *J. Atmospheric Terrest. Phys.* **34**, 1745.
Volland, H. and Mayr, H. G.: 1973, *Ann. Géophys.* **29**, 61.
Volland, H. and Mayr, H. G.: 1974, *Radio Sci.* **9**, 263.
von Zahn, U.: 1970, *J. Geophys. Res.* **75**, 5517.
von Zahn, U.: 1975, in B. M. McCormac (ed.) *Atmospheres of Earth and the Planets*, D. Reidel Publishing Company, Dordrecht, Holland, p. 133.
von Zahn, U., Fricke, K. H. and Trinks, H.: 1973, *J. Geophys. Res.* **78**, 7560.
Waldteufel, P.: 1970, *Planetary Space Sci.* **18**, 741.
Waldteufel, P.: 1970, Une étude par diffusion incohérente de la haute atmosphère neutre. Doctorat d'Etat, Université de Paris VI.
Waldteufel, P.: 1970, Thèse d'Etat, University of Paris VI.
Waldteufel, P. and Cogger, L.: 1971, *J. Geophys. Res.* **76**, 5322.
Walker, J. C. G. and McElroy, M. B.: 1966, *J. Geophys. Res.* **71**, 3779.
Walker, J. C. G. and Rees, R. H.: 1968, *Planetary Space Sci.* **16**, 459.
Warnock, J. M., McAfee, J. R. and Thomson, T. L.: 1970, *J. Geophys. Res.* **75**, 7272.
Watanabe, K. and Hinterreger, H. E.: 1962, *J. Geophys. Res.* **67**, 999.
Watt, T. M.: 1973, *J. Geophys. Res.* **78**, 2992.
Waynick, A. H.: 1974, *J. Atmospheric Terrest. Phys.* **36**, 2105.
Webber, W.: 1972, *J. Geophys. Res.* **77**, 5091.
Whipple, E. C. Jr.: 1965, *The Equilibrium Electric Potential of a Body in the Upper Atmosphere and Interplanetary Space*, Ph.D. thesis, The George Washington University; also published as NASA Tech. Note X–615–65–296.
Whipple, E. C., Warnock, J. M., and Winkler, R. H.: 1974, *J. Geophys. Res.* **79**, 179.
Whitehead, J. D.: 1961, *J. Atmospheric Terrest. Phys.* **20**, 49.

Whitehead, J. D.: 1967, *Space Res.* **7**, 90.

Wickwar, V. B.: 1971, *Photoelectrons from the Magnetic Conjugate Point Studied by Means of 6300Å Predawn Enhancement and the Plasma Line Enhancements*, Ph.D. Thesis, Rice University, Houston, Texas.

Wickwar, V. B.: 1975, in B. M. McCormac (ed.) *Atmospheres of Earth and the Planets*. D. Reidel Publishing Company, Dordrecht, Holland, p. 111.

Wilcox, J. H. and Ness, N. F.: 1965, *J. Geophys. Res.* **70**, 5793.

Wilson, J. W. and Garside, G.: 1968, *Planetary Space Sci.* **16**, 257.

Wilkes, M. V.: 1954, *Proc. Phys. Soc. London* **67** Part 4, 304.

Witt, G.: 1969, *Space Res.* **9**, 157.

Woodman, R. F.: 1971, *J. Geophys. Res.* **76**, 178.

Woodman, R. F. and La Hoz, C.: 1976, *J. Geophys. Res.* **81**, 5447.

Wrenn, G. L.: 1969, *Proc. Inst. Elec. Engrs. London* **57**, 1072.

Yeh, K. C. and Liu, C. H.: 1972, *Theory of Ionospheric Waves*, Academic Press, New York and London.

Yngvesson, K. O. and Perkins, F. W.: 1968, *J. Geophys. Res.* **73**, 97.

Young, J. M., Carruthers, G. R., Holmes, J. C., Johnson, C. Y. and Patterson, N. P.: 1968, *Science* **160**, 990.

Young, J. M., Johnson, C. Y. and Holmes, J. C.: 1967, *J. Geophys. Res.* **72**, 1473.

Zbinden, P. A., Hidalgo, M. A., Eberhardt, P. and Geiss, J.: 1975, *Planetary Space Sci.* **23**, 1621.

Zimmerman, S. P.: 1966, *J. Geophys. Res.* **71**, 2439.

Zimmerman, S. P., Pereira, G. P., Murphy, E. A. and Theon, J.: 1973, *Space Res.* **13**, 209.

SUBJECT INDEX

GEOPHYSICS AND ASTROPHYSICS MONOGRAPHS

AN INTERNATIONAL SERIES OF FUNDAMENTAL TEXTBOOKS

1. R. Grant Athay, *Radiation Transport in Spectral Lines.* 1972, XIII + 263 pp.
2. J. Coulomb, *Sea Floor Spreading and Continental Drift.* 1972, X + 184 pp.
3. G. T. Csanady, *Turbulent Diffusion in the Environment,* 1973, XII + 248 pp.
4. F. E. Roach and Janet L. Gordon, *The Light of the Night Sky.* 1973, XII + 125 pp.
5. H. Alfvén and G. Arrhenius, *Structure and Evolutionary History of the Solar System.* 1975, XVI + 276 pp.
6. J. Iribarne and W. Godson, *Atmospheric Thermodynamics*, 1973, X + 222 pp.
7. Z. Kopal, *The Moon in the Post-Apollo Era,* 1974, VIII + 233 pp.
8. Z. Švestka, *Solar Flares,* 1976, XV + 376 pp.
9. A. Vallance Jones, *Aurora,* 1974, XVI + 301 pp.
10. C.-J. Allègre and G. Michard, *Introduction to Geochemistry,* 1974, XI + 142 pp.
11. J. Kleczek, *The Universe,* 1976, VII + 259 pp.
12. E. Tandberg-Hanssen, *Solar Prominences,* 1974, XVI + 155 pp.
13. A. Giraud and M. Petit, *Ionospheric Techniques and Phenomena,* 1978, XX + 264 pp.
14. E. L. Chupp, *Gamma Ray Astronomy,* 1976, XIII + 317 pp.
15. W. D. Heintz, *Double Stars,* 1978, IX + 174 pp.

J. Audouze and S. Vauclair, *Nuclear Astrophysics,* (Forthcoming).
M. M. Millán, *Dispersive Correlation Spectroscopy for In-Situ and Remote Pollution Monitoring,* (Forthcoming).